"Why do engineers prefer GLP FE/EIT study guides?"

Reader remarks from the GLP in-basket....

"I greatly appreciate you coming out with such an excellent *FE/EIT Discipline Review*. I walked into the exam feeling prepared and confident...and most importantly, I passed!"

—Bill Johnston, Raleigh, North Carolina

"After looking at several other books, I am convinced your *Fundamentals of Engineering* is the best review around. It is very well organized, thorough, and concise. An outstanding book."

—Paul Griesmer, Cleveland, OH

"This is a time and cost effective tool. I easily passed the FE on my first attempt."

—Omar Mureebe, Uniroyal Chemical Company, Middlebury, CT

"In your FE book you cut to the chase and don't embellish it... Yours is the only review that doesn't scare the pants off people. It's exactly what I want for teaching my review course."

—John Thorington

"I was talking to a student a few hours ago and he said the morning problems on the FE/EIT were like the sample problems in your book—verbatim. It's the best review I've ever seen. When we started using your review in 1994, our students' pass rates went up 20%! Now we will continue to use your book because we take professional licensing seriously."

—Dr. Gary Rogers, Virginia Military Institute

"Your FE study guide is very thorough, well-organized and easy to use. It was instrumental in helping me brush up on engineering material I had not seen for 5 years. I passed the FE the first time!"

—Brian Campbell, Project Engr, Trailmaster Corp, Ft. Worth, TX

"Thank you for the book! I think I passed the test because of it. Stuff I hadn't studied for more than 15 years came back to me clearly. The two practice exams are what helped me the most."

—Juan Alfaro, College Park, MD

"Three years out of school and all I needed to pass the FE exam, on the first try, was your review book! I have recommended it to everyone I know. I look forward to using your PE reference."

—David Robins, Syracuse, NY

"Your book is great! Very easy to use. Although I only had time to skim through it, it helped me review the very basics I needed to know (i.e., Chemistry, Econ). I passed without a problem! Thanks! I'm keeping it as a handy reference."

—Julie Volby, Civil Engineering Student, Univ of M, Mnpls, MN

"Your review material is perfect for the non-traditional 'night student.' It refreshes and restores many years of part-time study, without being overly exhaustive."

—Jerome Bobak, Designer Mechanical Engineer, Grand Island, NY

"This book provided all of the necessary information for passing the FE exam. I was able to review for and *pass* the exam in only 3 weeks!"

—Joseph Rozza, Recent Graduate, Orlando, FL

Important Information

FE examination date: ____________________ time: ________________

location: ____

____

Examination Board Address: ____

____

____

phone: ____

Application was requested on this date: ____

Application was received on this date: ____

Application was accepted on this date: ____

This book belongs to: ____

____

____

____

phone: ____

...From the Professors who know it best!...

•FE/EIT• DISCIPLINE REVIEW

2nd edition

Merle C. Potter, Phd, PE—Editor

For Civil, Mechanical, Electrical and Chemical DS Tests

Authors:

Thomas F. Wolff .... Geotechnical
Frank J. Hatfield .... Structural Analysis
Thomas L. Maleck .... Transportation
Francis McKelvey .... Airport Design
Gilbert Baladi .... Pavement Design
Susan J. Masten .... Environmental Engineering
Merle C. Potter .... Hydrology

Merle C. Potter .... Thermodynamics, Fluids, Vibrations
Craig W. Somerton .... Heat Transfer
Brian S. Thompson .... Machine Design
Brian F. Feeny .... Control Systems

H. Roland Zapp .... Electromagnetic Fields
Dennis Wiitanen .... Electrical Power
H. Roland Zapp .... Communications and Signal Processing
Bong Ho .... Solid State Devices
R. Lal Tummala .... Computer Engineering

Bruce W. Wilkinson .... Chemical Thermodynamics, Reaction Engineering, Flow Reactors, Distillation, Mass Transfer, Fluid Flow

published by

GREAT LAKES PRESS
P.O. Box 483
Okemos, MI 48805
Customer Service (314) 273-6016
www.glpbooks.com

International Standard Book Number 1-881018-32-6

All comments and inquiries should be addressed to:
Great Lakes Press
PO Box 550
Grover, MO 63040-0550
Phone (314) 273-6016
Fax 314-273-6086
www.glpbooks.com

Library of Congress
Cataloging-in-Publication Data

Printed in the USA by Braun-Brumfield, Inc. of Ann Arbor, Michigan.

10 9 8 7 6 5 4 3 2 1

Table of Contents

Preface

by M. C. Potter

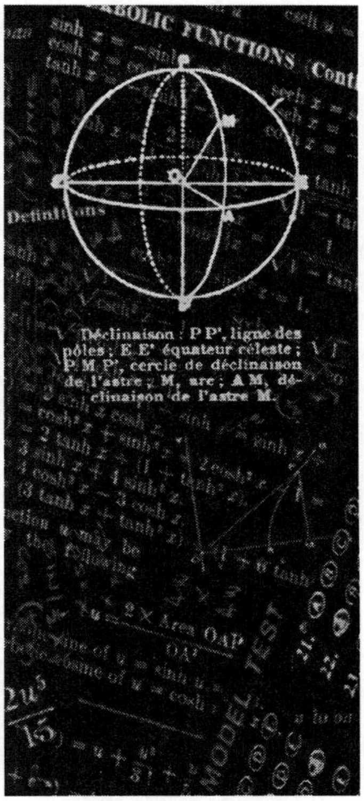

The powers that be decided that as of October 1996 the FE/EIT exam shall consist of a General engineering test during the morning session that covers twelve subject areas required in most engineering curricula, and an afternoon session that is discipline-specific. The afternoon session will consist of one of six discipline-specific (DS) tests: Chemical, Civil, Electrical, Industrial, Mechanical, or General. The General test is intended for those engineers who do not fall into any of the five specific disciplines. Each examinee will take the morning General test and one of the afternoon tests.

Since there are over 30 engineering disciplines, examinees in over 25 of those disciplines will be expected to take the General test during the afternoon session. Most of the state boards will recognize a "pass" on any of the afternoon tests by any examinee. Consequently, it may be advisable for you to elect the afternoon General test even though you are a Civil, Electrical, Mechanical, Chemical, or Industrial engineer. Check with your state board if you are uncertain of your state's position.

This book was written as a review for the afternoon test of the FE/EIT exam in Civil, Electrical, Mechanical, and Chemical engineering. Our strategy is to offer a short review of each subject of these four tests and illustrate each subject with at least one example problem which is as exam-like as possible; this has been done for all the subjects and primary equations included in the official NCEES Reference Handbook. Since the Handbook is the only book allowed into the examination, this review should cover most all problems tested on the afternoon session of the new FE/EIT exam.

Many of the afternoon questions can be answered by referring to equations, tables, or charts included in the first 81 pages of the NCEES Reference Handbook —material that is primarily intended for the General tests. We recommend that you use our *Fundamentals of Engineering Review*, a book written to help you prepare for the morning General session, and much of the afternoon DS tests, as well as the afternoon General test. The 7th edition has been prepared for the new exam format. Call 1-800-837-0201 to order a copy. We also offer HP48G and GX preprogrammed calculators and a companion 170-page *Jump-Start the HP48G/X* helpbook, additional aids to help you pass the FE/EIT exam.

The new test requires two books to prepare for, unless one chooses the General DS afternoon option. To offset the added expense, we offer a $10 discount/rebate with the purchase of both our *FE Review* and our *FE/EIT Discipline Review*. Call 1-800-837-0201 to order either copy or a set.

In this new second edition we have added 3 full 60-question practice tests, in Civil, Mechanical and Electrical engineering, with full solutions. These exams should help you decide if you should take the DS test in your area or select the General test option. For this edition, we have also added expanded reviews of the material in several areas.

We have included all the DS reviews in this one volume rather than create four separate volumes. This is to help the review course instructor who may not know how many engineers in each discipline will be present in the course. We have priced the book as if it were covering just one discipline—keeping it low-cost for its overall contents—since we realize that each engineer will be interested in only one section of the book.

Best wishes in your study and on your test day!

Dr. Merle C. Potter, PE
Okemos, Michigan

Passing the FE/EIT Discipline Specific Exam

A Brief Outline Of This Review

This book presents an efficient review of four discipline-specific subjects covered on the new format of the FE/EIT exam: Civil, Electrical, Mechanical and Chemical Engineering. GLP offers a companion book, *Fundamentals of Engineering*, 7th edition, which presents a succinct yet complete review of the material covered on the morning and afternoon general test. We strongly recommend that you use that review to prepare for the twelve subjects of the morning session which all test-takers must take. You may order a copy of the general test review by calling 1-800-837-0201.

This book contains:

- Short, succinct reviews of the important aspects of most of the discipline-specific subject areas. Some subject areas are presented with sample problems only.
- Sample problems that present the situational use of the more important equations presented in the NCEES Handbook.
- Complete solutions to each sample problem showing the details needed to arrive at a solution.
- Full 60-question practice tests in Civil, Mechanical, and Electrical engineering, with solutions.

Why We Created This FE Review

In 1983, we at Great Lakes Press developed an FE/EIT review with the cooperation of a team of colleagues in response to our experience in coordinating FE/EIT review courses. The kind of well-planned material that an engineer naturally desires to use did not exist at the time. The options were either to pick and choose from a large, encyclopedia-like book that covers almost every topic in engineering, or to use material that didn't even attempt to cover the topics tested on the exam. It was either feast or famine. And both options were often rather expensive! So we recruited popular lecturers from the university campus and prepared our first *Fundamentals of Engineering* review.

With the announcement of the new exam format, we pulled together a team of 15 professors to prepare this discipline-specific review.

On a daily basis we act as advocate for the test-taker in our involvement with engineering departments, associations, review courses and registration governing bodies. We are dedicated to making the entire licensing and registration process as reasonable as possible. Our present strategy for beating the new Discipline-Specific format is a prime example of our leadership. We've prepared this review focusing solely on the equations, figures, and tables presented in the NCEES Reference Handbook, the only aid allowed into the exam besides a calculator. There will be questions that may not require the use of the equations from the Handbook; these will be based on general information from the respective courses. Such general information may not be included in this review.

How To Become A Professional Engineer

To become registered as an engineer, a state may require that you:

1. Graduate from an ABET-accredited engineering program
2. Pass the *Fundamentals of Engineering* exam
3. Pass the *Principles and Practice of Engineering* (PE) exam after several years of engineering experience

Requirements vary from state to state, so you should obtain local guidelines and follow them carefully.

Why Get an Engineering License?

Registration is necessary if an engineer works as a consultant, and is highly recommended in certain industries—especially when one is, or hopes to be, in a management position.

How The FE Exam Is Scored

In the morning session, each of the 120 problems is worth one-half point. Thus a maximum score of 60 is possible.The afternoon session is different from the morning in that each problem is worth one point. With half the number of questions, the maximum possible score in the afternoon session is also 60.

The two-session total is 120 points, with both sessions having equal weight. A predetermined passing percentage is not established, nor is the exam graded on a curve. From recent history, a score of 60 (50% correct) would probably be a passing score. Each year the exact score is determined by statistical methods, so the score of 60 is an approximation.

Examination Format

The FE is composed of a four-hour general subject morning session, followed by a one-hour break, then a four-hour discipline-specific afternoon session of 60 more difficult questions. The two sessions have a total of 180 4-part multiple-choice questions covering the subject areas listed in the tables that follow.

Morning Session

Subject Area	*Approximate Number*
Chemistry	11
Computers	6
Dynamics	10
Economics	5
Electrical Circuits	12
Ethics	5
Fluids	8
Materials Science	8
Mathematics	24
Mechanics of Materials (Solids)	8
Statics	12
Thermodynamics	11
Total Questions:	120

Subject Lists of the Discipline-Specific Afternoon Tests

General • Chemistry, Computers, Dynamics, Economics, Electrical, Ethics, Fluids, Material Science, Mathematics, Mechanics of Materials (Solids), Statics, Thermodynamics

Civil • Construction, Environmental, Hydraulics, Hydrology, Numerical Methods, Legal, Soils, Foundations, Structural, Design, Surveying, Water Treatment, Waste Water

Mechanical • Controls, Computers, Systems, Energy conversion, Power plants, Turbomachines, Fluids, Heat transfer, Materials, Instrumentation, Design, HVAC, Solids, Thermodynamics

Electrical • Analog, Communication, Numerical Methods, Computer Hardware, Software, Controls, Systems, Digital, Electromagnetics, Instrumentation, Networks, Power, Signal Processing, Solid State Devices

Chemical • Reactions, Thermodynamics, Numerical Methods, Heat Transfer, Mass Transfer, Energy, Pollution, Controls, Design, Economics, Equipment, Safety, and Transport Phenomena

Industrial • Design, Economics, Statistics, Facilities, Cost Analysis, Computer Modeling, Ergonomics, Management, Systems, Manufacturing, Material Handling, Optimization, Production, Productivity, Queuing, Simulation, Quality Control, Quality Management, Performance

How to Use the NCEES Handbook

In the fall of 1996, the NCEES launched the new DS Test format, along with a new edition of the Handbook. A copy of this handbook will be mailed to you sometime after you register. Your first impression of this handbook may leave you overwhelmed! It has a tremendous amount of equations, figures, and information that may, at first glance, seem unfamiliar to you. Indeed, afternoon DS Test info is mixed in with morning General info. In our review we have used the NCEES Handbook to compile summaries of the more important equations from the most significant General subject areas, in fitting with the Morning Session and our recommended Afternoon Strategy—this is Appendix A at the back of this book. We have attempted to adapt our nomenclature throughout to that of the NCEES Handbook.

We strongly urge you to have the official NCEES Reference Handbook by your side the entire time you study and use a highlighter to identify the equations you most often use. Initially, you can use our Equation Summary Sheets and your own best judgment to highlight the key equations in the general part of the Handbook.

Familiarize yourself with the location of these equations so you can quickly locate them during the actual exam in the new handbook that will be given to you at the test site. If you do not prepare this way, you might become lost in the handbook during the exam.

Strategies for Study

The following strategies will help you prepare for the FE Exam:

1. Focus your review on the eight subjects that have the most exam questions: Chemistry, Dynamics, Electrical, Fluids, Math, Solids, Statics and Thermo.

2. *Quickly* review the remaining four major subject areas: Computers, Economics, Ethics and Materials Science.

3. Spend the majority of your time reviewing material with which you are <u>most</u> familiar.

4. You will receive a copy of the NCEES Handbook when you register for the exam or you can order a copy ($12) from GLP by calling 1-800-837-0201. You will not be allowed to bring this handbook (or any other printed material) with you into the exam. However, a clean copy of this same handbook will be given to you upon entering the exam site. The new handbook is particularly confusing. It includes about 80 pages that cover the General subjects (a.m. and p.m.) and 35 pages that cover the Discipline-Specific subjects. The trouble is that the a.m. and p.m. subjects are intermixed! Thus, you'll find Heat Transfer (for the ME DS Test) covered early in the pamphlet, hidden in with most of the General topics. If you elect to take the General Test, boldly mark the pages that cover the general subjects. In our book, we have provided most of the same equations that you will find in the NCEES Handbook for the major a.m. and General Test p.m. subjects. Then search through the handbook and highlight those important equations. This will allow you to familiarize yourself with the location of the equations you need most, and you will be able to quickly locate those equations in the Handbook during the exam. You will find this particularly helpful since the Handbook is filled with plenty of extraneous material. If you do not train yourself in this way, you may well find it difficult to quickly locate the appropriate

equations during the exam. Being familiar with the Handbook is perhaps the most important detail you should attend to.

Outline of A Good Study Program For Quick Progress

For a senior in an engineering college who has a busy schedule, we suggest 8 weeks as an ideal study period. During those 8 weeks, you must be willing to perform a fairly concentrated study. You should set aside blocks of 3 hours at least 2 days a week. In any case, we recommend that you study no less than 4 weeks for the FE exam. For those of you who have been away from this material for a while, base the length of your study period on the number of years you have been away from school and your own memory capability.

Perform an initial review during the first half of your selected study period (for engineering seniors, 4 weeks). But if you are trying to get through your review quickly, only the Practice Problems that have been starred (*) should be worked and studied in our *Fundamentals of Engineering*, 7th, edition, General Test Review.

Two days prior to exam day, review all subject areas briefly, using your highlighted handbook of equations and tables. Be sure you can quickly find the equations you will use most often in the *unmarked* booklet you will be given on exam day.

The day before the exam, relax and go to bed early. Do *not* cram or perform any panic studying. By now, you are as prepared as you can be for the exam.

The morning of your test, get up early and have a light, healthy breakfast (and maybe some coffee!). Arrive at the exam site at least 20 minutes early. You need to allow time for parking and getting settled—be sure to bring some change to pay for parking! During the one-hour lunch break, it is best to plan to meet with a friend who can help you relax and get refreshed for the Afternoon Session. If there are no restaurants nearby, bring a bag lunch and eat outside on the lawn somewhere. Get some fresh air. Do *not* spend the entire hour reviewing. Try not to talk to other test-takers about how it is going for them. This can easily induce either insecurity or false confidence. If you understand engineering principles and have prepared well, after the dust and sweat of the test day clears, you'll find you've passed!

Pacing Yourself During the FE Exam

The problems in the morning session are, for the most part, unrelated. Consequently, *two minutes* (on the average) can be spent on each problem. This makes fast recall *essential* as time does not allow you to contemplate various methods of solution. But each problem is only worth one-half point—so do not fall into the trap of spending too much time on any one problem from the morning session.

In the afternoon session, you can spend an average of 4 minutes per individual question. Many questions in the afternoon session are related to a common problem, divided into 2 to 6 sub-problems. Thus you may maintain a good pace and still spend up to 20-25 minutes solving a problem set that is more difficult for you.

Process of Elimination

Sometimes the best way to find the right answer is to look for the wrong ones and cross them out. On questions which are difficult for you, wrong answers are often much easier to find than right ones!

Answers are seldom given with more than three significant figures, and may be given with two significant figures. The choice *closest* to your own solution should be selected.

There is no penalty for *guessing*. Use the *process of elimination* when guessing at an answer. If only one answer is negative and three answers are positive, eliminate the one odd answer from your guess. Also, when in doubt, work backwards and eliminate those answers that you believe are untrue and then guess. By using a combination of methods, you greatly improve your odds of answering correctly.

Should I Guess?

Leave the last ten minutes of each session for making educated guesses. **Do not leave any answers blank** on your answer key. A guess *cannot* hurt you, only help you. Your score is based on the number of questions you answer correctly. An incorrect answer does not harm your score.

Place a question mark beside choices you are uncertain of, but seem correct. If time prevents you from reworking that problem, you will have at least identified your best guess.

Difficult Problems

If at first glance you know that a certain problem will require much time and is exceptionally difficult for you, make your best guess then *skip right over it.* Be sure to mark the problem in a unique manner (we suggest that you circle the problem number) so that if time permits, you may come back to it. (Note: it is not possible to return to the Morning Session problems in the Afternoon Session.)

When you are working through a problem and decide to move on due to difficulty, be sure to write down in your test booklet your notes and conclusions up to that point in case you have time to return to it. Then make your best guess and circle the problem number to return to if you have time.

If you feel you know how to work a difficult problem and could answer it with more time, identify it by circling the entire problem, not just the problem number—this identifies it as a 'most likely' candidate for your set aside ten minutes of 'guess' time.

Time-Saving Tips

Once you determine your answer, *always write the letter corresponding to the correct answer in the margin of the test booklet* beside the question. At the end of the page, you can then transfer all the answers from that page to the answer key at once. This will save you considerable time and help you maintain concentration as well!

Cross out choices that you have eliminated in your test booklet on the problems you will return to. Otherwise, you will have to reread them as you make your last-ditch deliberations.

Write Out Your Work

Feel free to write all over the writing space provided for you in your test booklet. Do not hesitate to work out a problem, no matter how simple it may be. Doing as much work as you can on paper will ease your mind and leave it less 'cluttered'—and it will help you if you need to return to a problem. You may think you're saving time, but your exam performance is *not* improved when you work problems in your head.

Bring a Calculator—Ideally, an HP-48G

You must take a silent calculator (it may be preprogrammed) into the exam. A calculator is essential when solving many problems. In fact, with the exception of a couple states, the premier engineering calculator, the HP-48G, is allowed into the exam (check with your state board). This calculator is a hand computer which has hundreds of basic equations and constants preprogrammed. If you need extra help with exam calculations, the HP-48G will be very useful. We at GLP offer this calculator and its higher-powered relative, the GX, for sale at substantial discounts. (48G $120, list $165. 48GX $230, list $350.) The GX, with its increased memory and removable card-slot, is of particular interest because cards are now made for it especially for the FE and PE exams—we have these cards available. We also offer a manual to guide you through the steps to using this calculator effectively for the FE Exam. (The manufacturer's manual is difficult to use for even basic operations.) Our engineering-oriented manual, called *Jump Start the HP-48*, is available for $19.95. Call us at 1-800-837-0201 for ordering or information, or browse our website with its Easy Order Form at www.glpbooks.com.

English vs. SI Units

Some questions can be worked using either English units or SI units in the morning session only. We recommend that all test takers prepare using only SI units, although some of our review material uses English units since that is what is used in the course. The answers from the array of choices will be the same using either set of units. The problems in this book are mostly in SI units. The afternoon Civil Engineering Test will use both sets of units in some of the subjects areas (e.g., Soils). A table of conversion factors is presented in Appendix B of this book.

Recommended Materials for FE/EIT Preparation

1. *FE/EIT Discipline Review*, 2nd ed., (this book) GLP; $34.95
2. Fundamentals of Engineering—General, 7th ed., GLP; $52.95
3. HP-48G/X Programmable Calculator; $120/$230
4. *Jump Start the HP-48*, 2nd ed., GLP; $19.95
5. NCEES FE/EIT Sample Exam; $28
6. NCEES Reference Handbook; $12

All these resources available from GLP by calling 1-800-837-0201.

State Boards of Registration Addresses and Phone Numbers

All State Boards of Registration administer the National Council of Engineering Examiners and Surveyors (NCEES) uniform examination.The dates of the exams cover a span of three days in mid-April and three days in late October. The specific dates are selected by each State Board. To be accepted to take the FE exam, an applicant must apply well in advance. For information regarding the specific requirements in your state, contact your state board's office. If incorrect here, your correct state board information can be obtained from the Executive Director of NCEES, P.O. Box 1686, Clemson, SC 29633-1686, phone (803) 654-6824. Any comments relating to the exam or the Reference Handbook should be addressed to NCEES at the above address.

Board Addresses

ALABAMA: State Board of Registration for Professional Engineers and Land Surveyors, P.O. Box 304451, Montgomery 36130-4451. Executive Secretary, Telephone: (334) 242-5568.

ALASKA: State Board of Registration for Architects, Engineers and Land Surveyors, Pouch D, Juneau 99811. Licensing Examiner, Telephone: (907) 465-2540.

ARIZONA: State Board of Technical Registration, 1951 W. Camelback Rd., Suite 250, Phoenix 85015. Executive Director, Telephone: (602) 255-4053.

ARKANSAS: State Board of Registration for Professional Engineers and Land Surveyors, P.O. Box 2541, Little Rock 72203. Secretary-Treasurer, Telephone: (501) 324-9085.

CALIFORNIA: State Board of Registration for Professional Engineers and Land Surveyors, 2535 Capitol Oaks Dr., Sacramento 95853. Executive Secretary, Telephone: (916) 263-2222.

COLORADO: State Board of Registration for Professional Engineers and Professional Land Surveyors, 1860 Broadway, Suite 1370, Denver 80202. Program Administrator, Telephone: (303) 894-7788.

CONNECTICUT: State Board of Registration for Professional Engineers and Land Surveyors, The State Office Building, Room G-3A, 165 Capitol Avenue, Hartford 06106. Administrator, Ph: (203) 566-3290.

DELAWARE: Delaware Association of Professional Engineers, 2005 Concord Pike, Wilmington 19803. Executive Secretary, Telephone: (302) 577-6500.

DISTRICT OF COLUMBIA: Board of Registration for Professional Engineers, 614 H Street, N.W., Room 910, Washington 20001. Executive Secretary, Telephone: (202) 727-7468.

FLORIDA: Department of Business and Professional Regulations, Board of Professional Engineers, Northwood Centre, 1940 N. Monroe Street, Tallahassee 32399-0755. Executive Director, Telephone: (904) 488-9912.

GEORGIA: State Board of Registration for Professional Engineers and Land Surveyors, 166 Pryor Street, SW, Atlanta 30303-3465. Executive Director, Telephone: (404) 656-3926.

GUAM: Territorial Board of Registration for Professional Engineers, Architects and Land Surveyors, Department of Public Works, Government of Guam, P.O. Box 2950, Agana 96911. Chairman, Telephone: (671) 646-3115.

HAWAII: State Board of Registration for Professional Engineers, Architects, Land Surveyors and Landscape Architects, P.O. Box 3469, (1010 Richards Street), Honolulu 96801. Executive Secretary, Telephone: (808) 586-2693.

IDAHO: Board of Professional Engineers and Land Surveyors, 600 S. Orchard, Suite A, Boise 83705. Executive Secretary, Telephone: (208) 334-3860.

ILLINOIS: Department of Registration and Education, Professional Engineers' Examining Committee, 320 West Washington, 3rd Floor, Springfield 62786. Unit Manager, Telephone: (217) 782-0458.

INDIANA: Indiana Professional Licensing Agency, 302 W. Washington St., Room E034, Indianapolis 46204. Executive Director, Telephone: (317) 232-2980.

IOWA: State Board of Engineering Examiners, Capitol Complex, 1918 S.E. Hulsizer, Ankeny 50021. Executive Secretary, Telephone: (515) 281-5602.

KANSAS: State Board of Technical Professions, 900 Jackson, Room 507, Topeka 66612. Executive Secretary, Telephone: (913) 296-3053.

KENTUCKY: State Board of Registration for Professional Engineers and Land Surveyors, 160 Democrat Dr., Frankfort 40601. Executive Director, Telephone: (502) 573-2680.

LOUISIANA: State Board of Registration for Professional Engineers and Land Surveyors, 1055 St. Charles Avenue, Suite 415, New Orleans 70130. Executive Secretary, Telephone: (504) 568-8450.

MAINE: State Board of Registration for Professional Engineers, State House, Station 92, Augusta 04333. Secretary, Telephone: (207) 287-3236.

MARYLAND: State Board of Registration for Professional Engineers, 501 St. Paul Place, Room 902, Baltimore 21202. Executive Secretary, Telephone: (410) 659-6322.

MASSACHUSETTS: State Board of Registration of Professional Engineers and of Land Surveyors, Room 1512, Leverett Saltonstall Building, 100 Cambridge Street, Boston 02202. Secretary, Ph: (617) 727-9957.

MICHIGAN: Board of Professional Engineers, P.O. Box 30018, (611 West Ottawa), Lansing 48909. Administrative Secretary, Telephone: (517) 335-1669.

MINNESOTA: State Board of Registration for Architects, Engineers, Land Surveyors and Landscape Architects, 133 7th St. E., St. Paul 55101. Executive Secretary, Ph: (612) 296-2388.

MISSISSIPPI: State Board of Registration for Professional Engineers and Land Surveyors P.O. Box 3, Jackson 39205. Executive Director, Telephone: (601) 359-6160.

MISSOURI: Board of Architects, Professional Engineers and Land Surveyors, P.O. Box 184, Jefferson City 65102. Executive Director, Telephone: (314) 751-0047.

MONTANA: State Board of Professional Engineers and Land Surveyors, Department of Commerce, 111 N. Jackson, P.O. Box 200513, Helena 59620-0513. Administrative Secretary, Telephone: (406) 444-4285.

NEBRASKA: State Board of Examiners for Professional Engineers and Architects, P.O. Box 94751, Lincoln 68509. Executive Director, Telephone: (402) 471-2021.

NEVADA: State Board of Registered Professional Engineers and Land Surveyors, 1755 East Plum Lane, Ste. 135, Reno 89502. Executive Secretary, Telephone: (702) 688-1231.

NEW HAMPSHIRE: State Board of Professional Engineers, 57 Regional Drive, Concord 03301. Executive Secretary, Telephone: (603) 271-2219.

NEW JERSEY: State Board of Professional Engineers and Land Surveyors, P.O. Box 45015, Newark 07101. Executive Secretary-Director, Telephone: (201) 504-6460.

NEW MEXICO: State Board of Registration for Professional Engineers and Land Surveyors, 1010 Marquez Pl., Santa Fe 87501. Secretary, Telephone: (505) 827-7561.

NEW YORK: State Board for Engineering and Land Surveying, The State Education Department, Cultural Education Center, Madison Avenue, Albany 12230. Executive Secretary, Telephone: (518) 474-3846.

NORTH CAROLINA: State Board of Professional Engineers and Land Surveyors, 3620 Six Forks Rd., Suite 300, Raleigh 27609. Executive Secretary, Telephone (919) 781-9499

NORTH DAKOTA: State Board of Registration for Professional Engineers and Land Surveyors, P.O. Box 1357, Bismarck 58502. Executive Secretary, Telephone: (701) 258-0786.

OHIO: State Board of Registration for Professional Engineers and Surveyors, 77 S. High St., 16th Fl., Columbus 43266-0314. Executive Secretary, Telephone: (614) 466-8999.

OKLAHOMA: State Board of Registration for Professional Engineers and Land Surveyors, Oklahoma Engineering Center, Room 120, 201 N.E. 27th Street, Oklahoma City, 73105. Executive Secretary, Telephone: (405) 521-2874.

OREGON: State Board of Engineering Examiners, Department of Commerce, 750 Front St., NE, Suite 240, Salem 97310. Executive Secretary, Telephone: (503) 378-4180.

PENNSYLVANIA: State Registration Board for Professional Engineers, P.O. Box 2649, Harrisburg 17120-264g. Administrative Assistant, Telephone: (800) 877-3926.

PUERTO RICO: Board of Examiners of Engineers, Architects, and Surveyors, Box 3271, San Juan 00904. Director, Examining Boards, Telephone: (809) 722-2122.

RHODE ISLAND: State Board of Registration for Professional Engineers and Land Surveyors, 10 Orms St., Suite 324, Providence 02904. Administrative Assistant, Telephone: (401) 277-2565.

SOUTH CAROLINA: State Board of Registration for Professional Engineers and Land Surveyors, P.O. Drawer 50408, Columbia 29250. Agency Director, Telephone: (803) 734-9166.

SOUTH DAKOTA: State Commission of Engineering, Architectural Examiners and Land Surveying, 2040 West Main Street, Suite 304, Rapid City 57702-2447. Executive Secretary, Telephone: (605) 394-2510.

TENNESSEE: State Board of Architectural and Engineering Examiners, 3rd Fl. Volunteer Plaza, 500 James Robertson Pkwy, Nashville 37243-1142. Administrator, Telephone: (615) 741-3221.

TEXAS: State Board of Registration for Professional Engineers, P.O.Drawer 18329, Austin 78760-8329. Executive Director, Telephone: (512) 440-7723.

UTAH: Division of Occupational and Professional Licensing, P.O. Box 45805, Salt Lake City 84145-0805. Director, Ph: (801) 530-6628.

VERMONT: State Board of Registration for Professional Engineers, Division of Licensing and Registration, Pavilion Building, Montpelier 05602. Executive Secretary, Telephone: (802) 828-2875.

VIRGINIA: State Board of Architects, Professional Engineers, Land Surveyors and Certified Landscape Architects, 3600 West Broad St., Seaboard Building, 5th Floor, Richmond 23230-4917. Assistant Director, Telephone: (804) 367-8506.

VIRGIN ISLANDS: Board for Architects, Engineers and Land Surveyors, Bldg 1, Sub-Base, Rm 205, St. Thomas 00801. Secretary, Telephone: (809) 774-3130.

WASHINGTON: State Board of Registration for Professional Engineers and Land Surveyors, P.O. Box 9649, Olympia 98504. Executive Secretary, Telephone: (206) 753-3634.

WEST VIRGINIA: State Board of Registration for Professional Engineers, 608 Union Building, Charleston 25301. Executive Director, Telephone: (304) 348-3554.

WISCONSIN: State Examining Board of Professional Engineers, P.O. Box 8935, Madison 53708. Administrator, Telephone: (608) 266-1397.

WYOMING: State Board of Examining Engineers, Herschler Building, Room 4135, Cheyenne 82002. Secretary-Accountant, Telephone: (307) 777-6156.

The D/S Review for Civil Engineering

edited by M. C. Potter

Overview of the CE/DS Topics

Chapters

What You Need to Know About the NCEES Handbook

Most of the problems on the Civil Engineering test of the afternoon session can be worked using pages 85 to 95 of the NCEES Handbook. Many of the problems, however, on the afternoon Civil Engineering test will come from the material in pages 1 to 81 of the NCEES Handbook, material used for the General tests of the morning and afternoon sessions. Since all engineers must take the General morning exam, it is assumed that you are familiar with that material. You must realize, however, that you will have to make reference to equations, tables, and charts located in those first 81 pages of the NCEES Handbook when you are working problems in the afternoon Civil Engineering session. The general engineering sections that you should be particularly familiar with are the sections on Mathematics, Statics, Mechanics of Materials, and Fluid Mechanics. A review of general engineering material is found in our *Fundamentals of Engineering*, a review book written specifically for the FE/EIT exam.

Do I Have to Take the DS Test?

If, after you study this Civil Engineering review section, and take the Civil Engineering Discipline Practice Exam, you feel that you could more easily pass the General exam of the afternoon session, you should register to take the afternoon General test. If you have already registered for the Civil Engineering afternoon test, you may contact your state board and request a change. Make sure your state board recognizes a *pass* on the General afternoon test for your particular discipline before you make that decision; the vast majority of the state boards will recognize a pass on any of the afternoon tests by any registrant.

This Review is Focused on the NCEES Handbook

In order to provide a very efficient, yet effective review, this DS Review has been designed to review only the material associated with the formulas and figures presented in the NCEES Handbook. The Handbook is the only material allowed in the exam. Thus the Handbook makes an efficient base to study around. It is assumed that you have a copy of the NCEES Handbook dated 1996 or later. If you do not, you can obtain one from Great Lakes Press by calling 1-800-837-0201.

GLP Offers a Concise Review for the A.M. Session & Other Study Aides

We also have an effective review for the General tests, both morning and afternoon, entitled *Fundamentals of Engineering*. It is used at hundreds of engineering schools across the nation. We also distribute the HP-48G and GX calculators, which are pre-programmed with many of the equations you'll need when working problems on the FE/EIT exam. Also, we've developed *Jump Start your HP-48G/X*, a 150-page book designed to efficiently help engineers get the most out of the HP48G/X. You can order these items by calling 1-800-837-0201.

1. Geotechnical Engineering

by Thomas F. Wolff

Geotechnical engineering may be subdivided into two broad areas of subject matter, each of which is often a separate undergraduate course. The first, *soil mechanics*, deals with the properties, classification, and stress-strain behavior (compressibility and strength) of soil materials and the movement of water through soils. The second, traditionally called *foundation engineering*, and more recently called *geotechnical design*, involves the application of the theory of soil mechanics to problems involving foundations, slopes and retaining structures.

The geotechnical "Definitions" cited on page 85 of the NCEES Reference Handbook are in some cases definitions and in some cases equations. They relate to commonly encountered geotechnical engineering problems; however, many of the variables used in the definitions are themselves not defined. Furthermore, some are valid only for special cases and cannot always be used for general cases. Hence, this chapter provides a brief review to define all variables and describe limitations of the equations.

1.1 Weight–Mass–Volume Problems

1.1.1 Definitions and Unit Conversions

Weight–volume and mass–volume problems involve determining quantities, ratios or proportions of soil components working from other known quantities, ratios or proportions. In an actual volume of soil, the solids, water and (if not saturated) air are arranged in a complex fashion as shown in Figure 1.1. To simplify visualization of the problem, a *phase diagram* may be drawn. In a phase diagram, shown in Figure 1.2, the solid particles are shown as a single contiguous mass at the bottom of the diagram. Above the solids is shown the water, and above that, the air. On the left side of the diagram are shown the total volume V, the volume V_s of solids, the volume V_w of water, the volume V_a of air, and the volume V_v of voids, which is the combined volume of water and air. On the right side are shown the total weight W, the weight W_S of solids, and the weight W_w of water. The weight of air is taken as zero.

Figure 1.1 The Soil-Water System

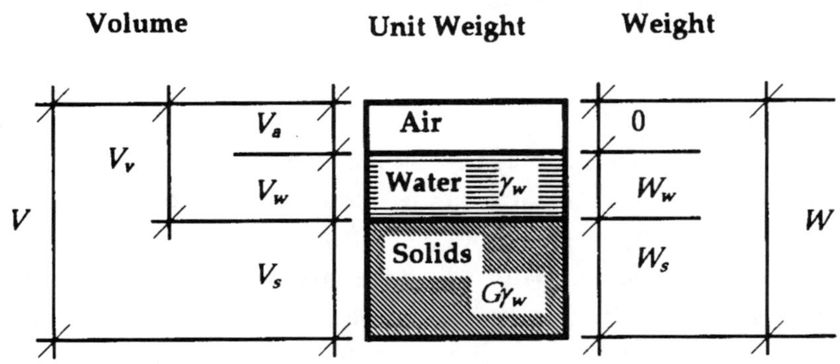

Figure 1.2 Phase Diagram

Observe that $V_v = V_w + V_a$, $V = V_s + V_v$, and $W = W_s + W_w$.

Relationships regarding soil and water volumes include the *void ratio e*, which is the ratio of the volume of voids to the volume of solids,

$$e = \frac{V_v}{V_s}$$

the *porosity* η, which is the ratio of the volume of voids to the total volume,

$$\eta = \frac{V_v}{V}$$

and the *degree of saturation S*, which is the ratio of the volume of water to the volume of voids and is usually expressed as a percentage,

$$S = \frac{V_w}{V_v} \times 100\%$$

The relationship involving soil and water weights is the *water content w* or *moisture content*, which is the ratio of the weight of water to the weight of solids,

$$w = \frac{W_w}{W_s} \times 100\%$$

Ratios of weights to volumes are termed *unit weights*. The *total unit weight* γ is the total weight of a soil mass (soil, water and air) divided by the total volume occupied, including the pore volume,

$$\gamma = \frac{W}{V}$$

The *saturated unit weight* γ_{sat} is the total unit weight obtained if the entire void volume were filled with water ($S = 100\%$ and $V_w = V_v$).

The *dry unit weight* γ_d is the ratio of the weight of solids to the total volume,

$$\gamma_d = \frac{W_s}{V}$$

The dry unit weight is to track the weight of soil solids in an entire occupied volume of solids and voids (e.g., the volume of a truck bed). For example,

$$W_s = \gamma_d V$$

The *specific gravity* G is the ratio of the unit weight of the soil solids to the unit weight of water,

$$G = \frac{\gamma_s}{\gamma_w}$$

hence, the *unit weight of solids* γ_s can be expressed as,

$$\gamma_s = G\gamma_w = \frac{W_s}{V_s}$$

For most of the minerals comprising soil solids, G is in the range 2.65 to 2.75; $G = 2.70$ is a common assumption when tests results are not available.

Problems involving SI units may be phrased in the context of masses and mass densities rather than weights and unit weights; however, unit weights must ultimately be obtained for stress calculations. Mass density is denoted by ρ. As weight W equals mass m times the acceleration of gravity g ($W = mg$), the unit weight values defined above are in fact the product of mass densities and g,

$$\gamma = \rho g$$
$$\gamma_d = \rho_d g$$

In the SI system, mass densities are commonly expressed as Mg/m^3, kg/m^3, or g/mL,

$$1 \text{ Mg/m}^3 = 1000 \text{ kg/m}^3 = 1 \text{ g/mL}$$

Recall the mass density and unit weight of water,

Mass density of water: $\rho_w = 1 \text{ Mg/m}^3 = 1000 \text{ kg/m}^3 = 1 \text{ g/mL}$

Unit weight of water: $\gamma_w = 9.807 \text{ MN/m}^3 = 9807 \text{ kN/m}^3 = 62.4 \text{ lb/ft}^3$

Example 1.1

Convert a total mass density of $\rho = 1900 \text{ kg/m}^3$ to a total unit weight for use in stress calculations.

(A) 19.0 kN/m^3

(B) 18.63 kN/m^3

(C) 1.90 kN/m^3

(D) 1.863 kN/m^3

Solution:

$$\gamma = \rho g = 1900 \text{ kg/m}^3 (9.807 \text{ m/s}^2) = 18\,630 \text{ kg/m}^2\text{s}^2$$
$$= 18\,630 \text{ N/m}^3 = 18.63 \text{ kN/m}^3.$$ The answer is B.

Example 1.2

Soil solids have a specific gravity of $G = 2.71$.

1. Find the unit weight of solids in SI units (kN/m^3).
   (A) 26.6 kN/m^3
   (B) 27.1 kN/m^3
   (C) 169.1 kN/m^3
   (D) 2.71 kN/m^3

2. Find the unit weight of solids in English units (lb/ft^3).
   (A) 26.6 lb/ft^3
   (B) 2.71 lb/ft^3
   (C) 169.1 lb/ft^3
   (D) 62.4 lb/ft^3

Solutions:

1. **A** $\gamma_s = G\gamma_w = 2.71(9.807\ kN/m^3) = 26.58\ kN/m^3$

2. **C** $\gamma_s = G\gamma_w = 2.71(62.4\ lb/ft^3) = 169.10\ lb/ft^3$

Example 1.3

A soil mass has a total unit weight of γ=130 lb/ft^3. Find the total unit weight in SI units (kN/m^3).
(A) 13.0 kN/m^3
(B) 4.04 kN/m^3
(C) 2.08 kN/m^3
(D) 20.4 kN/m^3

Solution:

Although a chain of conversion factors could be used, conversions can be simplified by taking the ratio of the unit weight of water in SI units to that in English units as an equality. (It's the same water). Thus,

$$\gamma = 130\ lb/ft^3\ [(9.807\ kN/m^3)/(62.4\ lb/ft^3)] = 20.43\ kN/m^3$$

The answer is D.

1.1.2 Solving Problems with Known Quantities

Although weight–volume problems can be worked directly with equations, errors are reduced and checks are provided if such problems are worked by drawing a phase diagram and performing the following steps.

- Fill in the known weights, volumes, and unit weights.
- Multiply known volumes by the corresponding unit weights to obtain weights. Divide known weights by the corresponding unit weight to obtain volumes. Use at least four significant digits to ensure that the results will balance when values are summed in the vertical direction and multiplied or divided in the horizontal direction.
- Where values of some relationships (γ, e, w , S, etc.) are given, use the definitions to obtain additional weights or volumes. Continue multiplication and division across the diagram and addition and subtraction up and down the sides until all weights and volumes are known and all quantities balance.
- Calculate the required values from the completed and checked diagram.

Example 1.4

A mold having a volume of 0.10 ft^3 was filled with moist soil. The weight of the soil in the mold was found to be 12.00 lb. The soil was oven dried and the weight after drying was 10.50 lb. The specific gravity of solids was known (or assumed) to be 2.70.

1. Find the water content.
   (A) 12.5%
   (B) 14.3%
   (C) 38.6%
   (D) 24.0%

2. Find the void ratio.
   (A) 0.604
   (B) 0.386
   (C) 0.143
   (D) 0.377

3. Find the porosity.
   (A) 0.604
   (B) 0.386
   (C) 0.143
   (D) 0.377

4. Find the degree of saturation.
   (A) 14.3%
   (B) 100%
   (C) 63.8%
   (D) 24.0%

5. Find the total unit weight.
   (A) 120 lb/ft^3
   (B) 105 lb/ft^3
   (C) 165 lb/ft^3
   (D) 62.4 lb/ft^3

6. Find the dry unit weight.
   (A) 120 lb/ft^3
   (B) 105 lb/ft^3
   (C) 165 lb/ft^3
   (D) 62.4 lb/ft^3

Solutions:

A phase diagram is drawn and the known quantities are entered (shown in bold). The remaining quantities are then calculated as follows:

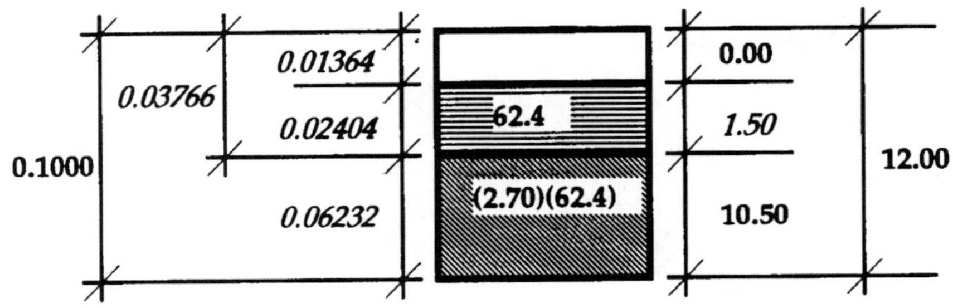

The weight and volume of water are, respectively,

$$W_w = W - W_s = 12.00 - 10.50 = 1.50 \text{ lb}$$

$$V_w = W_w / \gamma_w = 1.50 / 62.4 = 0.02404 \text{ ft}^3$$

The volume of solids is

$$V_s = \frac{W_s}{G\gamma_w} = \frac{10.50}{2.70 \times 62.40} = 0.06232 \text{ ft}^3$$

The volume of air is

$$V_a = V - V_w - V_s = 0.10000 - 0.02404 - 0.06232 = 0.01364 \text{ ft}^3$$

The volume of voids is

$$V_v = V_w + V_a = 0.02402 + 0.01364 = 0.03766 \text{ ft}^3$$

The phase diagram is now complete and all the desired solutions can be obtained:

1. **B** $w = W_w/W_s = 1.50 / 10.50 = 0.143 = 14.30\%$
2. **A** $e = V_v/V_s = 0.03766 / 0.06232 = 0.604$

3. **D** $\eta = V_v/V = 0.03766 / 0.1000 = 0.377$

4. **C** $S = (V_w/V_s)\ 100\% = (0.02404 / 0.03768)\ (100\%) = 63.8\%$

5. **A** $\gamma = W / V = 12.00 / 0.100 = 120\ \text{lb/ft}^3$

6. **B** $\gamma_s = Ws/V = 10.50 / 0.100 = 105\ \text{lb/ft}^3$

Example 1.5

The soil in Example 1.4 becomes saturated by accumulating water. Assuming it retains the same total volume,

1. Find the saturated water content.
   (A) 19.6%
   (B) 37.6%
   (C) 22.4%
   (D) 60.4%

2. Find the saturated unit weight.
   (A) 128.5 lb/ft^3
   (B) 105 lb/ft^3
   (C) 120 lb/ft^3
   (D) 165 lb/ft^3

3. Find the dry unit weight.
   (A) 128.5 lb/ft^3
   (B) 105 lb/ft^3
   (C) 165 lb/ft^3
   (D) 120 lb/ft^3

Solutions:

After saturation, $V_w = 0.03768\ \text{ft}^3$. Taking $W_w = V_w\gamma_w$ the new weight of water is (0.03768)(62.4) = 2.350 lb. The phase diagram is modified as shown and solutions follow.

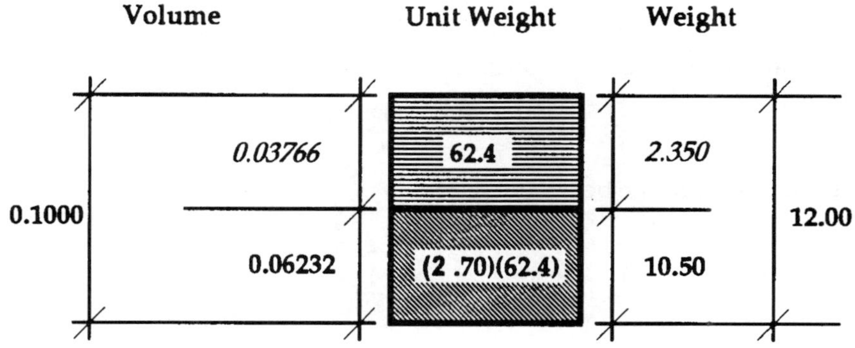

1. **C** $w = \frac{W_w}{W_s} = \frac{2.350}{10.50} = 0.2239$ or 22.4%

2. **A** $\gamma = \frac{W}{V} = \frac{10.50 + 2.35}{0.100} = 128.5\ \text{lb/ft}^3$

3. **B** $\gamma_d = \frac{W_s}{V} = \frac{10.50}{0.100} = 105.0\ \text{lb/ft}^3$

Note that the dry unit weight does not change with an addition of water at total constant volume.

1.1.3 Solving Problems Involving Only Relationships

For problems where only relationships (w, e, S, etc.) are given, only relationships may be calculated. The phase diagram approach may still be used by assuming a value for one weight or volume. While any quantity can be assumed, certain assumptions greatly simplify calculations:

- If w is given, assume $W_s = 1.00$ or 100.00 (lb or kg); then $W_w = w$ or $100w$.
- If e is given, assume $V_s = 1.000\ \text{ft}^3$ or m^3; then $V_v = e$.
- If γ is given, assume $V = 1.000\ \text{ft}^3$ or m^3; then $W = \gamma$.
- If γ_d is given, assume $V = 1.000\ \text{ft}^3$ or m^3; then $W_s = \gamma_d$.

It is recommended that weight-volume problems be solved using phase diagrams rather than only formulas; nevertheless, a very useful equation that relates four different relationships is

$$Se = wG$$

For saturated soils (S = 100%) this becomes

$$e = wG$$

The relationships between the void ratio and porosity are

$$e = \frac{\eta}{1-\eta}$$

and

$$\eta = \frac{e}{1+e}$$

The total unit weight can be expressed as:

$$\gamma = \frac{(G + Se) \times \gamma_w}{1+e}$$

The dry unit weight can be obtained as:

$$\gamma_d = \frac{G\gamma_w}{1+e}$$

Example 1.6

A soil has a water content of 20 percent, a void ratio of 0.800, and a specific gravity of 2.65.

1. Find the degree of saturation.
   (A) 80.0%
   (B) 20.0%
   (C) 66.3%
   (D) 53.0%

2. Find the porosity.
   (A) 0.444
   (B) 0.800
   (C) 0.530
   (D) 0.500

3. Find the total unit weight.
   (A) 165.4 lb/ft^3
   (B) 198.4 lb/ft^3
   (C) 110.2 lb/ft^3
   (D) 91.9 lb/ft^3

4. Find the dry unit weight.
   (A) 165.4 lb/ft^3
   (B) 91.9 lb/ft^3
   (C) 110.2 lb/ft^3
   (D) 198.4 lb/ft^3

Solutions:

Assume that the volume of solids V_s = 1.000 ft^3; the phase diagram is drawn as follows.

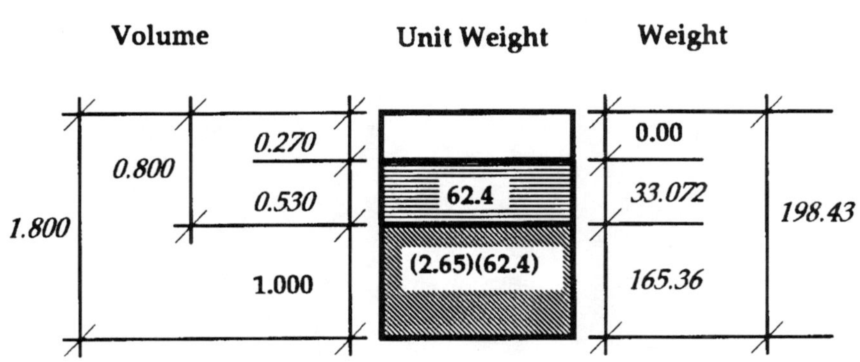

Remaining quantities are calculated as:

$$V_v = eV_s = (0.800)(1.000) = 0.800 \text{ ft}^3$$

$$W_s = V_s G\gamma_w = (1.000)(2.65)(62.4) = 165.36 \text{ lb}$$

$$W_w = wW_s = (0.20)(165.36) = 33.072 \text{ lb}$$

$$V_w = W_w/\gamma_w = (33.072)/(62.4) = 0.530 \text{ ft}^3$$

$$V_a = V_v - V_w = 0.800 - 0.530 = 0.270 \text{ ft}^3$$

At this point, the weights and volumes of all components are known for the assumed 1.000 ft^3 of solids and the required solutions can be calculated:

1. **C** $S = \dfrac{V_w}{V_v} \times 100\% = \dfrac{0.530}{0.800} \times 100\% = 66.3\%$

   Alternatively,

$$S = \frac{wG}{e} = \frac{(.20)(2.65)}{0.800} = 0.6625 \text{ or } 66.3\%$$

2. **A** $\eta = \dfrac{V_v}{V} = \dfrac{0.800}{1.000 + 0.800} = 0.444$

   Alternatively,

$$\eta = \frac{e}{1+e} = \frac{0.800}{1.800} = 0.444$$

3. **C** $\gamma = \dfrac{165.36 + 33.072}{1.800} = 110.2 \text{ lb/ft}^3$

   Alternatively,

$$\gamma = \frac{(G + Se)\gamma_w}{1+e}$$

$$= \frac{(2.65 + (0.6625 \times 0.800))62.4}{1.800} = 110.2 \text{ lb / ft}^3$$

4. **B** $\gamma_d = \dfrac{165.36}{1.800} = 91.9 \text{ lb/ft}^3$

   Alternatively,

$$\gamma_d = \frac{G\gamma_w}{1+e} = \frac{(2.65)(62.4)}{1.800} = 91.9 \text{ lb/ft}^3$$

1.2 Relative Density

The density of sands and gravels is often specified and measured in terms of the *relative density* D_d (the notation in the NCEES Handbook; note that many texts use D_r or I_D). The term is a misnomer as it deals with void ratio, not density. It scales the actual void ratio as a fraction of the range between the maximum and minimum void ratios:

$$D_d = \frac{e_{\max} - e}{e_{\max} - e_{\min}} \times 100\%$$

To determine the relative density using actual, minimum, and maximum dry unit weight values rather than void ratios, the above equation can be manipulated to obtain:

$$D_d = \frac{1/\gamma_{\min} - 1/\gamma_d}{1/\gamma_{\min} - 1/\gamma_{\max}} \times 100\%$$

Example 1.7

Tests are made to determine the maximum and minimum void ratios for a sand, and it is found that $e_{min} = 0.500$ and $e_{max} = 0.700$. A well-compacted sample in the field is found to have a void ratio of $e = 0.550$. What is relative density?

(A) 25%

(B) 55%

(C) 100%

(D) 75%

Solution:

$$D_d = \frac{e_{\max} - e}{e_{\max} - e_{\min}} \times 100\% = \frac{0.70 - 0.55}{0.70 - 0.50} \times 100\% = 75\%$$

The answer is **D**.

Example 1.8

Laboratory tests on a sand indicate a minimum dry unit weight of 16.0 kN/m^3 and a maximum dry unit weight of 18.0 kN/m^3. It is specified to be compacted to at least 60% relative density. What is the minimum acceptable unit weight that will pass the specifications?

(A) 17.20 kN/m^3

(B) 17.14 kN/m^3

(C) 10.80 kN/m^3

(D) 9.60 kN/m^3

Solution:

Substituting in $D_d = \dfrac{1/\gamma_{\min} - 1/\gamma_d}{1/\gamma_{\min} - 1/\gamma_{\max}} \times 100\%$,

$$60\% = \frac{0.06250 - 1/\gamma_d}{0.06250 - 0.05556} \times 100\%$$

$$1/\gamma_d = 0.0625 - 0.600 \times 0.00694 = 0.05834$$

$$\gamma_d = 17.14 \text{ kN/m}^3$$

Note that four or more significant digits must be retained to preserve accuracy as the problem involves a ratio of differences of inverses, a very numerically sensitive problem. The answer is **B**.

1.3 Grain-Size Characteristics of Soils

The *grain-size distribution* of sands and gravels is determined by a *sieve analysis* (or *mechanical analysis*). Opening sizes for commonly used standard sieves are shown in Table 1.1. A soil sample is passed over the sieves and the sieves are shaken until all particles have passed down to the sieve that retains them. The cumulative weight of all material larger than each opening size is divided by the total sample weight to obtain the *percentage retained* or the *percent coarser;* the cumulative weight of all material smaller than that size divided by the total sample weight is the *percentage passing* or the *percent finer.* The results are plotted as the percent finer (linear scale) versus the sieve opening or grain-size (log scale). The plot is referred to as a *grain size curve,* examples of which are presented in Figure 1.3.

Table 1.1 Opening Sizes of Standard Sieves

	Sieve Size	Opening
	1.5 in	37.5 mm
	1	25
	0.75	19
	0.5	12.5
No.:	4	4.75 mm
	10	2.00
	20	0.850
	40	0.425
	70	0.212
	100	0.150
	200	0.075

Certain sizes of interest may be read from the grain size curve. The D_{10} size, sometimes called the *effective size,* is the grain diameter for which 10 percent of the sample (by weight) is finer. The D_{50} size, called the *median grain size,* is the grain diameter for which 50 percent of the sample (by weight) is finer. In general, the notation D_{xx} refers to the grain diameter for which *xx* percent of the sample is finer by weight.

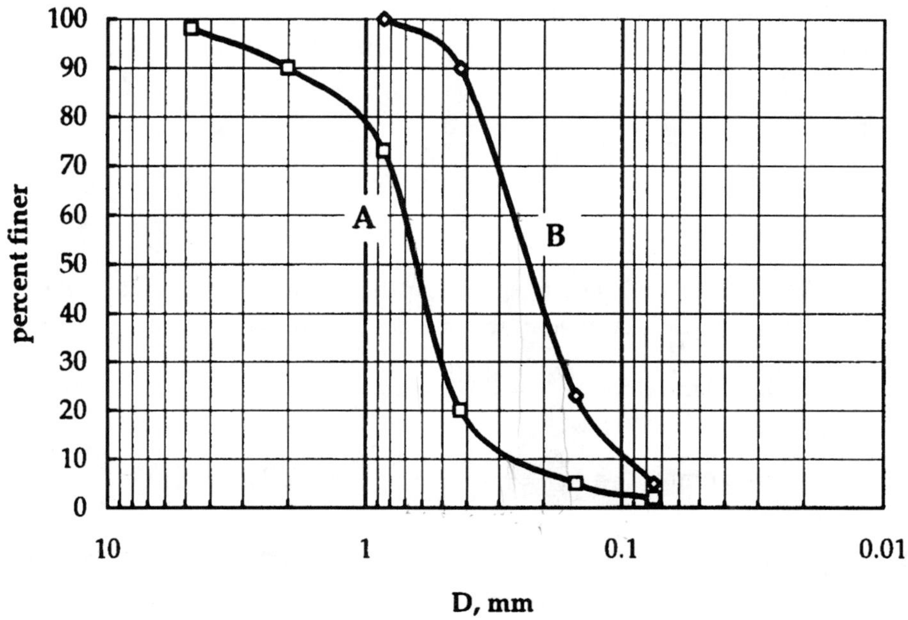

Figure 1.3 Grain Size Curves

Two parameters are used to describe the shape of the grain-size curve. The *coefficient of uniformity* c_u is:

$$c_u = \frac{D_{60}}{D_{10}}$$

It provides a measure of the slope of the curve. The *coefficient of curvature* or *coefficient of gradation* c_c is:

$$c_c = \frac{D_{30}^{\ 2}}{D_{60}D_{10}}$$

It provides a measure of the smoothness of the curve.

Example 1.9

For the grain size curve A in Figure 1.3,

1. Determine the effective grain size, D_{10}.
   (A) 0.12 mm
   (B) 0.41 mm
   (C) 10.0 mm
   (D) 0.28 mm

2. Determine the coefficient of uniformity.
   (A) 2.5
   (B) 0.40
   (C) 0.70
   (D) 25

3. Determine the coefficient of curvature.
   (A) 2.5
   (B) 0.75
   (C) 1.33
   (D) 0.51

Solutions:

1. **D** Start at the 10% passing point on the vertical axis, move horizontally across to Curve A, then down to read $D_{10} = 0.28$ mm

2. **A** $D_{60} = 0.70$ mm, $c_u = \dfrac{D_{60}}{D_{10}} = \dfrac{0.70}{0.28} = 2.5$

3. **C** $D_{30} = 0.51$ mm, $c_c = \dfrac{D_{30}^2}{D_{60}D_{10}} = \dfrac{0.51^2}{(0.70)(0.28)} = 1.33$

Example 1.10

Refer to the grain size curve B in Figure 1.3.

1. Determine the effective grain size, D_{10}.
   (A) 0.019 mm
   (B) 0.27 mm
   (C) 1 mm
   (D) 0.095 mm

2. Determine the coefficient of uniformity.
   (A) 2.50
   (B) 2.84
   (C) 0.35
   (D) 0.27

3. Determine the coefficient of curvature.
   (A) 1.13
   (B) 0.17
   (C) 0.88
   (D) 5.88

Solutions:

1. **D** Start at the 10% passing point on the vertical axis, move horizontally across to Curve B, then down to read $D_{10} = 0.095$ mm

2. **B** $D_{60} = 0.27$ mm, $c_u = \dfrac{D_{60}}{D_{10}} = \dfrac{0.27}{0.095} = 2.84$

3. **A** $D_{30} = 0.17$ mm, $c_c = \dfrac{D_{30}^2}{D_{60}D_{10}} = \dfrac{0.17^2}{(0.27)(0.095)} = 1.13$

1.4 Atterberg Limits and Plasticity

Atterberg limits are specific water contents at which soil behavior changes. The upper limit of plasticity, above which soil behaves as a liquid, and below which soil behaves as a plastic solid, is the *liquid limit,* denoted LL (or w_L). The lower limit of plastic behavior, above which the soil behaves as a plastic solid, and below which the soil behaves as a brittle solid, is the *plastic limit,* denoted PL (or w_P). The water content below which the soil no longer reduces in volume with a reduction in water content is the shrinkage limit SL or w_S. Although Atterberg limits are water contents (decimals or percentages), they are by convention expressed as integers values (implying a percentage).

The *plasticity index,* denoted PI (or I_P) is the difference of the liquid limit and the plastic limit:

$$PI = LL - PL$$

The *liquidity index,* denoted LI, is a measure of the natural water content relative to the plastic limit and the liquid limit:

$$LI = \frac{w - PL}{LL - PL}$$

The shrinkage index, denoted SI, is the difference of the plastic limit and the shrinkage limit:

$$SI = PL - SL$$

Example 1.11

A soil has a natural water content of 40%. Its liquid limit is 50, the plastic limit is 30 and the shrinkage limit is 20.

1. Determine the plasticity index.
   (A) 20
   (B) 30
   (C) 10
   (D) 80

2. Determine the liquidity index.
   (A) 2.00
   (B) 0.75
   (C) 0.50
   (D) 0.30

3. Determine the shrinkage index.
   (A) 20
   (B) 30
   (C) 80
   (D) 10

Solutions:

1. **A** $PI = LL - PL = 50 - 30 = 20$

2. **C** $LI = \dfrac{w - PL}{LL - PL} = \dfrac{40 - 30}{50 - 30} = 0.50$

3. **D** $SI = PL - SL = 30 - 20 = 10$

1.5 Stresses in Soil

Normal stresses (σ) in a soil mass are carried partly by the soil skeleton as effective stress (σ') and partly by pore water pressure (u). Shear stresses (τ) must be carried entirely by the soil skeleton since water cannot sustain shear stress. Terzaghi's *effective stress principle* states that:

$$\sigma = \sigma' + u$$

The vertical normal stress σ'_v on a horizontal plane may be calculated by dividing the overlying total weight of the soil by the area over which the soil acts:

$$\sigma = \frac{W}{A}$$

The NCEES Handbook uses the general equation $\sigma = P/A$. As the weight of soil is the unit weight times the volume, or $W = \gamma h_s A$, the above equation can be written

$$\sigma = \Sigma\, \gamma_s h_s$$

where the summation sign denotes that the unit weight and height of each soil layer must be treated separately.

The pore pressure, in conditions of a static water table, is the unit weight of water times the height of water above the plane on which stresses are calculated:

$$u = \gamma_w h_w$$

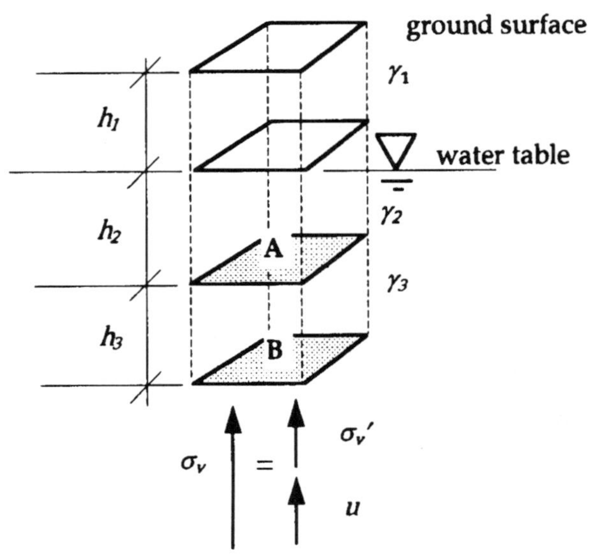

Figure 1.4 Column Through Soil Mass

Example 1.12

In Figure 1.4, assume $h_1 = 3$ ft, $\gamma_1 = \gamma_{moist} = 110$ lb/ft^3, $h_2 = 4$ ft, $\gamma_2 = \gamma_{sat} = 120$ lb/ft^3, $h_3 = 3$ ft, $\gamma_3 = \gamma_{sat} = 125$ lb/ft^3.

1. Determine the total vertical stress on plane A.
   (A) 770 lb/ft^2
   (B) 810 lb/ft^2
   (C) 840 lb/ft^2
   (D) 230 lb/ft^2

2. Determine the pore pressure on plane A.
   (A) 437 lb/ft^2
   (B) 187 lb/ft^2
   (C) 250 lb/ft^2
   (D) 62.4 lb/ft^2

3. Determine the effective vertical stress on plane A.
   (A) 560 lb/ft^2
   (B) 373 lb/ft^2
   (C) 623 lb/ft^2
   (D) 810 lb/ft^2

4. Determine the total vertical stress on plane B.
   (A) 1250 lb/ft^2
   (B) 1100 lb/ft^2
   (C) 1200 lb/ft^2
   (D) 1185 lb/ft^2

5. Determine the pore pressure on plane B.
   (A) 624 lb/ft^2
   (B) 437 lb/ft^2
   (C) 62.4 lb/ft^2
   (D) 187 lb/ft^2

6. Determine the effective vertical stress on plane B.
   (A) 561 lb/ft^2
   (B) 1185 lb/ft^2
   (C) 748 lb/ft^2
   (D) 626 lb/ft^2

Solutions:

1. **B** $\sigma_v = \Sigma\, \gamma_s\, h_s = (110)(3) + (120)(4) = 810\ \text{lb/ft}^2$
2. **C** The water table is 4 ft above the plane, hence, $u = \gamma_w h_w = (62.4)(4.0) = 249.6\ \text{lb/ft}^2$
3. **A** $\sigma'_v = \sigma_v - u = 810 - 249.6 = 560.4\ \text{lb/ft}^2$
4. **D** $\sigma_v = \Sigma\, \gamma_s\, h_s = (110)(3) + (120)(4) + (125)(3) = 1185\ \text{lb/ft}^2$
5. **B** $u = \gamma_w h_w = (62.4)(4.0 + 3.0) = 436.8\ \text{lb/ft}^2$
6. **C** $\sigma'_v = \sigma_v - u = 1185 - 436.8 = 748.2\ \text{lb/ft}^2$

Example 1.13

In Figure 1.4, assume $h_1 = 2$ m, $\gamma_1 = \gamma_{moist} = 18\ \text{kN/m}^3$, $h_2 = 3$ m, $\gamma_2 = \gamma_{sat} = 19\ \text{kN/m}^3$, $h_3 = 2$ m, $\gamma_3 = \gamma_{sat} = 20\ \text{kN/m}^3$.

1. Determine the total vertical stress on plane A.
   (A) 95 kN/m²
   (B) 90 kN/m²
   (C) 93 kN/m²
   (D) 18 kN/m²

2. Determine the pore pressure on plane A.
   (A) 29.4 kN/m²
   (B) 49.1 kN/m²
   (C) 9.8 kN/m²
   (D) 93.0 kN/m²

3. Determine the effective vertical stress on plane A.
   (A) 93.0 kN/m²
   (B) 43.9 kN/m²
   (C) 0 kN/m²
   (D) 63.6 kN/m²

4. Determine the total vertical stress on plane B.
   (A) 140 kN/m²
   (B) 133 kN/m²
   (C) 126 kN/m²
   (D) 20 kN/m²

5. Determine the pore pressure on plane B.
   (A) 68.7 kN/m^2
   (B) 19.6 kN/m^2
   (C) 49.1 kN/m^2
   (D) 9.8 kN/m^2

6. Determine the effective vertical stress on plane B.
   (A) 83.9 kN/m^2
   (B) 64.3 kN/m^2
   (C) 113.4 kN/m^2
   (D) 71.3 kN/m^2

Solutions:

1. **C** $\sigma_v = \Sigma\, \gamma_s h_s = (18)(2) + (19)(3) = 93.0\ kN/m^2$ (or 93.0 kPa)
2. **A** The water table is 3 m above the plane, hence,
$$u = \gamma_w h_w = (9.81)(3.0) = 29.4\ kN/m^2$$
3. **D** $\sigma'_v = \sigma_v - u = 93.0 - 29.4 = 63.6\ kN/m^2$
4. **B** $\sigma_v = \Sigma\, \gamma_s h_s = (18)(2) + (19)(3) + (20)(2) = 133\ kN/m^2$
5. **C** $u = \gamma_w h_w = (9.81)(3.0 + 2.0) = 49.1\ kN/m^2$
6. **A** $\sigma'_v = \sigma_v - u = 133.0 - 49.1 = 83.9\ kN/m^2$

1.6 Water Flow in Soil

1.6.1 Flow through Uniform Sections

The flow rate Q through a soil mass is given by *Darcy's Law*:

$$Q = vA = kiA$$

where Q is the flow (volume/time), v is the discharge velocity, k is the *coefficient of permeability* (length/time), i is the *hydraulic gradient*, and A is the cross-sectional area. The hydraulic gradient is the total head loss divided by the length of flow path over which that head loss occurs. Rearranging, the coefficient of permeability can be measured by creating a known gradient through a sample of known cross sectional area, and measuring the flow:

$$k = \frac{Q}{iA}$$

Example 1.14

The tube shown is circular with a uniform diameter D and contains a soil sample of length L. A constant head loss H is maintained across the sample by pouring in water at point A, and permitting water to overflow at point B.

Given: $D = 2$ cm, $H = 10$ cm, $L = 20$ cm, and $k = 0.1$ cm/s.

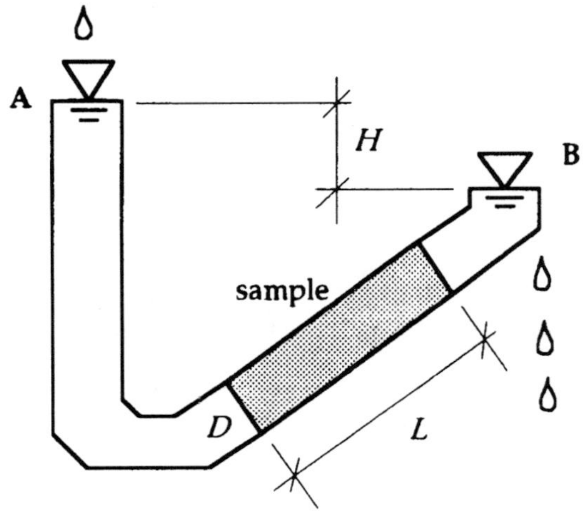

1. What is the cross-sectional area of the tube?
   (A) 12.57 cm^2
   (B) 3.14 cm^2
   (C) 4.0 cm^2
   (D) 2.0 cm^2

2. What is the gradient?
   (A) 0.5
   (B) 1.0
   (C) 10.0
   (D) 2.0

3. What is the discharge velocity?
   (A) 0.1 cm/s
   (B) 20 cm/s
   (C) 0.25 cm/s
   (D) 0.05 cm/s

4. What is the flow rate?
   (A) 0.629 cm^3/s
   (B) 0.314 cm^3/s
   (C) 0.157 cm^3/s
   (D) 0.079 cm^3/s

5. How long will it take to pass 1 liter of water through the sample?
   (A) 0.157 s
   (B) 1.77 hr
   (C) 1.77 min
   (D) 1.77 day

Solutions:

1. **B** $A = \pi D^2/4 = (3.14)(2)^2/4 = 3.14\ \text{cm}^2$
2. **A** $i = H/L = 10/20 = 0.5$
3. **D** $v = ki = (0.1)(0.5) = 0.05\ \text{cm/s}$
4. **C** $Q = kiA = (0.1)(0.5)(3.14) = 0.157\ \text{cm}^3\text{/s}$
5. **B** $t = (1000\ \text{cm}^3)\ /(0.157\ \text{cm}^3\text{/s}) = 6369\ \text{sec} \approx 1.77\ \text{hr}$

Example 1.15

Refer to the figure used in the previous problem. Given: $D = 5$ cm, $L = 20$ cm, $H = 40$ cm, and $Q = 1.96\ \text{cm}^3\text{/sec}$.

1. What is the cross-sectional area of the tube?
   (A) 78.54 cm^2
   (B) 25 cm^2
   (C) 19.64 cm^2
   (D) 5 cm^2

2. What is the gradient?
   (A) 0.5
   (B) 2
   (C) 1.0
   (D) 4.0

3. What is the coefficient of permeability?
   (A) 0.05 cm/s
   (B) 0.10 cm/s
   (C) 0.025 cm/s
   (D) 1.96 cm/s

4. What is discharge velocity?
   (A) 0.2 cm/s
   (B) 0.1 cm/s
   (C) 0.4 cm/s
   (D) 1.96 cm/s

Solutions:

1. **C** $A = \pi D^2/4 = (3.14)(5)^2/4 = 19.64 \text{ cm}^2$
2. **B** $i = H/L = 40/20 = 2.0$
3. **A** $k = Q/(iA) = 1.96/[(2)(19.64)] = 0.05 \text{ cm/s}$
4. **B** $v = ki = (Q/iA)(i) = Q/A = 1.96/19.64 = 0.1 \text{ cm/s}$

1.6.2 Flow through Non-uniform Sections

For nonuniform flow regimes, flows and gradients can be determined from a *flow net,* a graphical solution of Laplace's equation obtained by trial-and-error sketching or computer solution. A flow net consists of a family of *equipotential lines,* which are contours of equal total head, and *flow lines,* which are curves describing selected flow paths of water particles. If a flow net is constructed for a homogeneous, isotropic flow regime such that the equipotiential lines and flow lines intersect at right angles and form curvilinear squares, the flow through a unit depth (into the paper) cross section can be expressed as:

$$Q = kH\frac{N_f}{N_e}$$

where H is the total head loss across the section, N_f is the number of flow channels, and N_e is the number of equipotential drops.

Example 1.16

A sheetpile is driven in a shallow lake and the area to one side is pumped down as shown. The foundation is sand underlain by impervious rock. A flow net is drawn as shown. $H = 5$ ft and $k = 0.2$ ft/min.

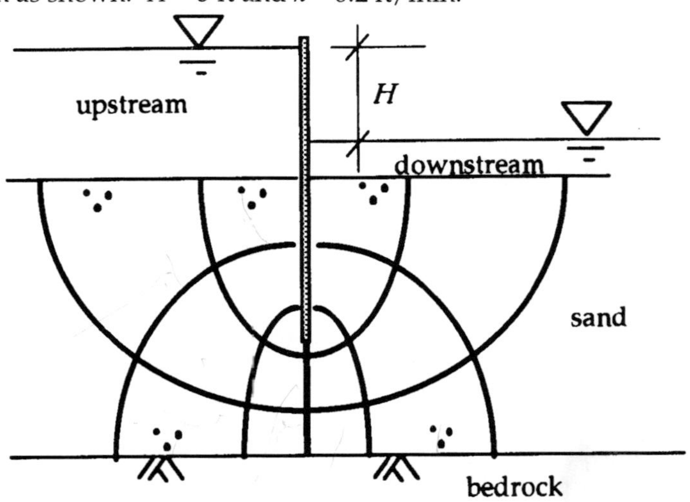

1. What is the flow per lineal foot of sheetpile?
   (A) 0.5 ft^3/min
   (B) 1.0 ft^3min
   (C) 2.0 ft^3/min
   (D) 1.5 ft^3/min

2. If the sheetpile and the excavation are 50 ft long (perpendicular to the paper), what capacity of pump is necessary to handle the flow?
   (A) 25 ft^3/min
   (B) 5 ft^3/min
   (C) 50 ft^3/min
   (D) 100 ft^3/min

3. If the sheetpile is extended upward to permit doubling H, what happens to the total flow?
   (A) it drops by half
   (B) it doubles
   (C) it stays the same
   (D) it increases by 41%

4. If the sheetpile is extended downward to cut through more of the sand, what happens to the flow?
   (A) it increases
   (B) it decreases
   (C) it stays the same
   (D) solution cannot be determined.

Solutions:

The number of flow channels N_f is three: one is between the sheet pile and the upper drawn flow line, the second is between the two drawn flow lines, and the third is between the lower drawn flow line and the bedrock. As water flows along any flow line from upstream to downstream, it traverses 6 squares or equipotential drops; $N_e = 6$.

1. **A** $Q = kH(N_f/N_e) = (0.2)(5)(3/6) = 0.5\ ft^2/min = 0.5\ ft^3/min/ft$ of length.

2. **A** $Q_{total} = Q \times$ length $= (0.5\ ft^3/min/ft)(50\ ft) = 25\ ft^3/min$.

3. **B** As the flow equation is linear in H, doubling H doubles the flow.

4. **B** For any constant number of flow lines N_f, deepening the sheet pile increase the resistance to flow, increasing N_e and decreasing the flow.

1.7 Settlement of Saturated Clay

1.7.1 Void Ratio — Pressure Relationship

Increasing the effective stress on a soil causes a volume reduction as soil solids are pushed closer together and water is expelled from the pores. For low-permeability soils such as clay, the volume change is time-dependent and the process is termed *consolidation*. The parameters used to calculate the rate and amount of consolidation can be measured in a *consolidation test*. A cylindrical sample of soil is confined in a rigid ring and submerged in water with a porous stone above and below the soil. A load is applied and the reduction in soil height is measured as a function of time. When compression has essentially stopped, another load is applied and the process is repeated.

The test results are analyzed by plotting the final void ratio e under each load on a linear scale versus the applied pressure p (actually an effective vertical stress σ'_v) plotted on a log scale. The plot, often termed an *e-log p curve*, is shown in Figure 1.5.

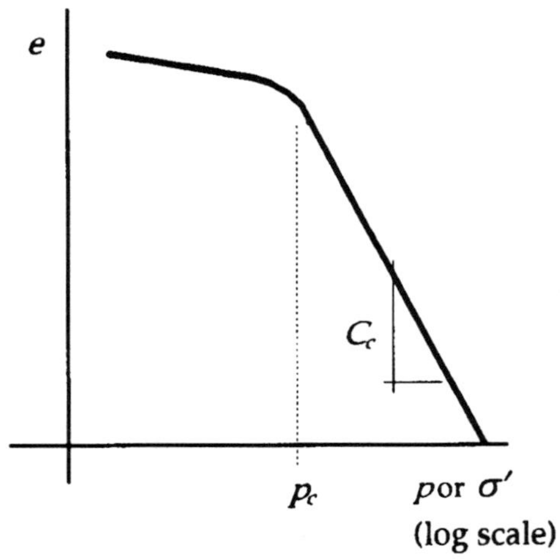

Figure 1.5 e **- log** p **Consolidation Curve**

The *preconsolidation pressure* p_c, or maximum effective stress the soil has been subjected to in the past, corresponds to the break in the curve. At pressures below the preconsolidation pressure, the curve has a relatively flat slope, and is termed the *recompression curve*. The soil in this stress range is *overconsolidated*. Above the preconsolidation pressure, the curve is steeper and is referred to as the *virgin consolidation curve*. In this range, the soil is said to be *normally consolidated*.

The virgin curve can be approximated as a straight line with a slope of C_c, termed the *compression index*. The compression index is

$$C_c = \frac{\Delta e}{\Delta \log_{10} p} = \frac{e_2 - e_1}{\log_{10} p_2 - \log_{10} p_1}$$

Although the compression index is always negative, the sign is not usually shown. The *recompression index* C_r can be calculated in the same manner using two points on the recompression curve.

The compression index of normally consolidated clays can be estimated from

$$C_c = 0.009(LL - 10)$$

where LL is the liquid limit expressed as a percentage.

Example 1.17

A normally consolidated clay layer has a liquid limit of 45. The effective stress at the midpoint of the layer is 2000 lb/ft^2.

1. Estimate the compression index.
   (A) 0.405
   (B) 0.004
   (C) 0.315
   (D) 0.003

2. Placement of a sand fill on the surface over a wide area causes a vertical stress increase of 800 lb/ft^2. How much will the void ratio at the midpoint of the layer decrease due to the stress increase?
   (A) 0.141
   (B) 0.125
   (C) 1.160
   (D) 0.315

Solutions:

1. **C** $C_c = 0.009(45\text{–}10) = 0.315$
2. **A** $\Delta e = C_c\,(\Delta \log_{10} p) = 0.315\,[(\log_{10} 2800 - \log_{10} 2000)]$
   $= 0.315 \log_{10}(2800/2000) = 0.141$

Example 1.18

The pressure is doubled on a soil for which $C_c = 0.30$ and the initial void ratio e_i is 0.800 (note: the NCEES Handbook uses e_i; most texts use e_o). What is the void ratio after the soil consolidates under the pressure increase?
   (A) 0.090
   (B) 0.890
   (C) 0.830
   (D) 0.710

Solution:

1. **D** $\Delta e = C_c\,(\Delta \log_{10} p) = 0.30\,(\log_{10} 2p - \log_{10} p) = 0.30 \times \log_{10} 2 = 0.090$
   $e_{final} = e_i - \Delta e = 0.800 - 0.090 = 0.710$

1.7.2 Magnitude of Settlement

The settlement of a clay layer can be calculated by determining the vertical strain ε_v as a function of depth and integrating the strain with respect to depth. As shown in Figure 1.6, if consolidation causes a void ratio reduction of Δe, the vertical strain ε_v is

$$\varepsilon_v = \frac{\Delta e}{1+e_i}$$

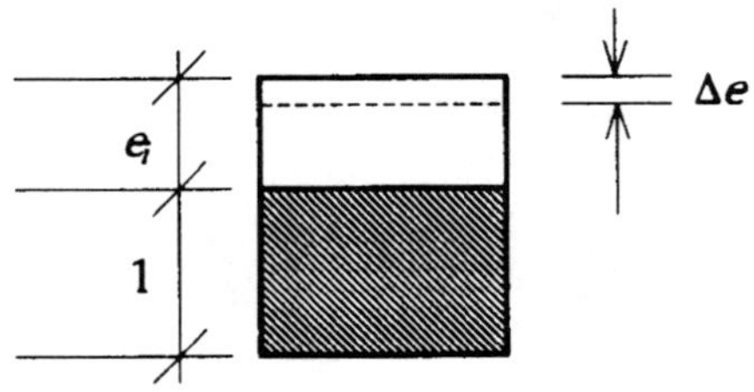

Figure 1.6 Void Ratio Reduction due to Consolidation

For a sufficiently thin layer of thickness H, the strain may be nearly constant, and the settlement ΔH can be calculated as the strain times the layer thickness:

$$\Delta H = H\varepsilon_v$$

Substituting in the previous equations for Δe, the following expression, found in the NCEES Handbook, is obtained:

$$\Delta H = H\frac{C_c}{1+e_i}\log_{10}\frac{p_i+\Delta p}{p_i}$$

Note that this equation is accurate only when

- the soil is normally consolidated, (consolidation occurs along the virgin curve, with constant slope C_c)
- the layer thickness H is sufficiently small such that the initial pressure p_i and the pressure change Δp are good averages over the layer

Example 1.19

A 5 ft thick normally consolidated clay layer is located between two sand layers. Assume that e_i = 0.900 and C_c = 0.35. A surface load increases the existing average effective stress in the layer from 1000 lb/ft^2 by an additional 500 lb/ft^2.

1. Calculate the strain in the clay layer.
   (A) 0.0620
   (B) 0.0326
   (C) 0.1630
   (D) 0.3500

2. Calculate the settlement of the clay layer.
   (A) 1.96 in
   (B) 0.196 in
   (C) 0.033 in
   (D) 0.263 in

Solutions:

1. **B** $\Delta e = C_c\,(\Delta \log_{10} p) = 0.35\,(\log_{10} 1500 - \log_{10} 1000)$

   $= 0.35 \log_{10}(1500/1000) = 0.062$

   $\varepsilon_v = \Delta e/(1+e_i) = 0.062/(1.900) = 0.0326$

2. **A** $\Delta H = H\,\varepsilon_v = 5(0.0326) = 0.163 \text{ ft} = 1.96 \text{ in}$

   Note that this is the same solution that would be obtained using the longer, combined equation above; it has merely been broken down to show Δe and ε_v.

1.7.3 Time Rate of Settlement

Applying a stress increase to a clay soil induces excess pore pressures, which dissipate with time in accordance with the theory of consolidation. For one-dimensional consolidation of a uniform clay layer, the average percent consolidation $U_{avg}(t)$ which has occurred by time t can be expressed as

$$U_{avg}(t) = f(T)$$

where

$$T = \frac{C_v t}{H_{dr}^2}$$

and

- T is a non-dimensional time factor from which $U_{avg}(t)$ can be determined.
- C_v is the coefficient of consolidation, with units of length^2/time.
- H_{dr} is the maximum length of drainage path in the clay layer. For a *doubly drained* layer (e.g., sand above and below the clay), it is one-half the clay layer thickness. For a *singly drained* layer (e.g., impervious shale under the clay), it is the same as the clay layer thickness. The subscript is used to emphasize that H_{dr} is not always equal to the layer thickness H (the NCEES Handbook uses the notation H in both cases).

The relationship $U_{avg} = f(t)$ is tabulated in Table 1.2.

Table 1.2 Relationship between U_{avg} and T

U_{avg}, %	T	U_{avg}, %	T
0 %	0.000	70	0.403
10	0.008	75	0.477
20	0.031	80	0.567
30	0.071	85	0.684
40	0.126	90	0.848
50	0.197	95	1.129
55	0.238	99	1.781
60	0.287	100	∞
65	0.340		

The settlement $S(t)$ at any time t is equal to the total calculated settlement multiplied by the degree of consolidation at time t.

Example 1.20

A 10 ft thick clay layer is drained by sand layers at the top and bottom. It is predicted to settle a total of 2.0 inches under the stress increase due to a foundation load. The coefficient of consolidation is 0.03 ft^2/day.

1. The time required for 50% of the total settlement to occur is most nearly:
   (A) 640 days
   (B) 320 days
   (C) 160 days
   (D) 80 days

2. The time required for 90% of the total settlement to occur is most nearly:
   (A) 700 days
   (B) 2800 days
   (C) 1400 days
   (D) 350 days

3. Determine the settlement at the end of one year.
   (A) 0.73 in
   (B) 1.45 in
   (C) 2.00 in
   (D) 0.37 in

Solutions:

1. **C** As the 10 ft thick clay layer is bounded by a sand deposit on top and bottom, it is doubly drained and $H_{dr} = 10/2 = 5$ ft. For 50% consolidation, $T = 0.197$. Hence, $t_{50} = T_{50}H_{dr}^2/C_v = (0.197)(5^2)/0.03 = 164$ days.
2. **A** $t_{90} = T_{90}H_{dr}^2/C_v = (0.848)(5^2)/0.03 = 706$ days.
3. **B** $T = C_v t/H_{dr}^2 = (0.03)(365)/5^2 = 0.438$. For this value, $U_{avg} = 72.5$ %. $S(t) = 72.5\%(2 \text{ in}) = 1.45$ in.

1.8 Shear Strength of Soils

1.8.1 The Direct Shear Test

The shear strength of soil τ_f on any plane is a function of the normal stress on the plane. A simple test to measure shear strength is the *direct shear test,* illustrated on the left side of Figure 1.7. The soil is placed in a split box, and a vertical normal force V is applied. The horizontal force H needed to shear the soil is measured. If V and H are divided by the sample area A the normal stress σ and the shear stress τ_f are obtained. Plotting τ_f vs σ for samples tested at several normal stresses, as shown on the right of the figure, provides the strength parameters c and ϕ. The parameter c is commonly termed the *cohesion* and the parameter ϕ is termed the *angle of internal friction,* or simply the *friction angle.*

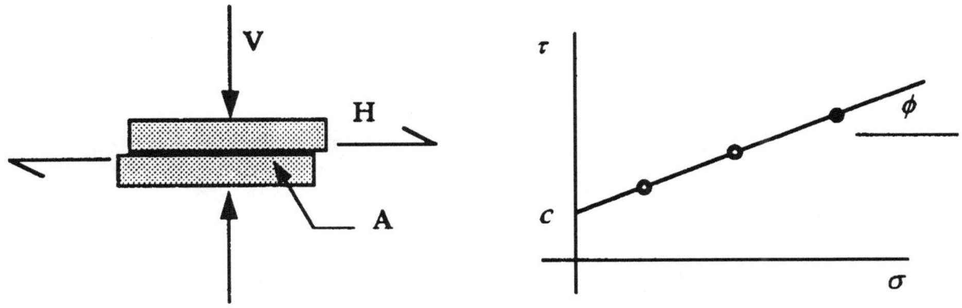

Figure 1.7 Direct Shear Test

The strength of soil along the failure plane is represented by the *Mohr-Coulomb strength equation*:

$$\tau_f = c + \sigma \tan \phi$$

where σ is the normal stress on the failure plane. Soil strength parameters may be defined in terms of *total stresses,* as written above, or in terms *of effective stresses,* as

$$\tau_f = c' + \sigma' \tan \phi'$$

The total stress representation is often used for short-term or undrained loading of saturated clays, for which $\phi = 0$, hence

$$\tau_f = c$$

For cohesionless sands, $c = 0$, and drained shear conditions (modeled with effective stress parameters) will usually prevail due to the relatively high permeability, hence:

$$\tau_f = \sigma' \tan \phi'$$

Nevertheless, one may often find the equation above written without the prime superscripts. For example, for dry sands, the total and effective stresses are the same.

Example 1.21

A sled with a rough bottom is loaded to a weight of 200 lb and pulled across a beach. It takes a horizontal force of 125 lb to pull the sled. What are the strength parameters of the beach sand?

(A) $c = 125\ \text{lb/ft}^2$, $\phi = 0$
(B) $c = 0$, $\phi = 32°$
(C) $c = 125\ \text{lb/ft}^2$, $\phi = 32°$
(D) $c = 200\ \text{lb/ft}^2$, $\phi = 32°$

Solution:

As the material is sand, $c = 0$. The shear stress on the base of the sled is $\tau = 125/A$, where A is the area of the sled. The normal stress on the base of the sled is $\sigma = 200/A$. Applying the Mohr-Coulomb equation:

$$\tau_f = c + \sigma \tan \phi$$

$$\frac{125}{A} = 0 + \frac{200}{A} \tan \phi$$

$$\therefore \tan \phi = 125/200 = 0.625 \quad \text{and} \quad \phi = 32°$$

The answer is **B**.

Example 1.22

A sand for which $\phi = 35°$ is tested in direct shear in a 3 inch by 3 inch square shear box under a normal load of 200 lb.

1. Determine the normal stress on the sample.
   (A) 200 psi
   (B) 22.2 psi
   (C) 28.3 psi
   (D) 35 psi

2. Determine the shear stress on the sample at failure.
   (A) 15.6 psi
   (B) 22.2 psi
   (C) 70.0 psi
   (D) 28.3 psi

3. Determine the shear force on the sample at failure.
   (A) 22.2 lb
   (B) 70 lb
   (C) 200 lb
   (D) 140 lb

Solutions:

1. **B** $\sigma = P/A = 200/(3.0)^2 = 22.2$ psi

2. **A** $\tau_f = c + \sigma \tan \phi = 0 + 22.2 \tan 35° = 15.6$ psi

3. **D** $H = \tau A = (15.6)(9) = 140.4$ lb
   Alternatively, $H = V \tan \phi = 200 \tan 35° = 140.0$ lb (difference due to three-digit rounding).

1.8.2 The Unconfined Compression Test

The triaxial test permits more control over applied stress conditions than the direct shear test. A cylinder of soil is subjected to an all-around confining stress σ_3 and a deviator stress $(\sigma_1 - \sigma_3)$ applied vertically until failure (see Figure 1.8). The stress conditions on the sample at failure can be determined for all planes by plotting a Mohr's circle. If the test is repeated for soil samples at different confining stresses, the Mohr-Coulomb strength envelope can be obtained by drawing a line tangent to the failure circles as shown on the right side of the figure.

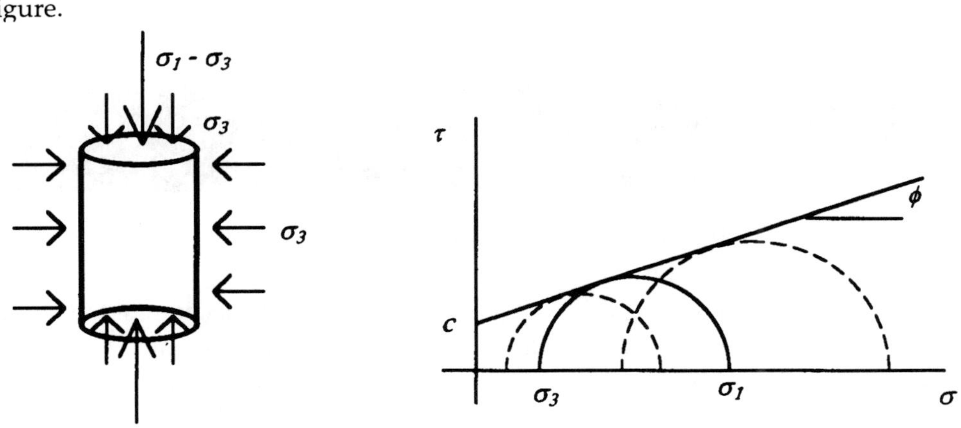

Figure 1.8 Triaxial Test, Mohr's Circles and Strength Envelope

For saturated clays where $\phi = 0$, it is convenient to test without any confining stress ($\sigma_3 = 0$). In the resulting *unconfined compression test*, the normal stress σ_1 required for shear is termed the *unconfined compressive strength* q_u and is equal to twice the cohesion:

$$q_u = 2c$$

Example 1.23

A 2.0 inch diameter by 4.0 inch high cylindrical sample of clay is tested in an unconfined compression test. The sample fails under a force of 200 lb.

1. What is the normal stress on the sample at failure?
   (A) 200 psi
   (B) 50.0 psi
   (C) 63.7 psi
   (D) 15.9 psi

2. What is the unconfined compressive strength of the sample?
   (A) 200 psi
   (B) 50.0 psi
   (C) 63.7 psi
   (D) 15.9 psi

3. What is the cohesion of the sample?
   (A) 31.8 psi
   (B) 63.7 psi
   (C) 127.4 psi
   (D) 200 psi

Solutions:

1. **C** $\sigma = P/A = 200/(\pi\ 2.0^2/4) = 63.67$ psi
2. **C** The unconfined compression strength q_u is equal to the normal stress at failure, 63.7 psi.
3. **A** $c = q_u/2 = 63.67/2 = 31.8$ psi

1.9 Bearing Capacity of Soils

The basis of bearing capacity calculations is shown in Figure 1.9. When the applied pressure on the base of a footing reaches the *ultimate bearing capacity* q_{ult}, shear failure occurs along a curved surface below and to the sides of the footing. In the Terzaghi bearing capacity theory, the soil above the footing is replaced by a surcharge pressure γD. Soil below the footing has the properties c, ϕ, and γ.

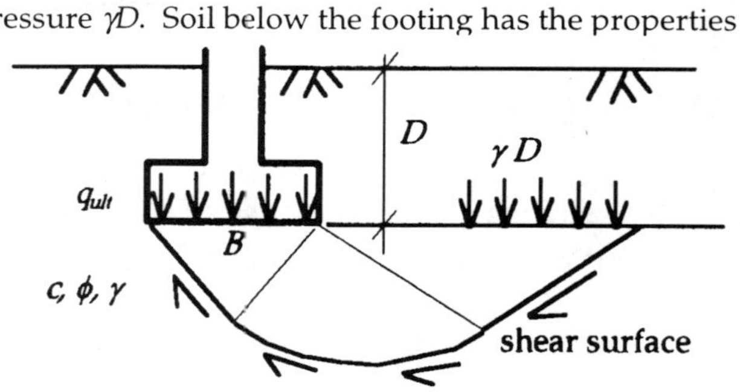

Figure 1.9 Bearing Capacity Failure under Footing

According to Terzaghi's theory, the ultimate bearing capacity q_{ult} for a concentrically loaded continuous footing of width B can be expressed as:

$$q_{ult} = cN_c + \gamma DN_q + 0.5\ B\gamma N_\gamma$$

where

c	is the cohesion of the soil below the base of the footing.
γD	is the vertical effective stress at the elevation of the footing base

(sometimes also denoted q).

B is the footing width.

γ is the unit weight of the soil below the footing base.

N_c, N_q and $N\gamma$ are bearing capacity factors which are functions of the friction angle ϕ of the soil below the footing. These functions are widely tabulated and plotted, and there are several versions, reflecting different assumptions in their derivation.

Refinements and extensions of the above equation are available for more general conditions.

Example 1.24

A continuous footing is founded 3 ft below the ground surface in a clay for which $\gamma = 125$ lb/ft^3, $c = 1200$ lb/ft^2, and $\phi = 0$. For $\phi = 0$, assume $N_c = 5.14$, $N_q = 1$ and $N\gamma = 0$.

1. Determine the ultimate bearing capacity.
   (A) 1200 lb/ft^2
   (B) 1575 lb/ft^2
   (C) 6168 lb/ft^2
   (D) 6540 lb/ft^2

2. If the factor of safety FS is to be at least 3, recommend an allowable bearing capacity q_a.
   (A) 2180 lb/ft^2
   (B) 19,350 lb/ft^2
   (C) 6540 lb/ft^2
   (D) 3600 lb/ft^2

Solutions:

1. **D** $q_{ult} = cN_c + \gamma DN_q + 0.5B\gamma N\gamma = (1200)(5.14) + (125)(3 \text{ ft})(1) + 0 = 6540$ lb/ft^2
2. **A** $q_a = q_{ult} / FS = 6450/3 = 2180$ lb/ft^2

Example 1.25

A continuous footing two feet wide is founded 3 ft below the ground surface in a uniform cohesionless sand. The soil parameters are $\gamma = 125$ lb/ft^3 , $c = 0$, and $\phi = 34°$. This corresponds to $N_q = 29.4$ and $N\gamma = 41.1$.

1. Determine the ultimate bearing capacity.
   (A) 11,040 lb/ft^2
   (B) 16,200 lb/ft^2
   (C) 5,160 lb/ft^2
   (D) 125 lb/ft^2

2. Determine the ultimate bearing force per lineal foot of foundation.
   (A) 16,200 lb/ft
   (B) 64,800 lb/ft
   (C) 32,400 lb/ft
   (D) 8,100 lb/ft

Solutions:

1. **B** $q_{ult} = cN_c + \gamma DN_q + 0.5B\gamma N_\gamma = 0 + (3)(125)(29.44) + (.5)(2)(125)(41.06) =$ 16,200 lb/ft^2
2. **C** $Q_{ult} = q_{ult} \times Area = q_{ult}(B)(1) = 16{,}200(2)(1) = 32{,}400$ lb/lineal ft

1.10 Earth Pressure and Retaining Walls

As soils possess shear strength, the horizontal stress in a soil can be greater or less than the vertical stress, and is dependent on the strain conditions. A small rotation of a retaining wall away from its backfill will reduce the horizontal earth pressure to its minimum value, the *active earth pressure* condition. Retaining walls are usually designed for active earth pressure conditions. For the special but common case of a frictionless vertical wall backfilled with a cohesionless material with a horizontal backfill surface, the *active earth pressure coefficient* K_a is

$$K_a = \tan^2(45 - \phi/2)$$

For dry sand with no water table, the active earth pressure p_a at any depth z is equal to

$$p_a = K_a\sigma'_v = K_a\,\gamma z$$

Knowing the active earth pressure at any point along a face of a wall, the total earth force on a wall can be obtained by integrating the pressure with respect to depth, or calculating the area of an earth pressure vs. depth diagram.

Example 1.26

The retaining wall shown is 8 ft high and retains a clean dry sand for which $\gamma = 120$ lb/ft^3 and $\phi = 32°$.

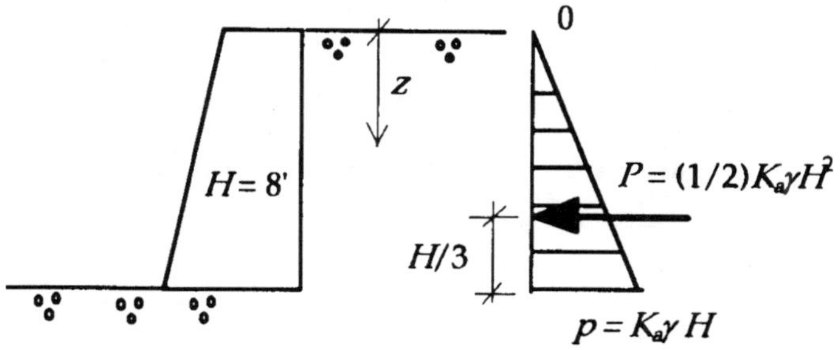

1. Calculate the active earth pressure at the base of the wall.
   (A) 960 lb/ft^2
   (B) 480 lb/ft^2
   (C) 295 lb/ft^2
   (D) 1180 lb/ft^2

2. Calculate the total earth force on the wall.
   (A) 960 lb/ft
   (B) 480 lb/ft
   (C) 295 lb/ft
   (D) 1180 lb/ft

3. Calculate the overturning moment per lineal foot of wall due to the active earth pressure.
   (A) 1180 ft-lb
   (B) 9440 ft-lb
   (C) 1280 ft-lb
   (D) 3140 ft-lb

Solutions:

1. **C** $K_a = \tan^2(45 - \phi/2) = \tan^2(45 - 32/2) = 0.307$
   $p_a = K_a\gamma z = (0.307)(120)(8) = 295 \text{ lb/ft}^2$

2. **D** The total active earth force is the integral of the earth pressure from $z = 0$ to 8′:
   $P_a = 0.5\, K_a\gamma H^2 = (0.5)(0.307)(120)(8)^2 = 1180$ lb/ft of wall

3. **D** The active force acts at the centroid of the earth pressure diagram; in this case, $H/3$. Thus $M_o = P_a \times H/3 = (1179)(8/3) = 3140$ ft-lb/ft of wall

Note: If a force is applied on the outside of the wall, pushing it toward the backfill, the horizontal pressure increases until it reaches the *passive earth pressure condition*, at which shear failure occurs and the pressure and force can no longer increase. The *passive earth pressure coefficient* is

$$K_p = \tan^2(45 + \phi/2)$$

Example 1.27

For the same wall shown in the previous example, assume a force is applied to the outside until passive earth pressure conditions occur.

1. Calculate the passive earth pressure at the base of the wall.
   (A) 3120 lb/ft^2
   (B) 295 lb/ft^2
   (C) 1560 lb/ft^2
   (D) 6240 lb/ft^2

2. Calculate the total passive earth force on the wall.
   (A) 6240 lb/ft
   (B) 3120 lb/ft
   (C) 12,480 lb/ft
   (D) 24,960 lb /ft

Solutions:

1. **A** $K_p = \tan^2(45 + 32/2) = 3.25$
   $p_a = K_p \gamma z = (3.25)(120)(8) = 3120 \text{ lb/ft}^2$

2. **C** The total active earth force is the integral of the earth pressure from $z = 0$ to 8′:
   $P_p = 0.5\, K_p \gamma H^2 = (0.5)(3.25)(120)(8)^2 = 12{,}480$ lb/ft of wall

1.11 Slope Stability

A simple approach to assessing the stability of a slope is to consider the statics on a wedge of soil such ABC shown in Figure 1.10.

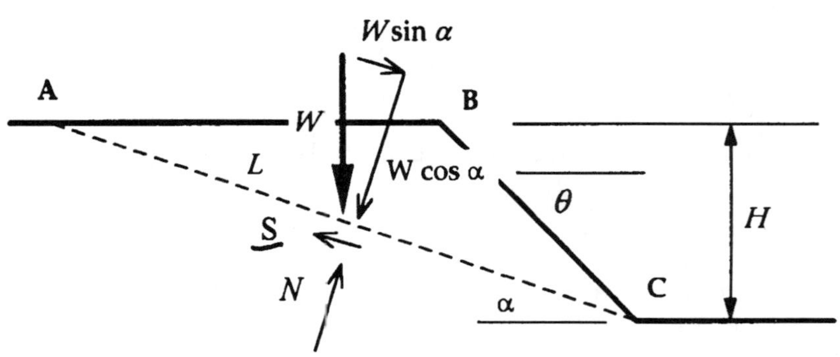

Figure 1.10 Free Body Considered for Slope Stability Analysis

If the wedge slides along a plane AC inclined at angle α, the applied shear force parallel to the plane is $W \sin \alpha$. The normal force on plane AC is $N = W \cos \alpha$. If the soil has strength parameters c and ϕ, the maximum shear force S that can be developed along plane AC is

$$S = cL + N \tan \phi = cL + W \cos\alpha \tan \phi$$

If the factor of safety against sliding is taken as the ratio of the available shear force to the applied shear force, one obtains the equation from the NCEES Handbook:

$$FS = (cL + W \cos \alpha \tan \phi)/W \sin \alpha$$

The critical angle α must be found by trial and error, and would be the value yielding the lowest factor of safety. In practice, non-planar surfaces, especially circular surfaces, are also considered, using more rigorous methods.

Example 1.28

Refer to Figure 1.9. Assume H = 20 ft, γ = 120 lb/ft^3, c = 500 lb/ft^2, and ϕ = 30°. Assume the slope angle θ = 26.6° (a 1 on 2 slope) and the failure surface angle α = 15°. Determine the factor of safety FS.

(A) 1.15

(B) 1.54

(C) 1.00

(D) 1.76

Solution:

B The length AB = $(H/\tan \alpha) - (H/\tan\theta)$ = (20/0.2679) – (20/0.5) = 34.65 ft

The weight of the soil wedge is

$$W = Area \times \gamma$$

$$= (0.5)(20)(34.65)(120) = 41{,}580 \text{ lb/ft}^2$$

The length $L = 20 / \sin \alpha = 77.27$ ft. Then

$$FS = (cL + W \cos \alpha \tan \phi) / W \sin \alpha$$

$$= [500 \times 77.27 + 41{,}580 \cos 15° \tan 30°]/41{,}580 \cos 15° = 1.54$$

2. Structural Analysis and Design

by Frank J. Hatfield

The material regarding structural analysis and design given on pages 85, 87, 88, 89 and 93 of the NCEES Handbook is an arbitrary sampling of particular methods and formulas. The following review is based on the hope that the examination itself will focus on the same specific topics as well as on basic concepts.

2.1 Analysis of Determinate Trusses

A truss is a structure in which all members are straight, all joints are hinges (i.e., cannot transmit a moment), and all loads are applied at joints. Therefore, there are no moments in the members, and member forces are directed axially. The virtual work formula for truss deflection is on page 85 of the Handbook. However, in the definitions of terms following the formula, the definitions of F_Q and F_p are erroneously interchanged.

Example 2.1

The figure shown represents a plane truss with a load.

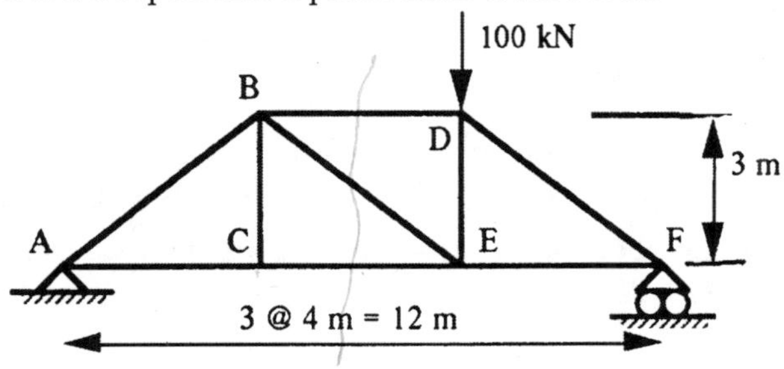

1. What is the reaction force at support F?
   (A) -33 kN
   (B) 0 kN
   (C) 33 kN
   (D) 67 kN

2. What is the force in member *BC* ?
   (A) 0 kN
   (B) 100 kN tension
   (C) 100 kN compression
   (D) 133 kN tension

3. What is the force in member *CE* ?
   (A) 0 kN
   (B) 44 kN tension
   (C) 44 kN compression
   (D) 89 kN tension

4. Assume that for all members, cross-sectional area is 900 mm^2 and modulus of elasticity is 200 GPa. What is the horizontal deflection at joint *F* ?
   (A) 0 mm
   (B) 1 mm
   (C) 2 mm
   (D) 4 mm

Solutions:

1. **D** Sum of moments about *A* is zero:

$$12\ R_F - 8\ \text{m} \times 100\ \text{kN} = 0.\quad R_F = 67\ \text{kN}.$$

2. **A** Cut members *AC, BC, CE* and investigate the free-body containing joint *C*. Sum of vertical forces is zero. $F_{BC} = 0$ kN.

3. **B** Cut members *BD, BE, CE* and investigate the free-body containing joints *D, E, F*. Sum of moments about *B* is zero:

$$8\ R_F - 4\ \text{m} \times 100\ \text{kN} - 3\ F_{CE} = 0.$$

Since $R_F = 67$ kN, then

$$F_{CE} = 44\ \text{kN}.$$

4. **D** Use virtual work. First compute all member forces caused by real load. Then remove real load and impose a virtual unit load horizontally at *F*. Compute member forces caused by the virtual load. $F_{QAC} = F_{QCE} = F_{QEF} = 1$; all other $F_Q = 0$:

$$\Delta = \Sigma\ F_Q\ F_p\ L\ /\ AE$$

$$= (1 \times 44\ \text{kN} \times 4\ \text{m} + 1 \times 44 \times 4 + 1 \times 89 \times 4) \times 1000/(900\ \text{mm}^2 \times 200\ \text{kN/mm}^2)$$

$$= 4\ \text{mm}$$

2.2 Analysis of Determinate Beams

A beam is a member subjected to lateral forces which cause internal shear and moment. The Handbook, on page 93, gives a table of fixed-end moments which suggests that the examination may include an indeterminate beam, and that it will be most conveniently analyzed by a pre-computer method called moment distribution. However, a determinate beam was chosen for the example that follows because in review it is most important to focus on basic concepts. The virtual work formula for beam deflection given on page 85 of the Handbook will be unfamiliar to many readers. It may be better to review a familiar method rather than learning the virtual work method. A definition of influence lines is given on page 85 of the Handbook.

Example 2.2

The figure shown represents a beam with an internal hinge at *B* and a moving load.

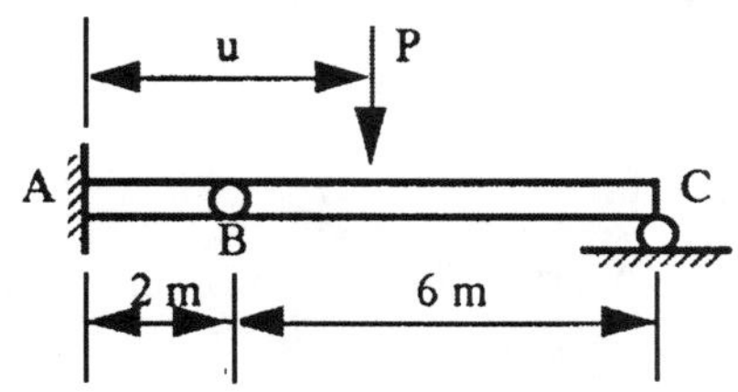

1. What is the ordinate of the influence line for the reaction force at *C* for u = 1.0 m?
   (A) 0
   (B) 0.25
   (C) 0.50
   (D) 1.0

2. What is the shear at *B* for P = 60 kN and u = 5 m ?
   (A) 0 kN
   (B) 30 kN
   (C) 40 kN
   (D) 60 kN

3. What is the maximum moment on segment *BC* for *P* = 60 kN ?
   (A) 0 kN-m
   (B) 10 kN-m
   (C) 60 kN-m
   (D) 90 kN-m

4. Assume the moment of inertia is 300×10^6 mm^4 and modulus of elasticity is 200 GPa. What is the angular rotation in radians at B on segment AB for $P = 60$ kN and $u = 2$ m ?
   (A) 0.001
   (B) 0.002
   (C) 0.004
   (D) 0.010

Solutions:

1. **A** Influence line is a graph of the value of R_C as a function of load position u, with $P = 1$. When $u \leq 2$, $R_C = 0$.

2. **B** Use segment BC as free-body. Sum of moments about C is zero:

$$3 \text{ m} \times 60 \text{ kN} - 6\ V_B = 0. \quad \therefore V_B = 30 \text{ kN}.$$

3. **D** Locate P at $u = 5$ m to maximize moment. Use right half of segment BC as free-body. Sum of moments about point at middle of BC is zero:

$$3\ R_C - M = 0.$$

Since $R_C = 30$ kN, then

$$M = 90 \text{ kN-m}.$$

4. **B** Use virtual work. First determine moment function for real load. $M = 60x - 120{,}000$ kN-mm for segment AB; $M = 0$ for segment BC. Then remove real load and impose a virtual unit moment at B. Determine the moment function for the virtual load; $m = 1$ for segment AB; $m = 0$ for segment BC:

$$\Delta = \int (mM/EI)\, dx$$

$$= \int 1\ (60x - 120{,}000)\, dx\ /\ (300 \times 10^6 \times 200)$$

$$= -120 \times 10^6\ /\ (300 \times 10^6 \times 200) = 0.002 \text{ radians}$$

Note: Δ can represent any type of deformation; here it represents rotation.

2.3 Design in Reinforced Concrete

The material on page 88 of the Handbook is selected from a specification published by the American Concrete Institute. A review of slabs, beams and columns is suggested.

Example 2.3

The figure shown represents the profile and cross-section of a reinforced concrete beam. Reinforcing is Grade 60 steel, and the 28-day compressive strength of the concrete is 3000 psi.

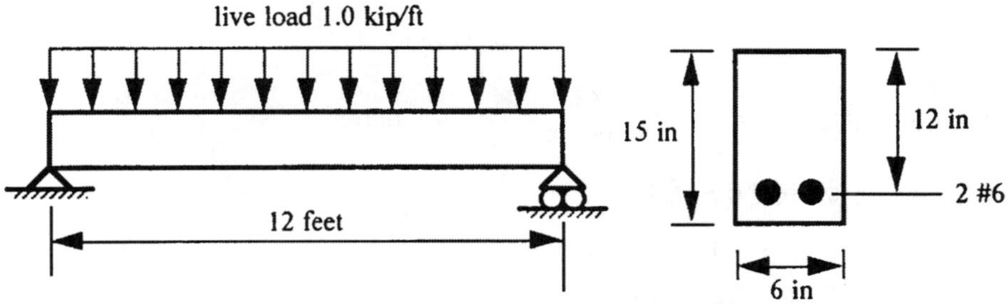

1. The beam is...
   (A) Over-reinforced
   (B) Under-reinforced
   (C) Balanced
   (D) Doubly reinforced

2. What is the design moment strength?
   (A) 20 kip-ft
   (B) 33 kip-ft
   (C) 41 kip-ft
   (D) 53 kip-ft

3. What is the required moment strength?
   (A) 15 kip-ft
   (B) 20 kip-ft
   (C) 24 kip-ft
   (D) 33 kip-ft

4. Shear reinforcement is...
   (A) not needed at the mid-span of the beam
   (B) not needed at the ends of the beam
   (C) not needed anywhere on the beam
   (D) needed for the full length of the beam

Solutions:

1. **B** $\rho = A_s / b\,d = 2 \times 0.44 \;/\; 6 \times 12 = 0.012$

$$\rho_b = \frac{0.85\,\beta f_c'}{f_y} \times \frac{87000}{87000 + f_y}$$

$$= \frac{0.85 \times 0.85 \times 3000}{60000} \times \frac{87000}{87000 + 60000} = 0.021$$

$\rho < \rho_b$ so beam is under-reinforced

2. **C** $a = \dfrac{A_s f_y}{0.85\, f_c' b} = \dfrac{2 \times 0.44 \times 60000}{0.85 \times 3000 \times 6} = 3.45$ inches

$$\phi\, M_N = \phi\, A_s f_y\, (d - a/2)$$

$$= 0.90 \times 2 \times 0.44 \times 60000\; (12 - 3.45 \;/\; 2)$$

$$= 488 \times 10^3 \text{ lb-in or } 41 \text{ kip-ft}$$

3. **D** $w_{\text{dead}} = (6/12)$ ft x $(15/12)$ ft x 150 pcf = 94 lb/ft

$$M_{\text{dead}} = w_{\text{dead}} L^2 \;/\; 8 = 0.094 \text{ kip/ft} \times (12 \text{ ft})^2 \;/\; 8 = 1.7 \text{ kip-ft}$$

$$M_{\text{live}} = w_{\text{live}} L^2 \;/\; 8 = 1.0 \text{ kip/ft} \times (12 \text{ ft})^2 \;/\; 8 = 18 \text{ kip-ft}$$

$$M_U = 1.4\, M_{\text{dead}} + 1.7\, M_{\text{live}} = 1.4 \times 1.7 + 1.7 \times 18 = 33 \text{ kip-ft}$$

4. **A** Shear requirement computed at distance *d* from support.

$$V_{\text{dead}} = 5 \text{ ft} \times 0.094 \text{ kip/ft} = 0.5 \text{ kip}$$

$$V_{\text{live}} = 5 \text{ ft} \times 1.0 \text{ kip/ft} = 5.0 \text{ kip}$$

$$V_U = 1.4\, V_{\text{dead}} + 1.7\, V_{\text{live}} = 1.4 \times 0.5 + 1.7 \times 5.0 = 9.2 \text{ kip}$$

$$V_C = 2\sqrt{f_c'}\, b\, d = 2\sqrt{3000} \times 6 \times 12 = 7890 \text{ lb or } 7.9 \text{ kips}$$

$$\phi\, V_C \;/\; 2 = 0.85 \times 7.9 \;/\; 2 = 3.4 \text{ kips}$$

Since $V_U > \phi\, V_C / 2$ shear reinforcement is needed near the supports. However V_U decreases with distance from the support, and is less than $\phi\, V_C \;/\; 2$ near midspan. Therefore, there is a section of the beam near the middle where shear reinforcement is not needed.

2.4 Design in Structural Steel

The material on pages 88 and 89 of the Handbook is selected from specifications published by the American Institute of Steel Construction. Both the ASD (allowable stress design) and the LRFD (load and resistance factor design) specifications are included. Since LRFD is the more modern methodology and is the one taught at nearly all colleges and universities, only it will be covered in the following example. The Handbook includes obsolete evaluations of the LRFD shear lag factor U for tension members and of the LRFD block shear design strength for bolted connections. These formulas are from the first edition LRFD specification and differ from those in the current second edition LRFD specification. For the purposes of passing the FE examination, it probably is safest to use the obsolete formulas given in the Handbook.

Example 2.4

The sketch shown represents the end of a double-angle steel tension member with holes for a bolted connection. Yield strength of the steel is 50 ksi, ultimate tensile strength is 65 ksi. The gross cross-sectional area of the two angles together is 2.88 in^2. Holes are punched 13/16 diameter for 3/4 inch bolts.

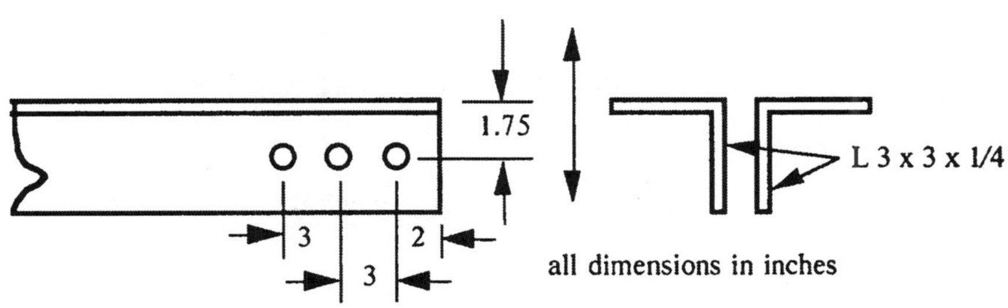

1. What is the design strength for yield?
   (A) 101 kips
   (B) 111 kips
   (C) 130 kips
   (D) 264 kips

2. What is the design strength for tension fracture?
   (A) 101 kips
   (B) 111 kips
   (C) 130 kips
   (D) 264 kips

3. What is the design strength for block shear fracture?
   (A) 101 kips
   (B) 111 kips
   (C) 130 kips
   (D) 264 kips

4. Assume that the critical design strength is 96 kips and that dead load is 40 kips. What is the maximum permissible live load?
   (A) 30 kips
   (B) 35 kips
   (C) 40 kips
   (D) 48 kips

Solutions: (LRFD)

1. **C** $\phi F_y A_g = 0.9 \times 50 \text{ ksi} \times 2.88 \text{ in}^2 = 130 \text{ kips}$

2. **A** $A_n = 2.88 - 2 \times 1/4 \text{ in} \times 7/8 \text{ in} = 2.44 \text{ in}^2$

   $A_e = U A_n = 0.85 \times 2.44 \text{ in}^2 = 2.07 \text{ in}^2$ (first edition LRFD)

   $\phi F_u A_e = 0.75 \times 65 \text{ ksi} \times 2.07 \text{ in}^2 = 101 \text{ kips}$

3. **B** Using first edition LRFD

   $A_{gt} = 2 \times 1/4 \text{ in} \times (3 - 1.75) \text{ in} = 0.625 \text{ in}^2$

   $A_{nv} = 2 \times 1/4 \times (8 - 2.5 \times 13/16) \text{ in} = 2.98 \text{ in}^2$

   $\phi (0.6\ F_u A_{nv} + F_y A_{gt}) =$

   $0.75 (0.6 \times 65 \text{ ksi} \times 2.98 \text{ in}^2 + 50 \times 0.625) = 111 \text{ kips}$

4. **A** $T_u = 1.2\ T_{\text{dead}} + 1.6\ T_{\text{live}}$

   $T_{\text{live}} = (T_u - 1.2\ T_{\text{dead}}) / 1.6 = (96 - 1.2 \times 40) / 1.6 = 30 \text{ kips}$

3. Transportation

by Thomas Maleck

Many of the questions relating to transportation can be answered by using the material found in the American Association of State Highway and Transportation Officials (AASHTO) "A Policy on Geometric Design of Highways and Streets, 1984". This guide is often referred to as the "GREEN BOOK". Additional help can be found in *Principles & Practice of Civil Engineering* by Merle C. Potter.

3.1 Braking Distance (stopping sight distance)

Please refer to chapter three of the AASHTO "A Policy on Geometric Design of Highways and Streets, 1984" to obtain the details needed to properly understand this material. See page 85 of the NCEES Reference Handbook for the equation used to calculate braking distance.

Example 3.1

A two lane highway has a grade of 2 percent. The design speed is 60 mph. Refer to the following table.

Design Speed (mph)	Assumed Speed for Condition (mph)	Brake Reaction		Coefficient of Friction f	Braking Distance on Level[a] (ft)	Stopping Sight Distance	
		Time (sec)	Distance (ft)			Computed[a] (ft)	Rounded for Design (ft)
50	44-50	2.5	161.3-183.3	0.30	215.1-277.8	376.4-461.1	400-475
55	48-55	2.5	176.0-201.7	0.30	256.0-336.1	432.0-537.8	450-550
60	52-60	2.5	190.7-220.0	0.29	310.8-413.8	501.5-633.8	525-650
65	55-65	2.5	201.7-238.3	0.29	347.7-485.6	549.4-724.0	550-725
70	58-70	2.5	212.7-256.7	0.28	400.5-583.3	613.1-840.0	625-850

[a]Different values for the same speed result from using unequal coefficients of friction.

1. What is the assumed driver reaction time?
   (A) 1.0 sec
   (B) 1.5 sec
   (C) 2.0 sec
   (D) 2.5 sec

2. What is the lowest allowable assumed speed used for determining stopping sight distance?
   (A) 52 mph
   (B) 55 mph
   (C) 57 mph
   (D) 60 mph

3. What is the assumed coefficient of friction?
   (A) 0.28
   (B) 0.29
   (C) 0.30
   (D) 0.31

4. What is the expected braking distance?
   (A) 405 ft
   (B) 425 ft
   (C) 445 ft
   (D) 465 ft

5. How far does the vehicle travel before the vehicle decelerates?
   (A) 220 ft
   (B) 230 ft
   (C) 240 ft
   (D) 250 ft

6. What is the required stopping sight distance?
   (A) 625 ft
   (B) 665 ft
   (C) 695 ft
   (D) 715 ft

Solutions:

1. **D** From the table, brake reaction time equals driver reaction time which equals 2.5 sec.

2. **A** From the table, assume speed for condition is 52-60 mph. For wet pavement 52 mph is assumed for design.

3. **B** From the table, coefficient of friction for a design speed of 60 mph is 0.29.

4. **C** $d = \dfrac{V^2}{2g(f \pm s)}$

   V = 60 mph, which = 88 fps, $g = 32.2$, $f = 0.29$, $s = -0.02$. s is negative because one direction of travel is going downhill.

   Therefore, $d = 445.36$ ft.

5. **A** Distance vehicle travels before braking is $T \times V$ where

   $T = 2.5$ sec (driver reaction)

   $V = 60$ mph = 88 fps

   Therefore, distance = 2.5(88) = 220 ft.

6. **B** Stopping sight distance = braking distance + distance vehicle travels before braking which equals questions No. 4 plus No. 5:

   445 + 220 = 665 ft

3.2 Sight Distance Over a Vertical Curve

The equations needed to solve the problem of sight distance over a vertical curve are found on page 85 of the NCEES Reference Handbook.

Example 3.2

A freeway has a 70 mph design speed. There is a 2% grade followed by a negative 3% grade. (Hint: this is a crest vertical curve.) Assume height of driver's eye to be 3.5 ft and object height to be 0.5 ft. Refer to the table given in Example 3.1, as well as the following table.

	Time(s)			Decision Sight Distance (ft)		
	Premaneuver					
Design Speed (mph)	Detection & Recognition	Decision & Response Initiation	Maneuver (Lane Change)	Summation	Computed	Rounded for Design
30	1.5-3.0	4.2-6.5	4.5	10.2-14.0	449- 616	450- 625
40	1.5-3.0	4.2-6.5	4.5	10.2-14.0	598- 821	600- 825
50	1.5-3.0	4.2-6.5	4.5	10.2-14.0	748-1,027	750-1,025
60	2.0-3.0	4.7-7.0	4.5	11.2-14.5	986-1,276	1,000-1,275
70	2.0-3.0	4.7-7.0	4.0	10.7-14.0	1,098-1,437	1,100-1,450

1. What is the required length of vertical curve needed to satisfy design stopping sight distance?
   (A) 1435 ft
   (B) 2720 ft
   (C) 4550 ft
   (D) 1670 ft

2. What is the required length of vertical curve needed to satisfy the minimum value for decision sight distance?
   (A) 1435 ft
   (B) 2720 ft
   (C) 4550 ft
   (D) 1670 ft

Solutions:

1. **B** From the table, stopping sight distance = 850 ft.

   For a crest vertical curve:

$$S < L \qquad L = \frac{AS^2}{100\left(\sqrt{2H_1} + \sqrt{2H_2}\right)^2}$$

$$S > L \qquad L = 2S - \frac{200\left(\sqrt{H_1} + \sqrt{H_2}\right)^2}{A}$$

   L = length of vertical curve
   A = algebraic difference in grade which equals $2-(-3)=5$
   S = sight distance requirement which equals 850 ft
   H_1 = height of driver's eye which equals 3.5 ft
   H_2 = height of object which equals 0.5 ft

   For $S > L$

$$L = (2)(850) - \frac{(200)\left(\sqrt{3.5} + \sqrt{0.5}\right)^2}{5}$$

$$= 1434.17 \text{ ft}$$

   False answer: S is <u>not</u> larger than L.

   For $S < L$

$$L = \frac{(5)(850)^2}{(100)\left(\sqrt{7.0} + \sqrt{1}\right)^2}$$

$$= 2717.90 \text{ ft}$$

   Correct answer: S is smaller than L.

2. **C** From the table, minimum value is 1,100 ft.

   A, H_1 and H_2 do not change.

   For $S > L$

$$L = (2)(1100) - \frac{(200)\left(\sqrt{3.5} + \sqrt{0.5}\right)^2}{5}$$

$$= 1934.17 \text{ ft}$$

   False answer: S is <u>not</u> larger than L.

   For $S < L$

$$L = \frac{(5)(1100)^2}{(100)\left(\sqrt{7} + \sqrt{1}\right)^2}$$

$$= 4551.78 \text{ ft}$$

   Correct answer: S is smaller than L.

3.3 Sight Distance for a Sag Vertical Curve

The sag vertical curve equations are also found on page 85 of the NCEES Handbook.

Example 3.3

A highway has a 55 mph design speed. There is a negative 1 percent grade followed by a 2 percent grade. Refer to the following table.

Design Speed (mph)	Assumed Speeds: Passed Vehicle (mph)	Assumed Speeds: Passing Vehicle (mph)	Minimum Passing Sight Distance (ft)
50	41	51	1,840
60	47	57	2,140
65	50	60	2,310
70	54	64	2,490

1. What is the required length of vertical curve needed to satisfy AASHTO stopping sight distance?
   (A) 320 ft
   (B) 585 ft
   (C) 200 ft
   (D) 1460 ft

2. What is the required length of vertical curve needed to satisfy AASHTO passing sight distance?
   (A) 320 ft
   (B) 585 ft
   (C) 200 ft
   (D) 1460 ft

3. What is the required length of vertical curve needed to satisfy AASHTO design sight distance?
   (A) 320 ft
   (B) 585 ft
   (C) 200 ft
   (D) 1460 ft

4. What is the required length of vertical curve to satisfy AASHTO requirements for comfort?
   (A) 320 ft
   (B) 585 ft
   (C) 200 ft
   (D) 1460 ft

Solutions:

1. **A** From the table in Example 3.1, stopping sight distance = 550 ft.

$$A = (-1) - 2 = -3 \text{ or } 3$$

For a sag vertical curve:

$$S > L \qquad L = 2S - \frac{400 + 3.55S}{A}$$

$$S < L \qquad L = \frac{AS^2}{400 + 3.55S}$$

For $S > L$

$$L = 2(550) - \frac{400 + 3.55(550)}{3}$$

$$= 315.83 \text{ ft}$$

Correct Answer: S is greater than L.

For $S < L$

$$L = \frac{3(550)^2}{400 + 3.55(550)}$$

$$= 385.76 \text{ ft}$$

False answer: S is not less than L.

2. **D** From the table given above, passing sight distance = 1950 ft for 55 mph.

A still equals 3.

For $S > L$

$$L = 2(1950) - \frac{400 + 3.55(1950)}{3}$$

$$= 1459.17 \text{ ft}$$

Correct answer: S is greater than L.

For $S < L$

$$L = \frac{3(1950)^2}{400 + 3.55(1950)}$$

$$= 1557.9 \text{ ft}$$

False answer: S is not smaller than L.

3. **B** From the table of Example 3.2, decision sight for 55 mph is 875 to 1150 ft. Using a minimum value of 875 for this example. A still equals 3.

For $S > L$

$$L = 2(875) - \frac{400 + 3.55(875)}{3}$$

$$= 581.25 \text{ ft}$$

Correct answer: S is greater than L.

For $S < L$

$$L = \frac{3(875)^2}{400 + 3.55(1950)}$$

$$= 313.67 \text{ ft}$$

False answer: S is not less than L.

4. C $L = \dfrac{AV^2}{46.5}$

$A = 3$ and $V = 55$ mph.

Therefore,

$$L = \frac{3(55)^2}{46.5} = 195.16 \text{ ft}$$

3.4 Vertical Curve Elevations

Refer to page 92 of the NCEES Reference Handbook for a figure and equations used for calculations involving vertical curves.

Example 3.4

An 800 ft vertical curve with equal legs is provided for a highway crest vertical curve. The grade of the back tangent is 2 % and the grade of the forward tangent is a negative 2 %. The elevation of the Point of Vertical Intersection (*PVI*) is 1200 ft.

1. What is the elevation of the Point of Vertical Curve (*PVC*)?
   (A) 1192 ft
   (B) 1208 ft
   (C) 1184 ft
   (D) 1216 ft

2. What is the elevation of the Point of Vertical Tangent (*PVT*)?
   (A) 1192 ft
   (B) 1208 ft
   (C) 1184 ft
   (D) 1216 ft

3. What is the Rate of Change of Grade?
   (A) –0.050
   (B) –0.500
   (C) 0.01
   (D) 0.10

4. What is the Tangent Offset at the *PVI*?
   (A) –0.40
   (B) –2.00
   (C) –4.00
   (D) 0.20

5. If the PVI is at station 11+00 and the PVC is at station 7+00, what is the Tangent Offset at station 8+00?
   (A) 0.25
   (B) 0.50
   (C) 1.00
   (D) –0.25

6. What is the Curve Elevation at station 8+00?
   (A) 1193.75 ft
   (B) 1194.00 ft
   (C) 1194.50 ft
   (D) 1195.00 ft

Solutions:

1. **A** $PVC = PVI - (g_1)L/2$

   $= 1200 - (2)8/2 = 1192$

   (L is in stations)

2. **A** $PVT = PVT + (g_2)L/2$

   $= 1200 + (-2)8/2 = 1192$

3. **B** Rate of Change:

   $$r = \frac{g_2 - g_1}{L} = \frac{(-2)-2}{8}$$
   $$= -0.500$$

4. **C** a = Parabola Constant

   $$a = \frac{g_2 - g_1}{2L} = \frac{(-2)-2}{2(8)}$$
   $$= -0.25$$

   E = Tangent Offset at PVI

   $$E = a\left(\frac{L}{2}\right)^2 = (-0.25)\frac{8^2}{2^2}$$
   $$= -4.00$$

5. **D** y = Tangent Offset

   $$y = ax^2 = (-0.25)(1)^2$$
   $$= -0.25$$

   (x is in stations)

6. **A** Curve Elevation $= Y_{PVC} + g_1 x + ax^2$

   $Y_{PVC} = 1192$ ft

   $g_1 = 2\ \%$

$x = 1$ station

$a = -0.25$

Therefore: Curve Elevation at station 8+00 equals:

$$1192 + 2(1) - 0.25(1)^2 = 1193.75$$

3.5 Horizontal Curves

The horizontal curve figure and equations are found on page 93 of the NCEES Handbook.

Example 3.5

A horizontal curve with a 1000 ft radius is to be provided for two perpendicular roads. The station of the *PI* is 12+00.00.

1. What is the Intersection of Angle?
   (A) 90°
   (B) 120°
   (C) 45°
   (D) 60°

2. What is the Degree of Curve?
   (A) 5.73°
   (B) 5.00°
   (C) 4.25°
   (D) 6.81°

3. What is the Tangent Distance?
   (A) 800 ft
   (B) 550 ft
   (C) 1000 ft
   (D) 1100 ft

4. What is the External Distance?
   (A) 411.8 ft
   (B) 414.2 ft
   (C) 618.2 ft
   (D) 411.2 ft

5. What is the Length of the Middle Ordinate?
   (A) 292 ft
   (B) 293 ft
   (C) 288 ft
   (D) 283 ft

6. What is the Length of Curve?
   (A) 1200.00 ft
   (B) 1242.81 ft
   (C) 1500.00 ft
   (D) 1570.80 ft

7. What is the stationing of the *PC*?
   (A) 17+70.80
   (B) 15+81.00
   (C) 2+00.00
   (D) 2+70.80

8. What is the stationing of the *PT*?
   (A) 17+70.80
   (B) 15+81.00
   (C) 2+00.00
   (D) 2+70.80

Solutions:

1. **A** Since the roads are perpendicular, the Angle of Intersection is 90°.

2. **A** The Degree of Curve $= \dfrac{5729.58}{R}$

   $R = 1000$ ft. Therefore,

$$D = \frac{5729.58}{1000} = 5.73°$$

3. **C** The Tangent Distance is

$$T = R\tan(I/2)$$

   $I = 90°$ (from question 1). Therefore,

$$T = (1000)\tan(90/2) = 1000.00 \text{ ft}$$

4. **B** The External Distance is

$$E = R[(\sec I/2) - 1]$$
$$= 1000[1.41421 - 1] = 414.21 \text{ ft}$$

5. **B** The length of the Middle Ordinate is

$$M = R - R\cos\frac{I}{2}$$

   $R = 1000.00$

   $I = 90°$

   Therefore: $M = 292.89$ ft

6. **D** The Length of Curve is

$$L = \frac{100I}{D}$$

$$= \frac{100(90)}{5.72958} = 1570.80 \text{ ft}$$

7. **C** Station of PC = station of PI – Tangent Distance:

$$PC = 12{+}00.00 - 1000.00 = 2{+}00.00$$

8. **A** Station of PT equals station of PC plus Length of Curve:

$$PT = PC + L$$

$$= 2 + 00.00 + 1570.80$$

$$= 17 + 70.80$$

3.6 Superelevations

To calculate the superelevation of a highway, refer to the material on page 86 of the NCEES Reference Handbook. The superelevation equation for railroads is also presented there.

Example 3.6

For the horizontal curve in Example 3.5, what is the minimum superelevation needed to provide for a 50 mph design speed? Refer to the following table.

Design Speed (mph)	Maximum e	Maximum f	Total (e + f)	Maximum Degree of Curve	Rounded Maximum Degree of Curve	Minimum Radius (ft)
20	.08	.17	.25	53.54	53.5	107
30	.08	.16	.24	22.84	22.75	252
40	.08	.15	.23	12.31	12.25	468
50	.08	.14	.22	7.54	7.5	764
60	.08	.12	.20	4.76	4.75	1,206
65	.08	.11	.19	3.85	3.75	1,528
70	.08	.10	.18	3.15	3.0	1,910

NOTE: In recognition of safety considerations, use of e_{max} = 0.04 should be limited to urban conditions.

(A) 0.165

(B) 0.04

(C) 0.140

(D) 0.027

Solution:

The maximum allowable lateral friction force is required to minimize the needed superelevation. Therefore, from the given table, the maximum allowable friction factor f is 0.140.

The required superelevation e is equal to

$$e + f = \frac{V^2}{gR}$$

Therefore:

$$e = \frac{V^2}{gR} - f$$

$V = 50 \text{ mph} = 73.33 \text{ ft/sec}$

$R = 1000 \text{ ft}$

$g = 32.2 \text{ ft/sec}^2$

$$e = \frac{73.33^2}{32.2(1000)} - 0.14 = 0.027$$

The answer is **D**.

3.7 Spirals

Spirals for highways are analyzed using the equation given on page 86 of the NCEES Reference Handbook. Railroad spirals are also analyzed using the equation given there.

Example 3.7

For the horizontal curve in Examples 3.5 and 3.6, what is the recommended length of spiral transition?

(A) 150 ft
(B) 200 ft
(C) 250 ft
(D) 300 ft

Solution:

The length of spiral is equal to:

$$L_S = 1.6\frac{V^3}{R}$$

$V = 50 \text{ mph}$

$R = 1000 \text{ ft}$

$$L_S = 1.6\frac{50^3}{1000} = 200 \text{ ft}$$

The answer is **B**.

4. Airport Design

by Francis McKelvey

The questions relating to airport layout and design can be answered by refering to pages 90 and 91 of the NCEES Handbook. We will present several of the more common problems encountered when designing an airport.

Example 4.1

The preliminary vertical profile for a transport airport runway centerline is given below. The specifications for the geometric design of the vertical profile of the centerline of a runway are presented on p. 91 of the Handbook.

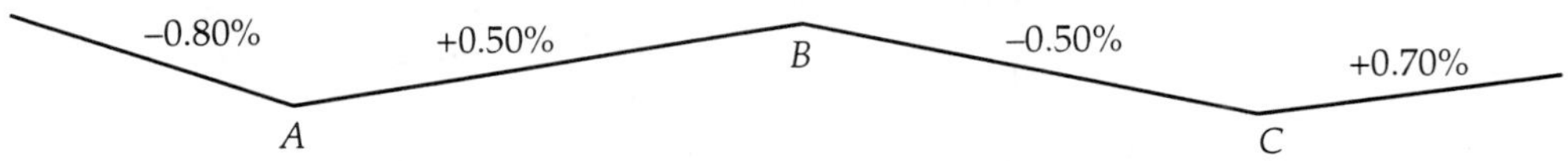

1. The minimum required length of a vertical curve at the point of intersection *A* is most nearly:
   (A) 300 ft
   (B) 600 ft
   (C) 1300 ft
   (D) 2000 ft

2. The minimum required length of a vertical curve at the point of intersection *B* is most nearly:
   (A) 100 ft
   (B) 500 ft
   (C) 1000 ft
   (D) 1500 ft

3. The minimum required length of a vertical curve at the point of intersection *C* is most nearly:
   (A) 200 ft
   (B) 500 ft
   (C) 700 ft
   (D) 1200 ft

4. The minimum required distance between point of intersection A and point of intersection B is most nearly:
   (A) 1000 ft
   (B) 1300 ft
   (C) 2000 ft
   (D) 2300 ft

5. The minimum required distance between point of intersection B and point of intersection C is most nearly:
   (A) 500 ft
   (B) 700 ft
   (C) 2200 ft
   (D) 2500 ft

Solutions:

1. **C** p.91 - Transport Airports: $L_1 = 1.30(1000) = 1300$ ft

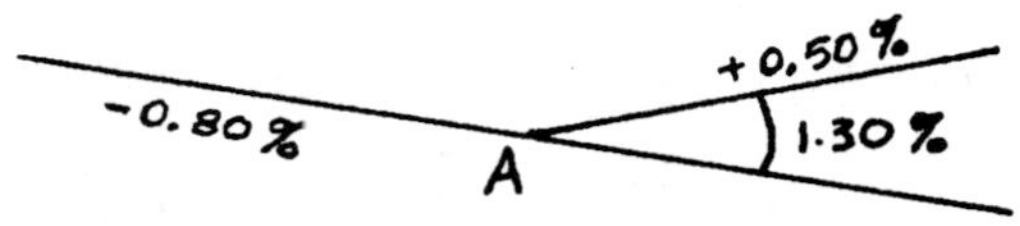

2. **C** p.91 - Transport Airports: $L_2 = 1.00(1000) = 1000$ ft

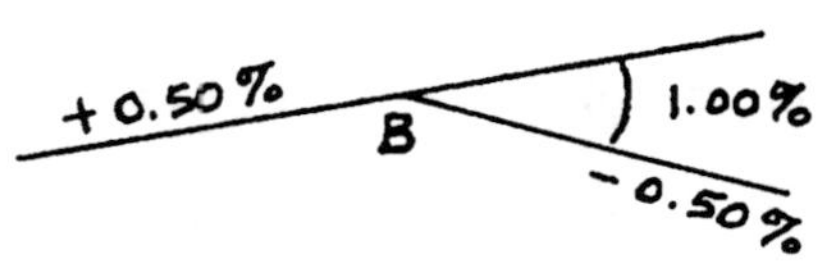

3. **D** p.91 - Transport Airports: $L_3 = 1.20(1000) = 1200$ ft

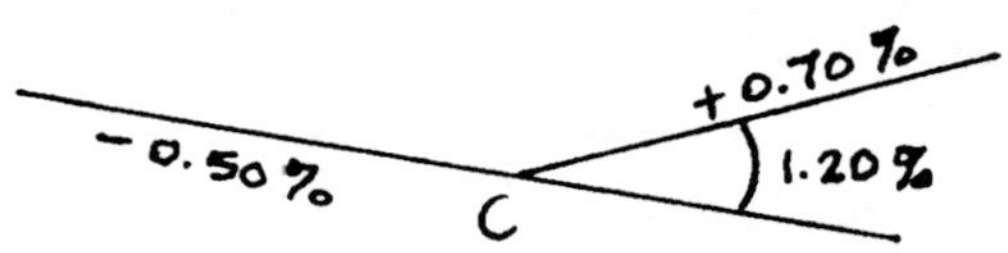

4. **D** p.91 - Transport Airports: $D = 1000(A + B)$
$= 1000(1.30 + 1.00) = 2300$ ft

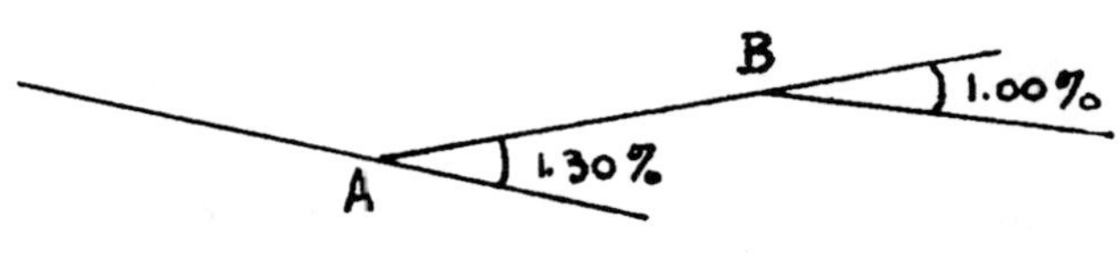

5. **C** p.91 - Transport Airports: $D = 1000(B + C)$
$= 1000(1.00 + 1.20) = 2200$ ft

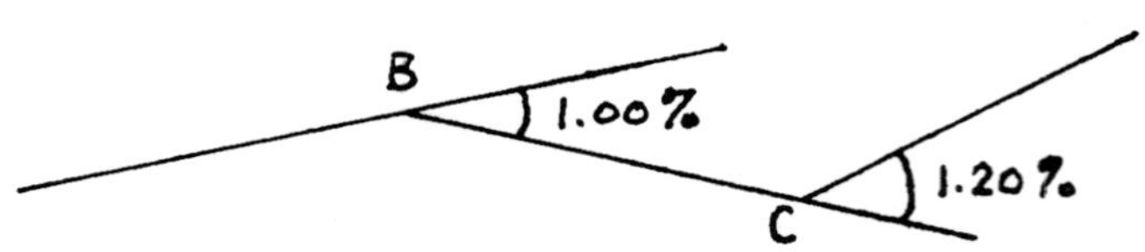

Example 4.2

The preliminary vertical profile for an utility airport runway centerline is given. The specifications for the geometric design of the vertical profile of the centerline of a runway are given on Page 91.

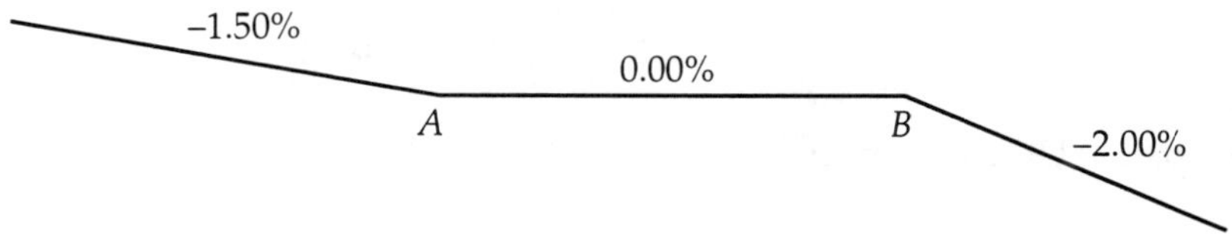

1. The minimum required length of a vertical curve at the point of intersection *A* is most nearly:
   (A) 150 ft
   (B) 300 ft
   (C) 450 ft
   (D) 600 ft

2. The minimum required length of a vertical curve at the point of intersection *B* is most nearly:
   (A) 100 ft
   (B) 200 ft
   (C) 300 ft
   (D) 600 ft

3. The minimum required distance between point of intersection *A* and point of intersection *B* is most nearly:
   (A) 150 ft
   (B) 200 ft
   (C) 675 ft
   (D) 875 ft

4. The maximum gradient of the runway centerline between point of intersection *A* and point of intersection *B* is most nearly:
   (A) 0.00%
   (B) +0.50%
   (C) +1.00%
   (D) +1.50%

Solutions:

1. **C** p.91 - Utility Airport: $L_1 = 300(1.50) = 450$ ft

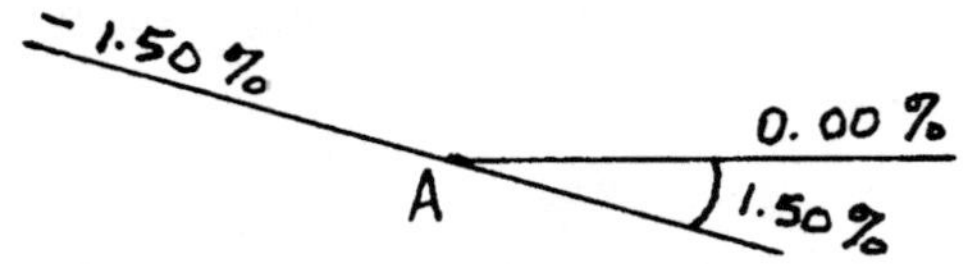

2. **D** p.91 - Utility Airports: $l_2 = 300(2.00) = 600$ ft

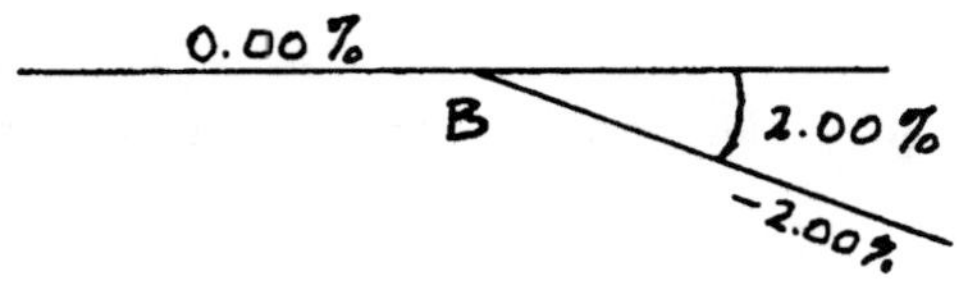

3. **D** p.91 - Utility Airports: $D = 250(A + B)$

$$= 250(1.50 + 2.00) = 875 \text{ ft}$$

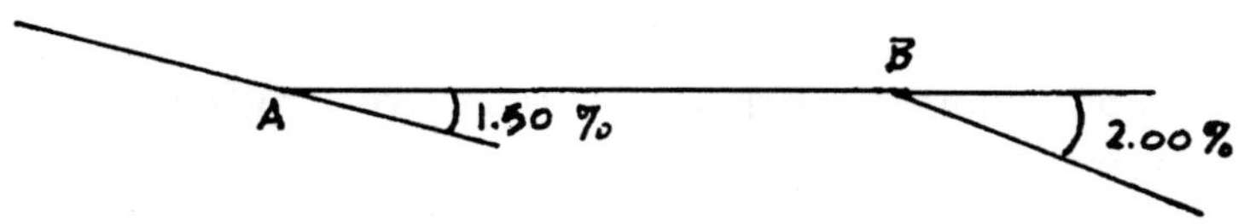

4. **A** From p.91 the maximum gradient = 2.00%. But also the maximum grade change = 2.00%.

$A + A$:

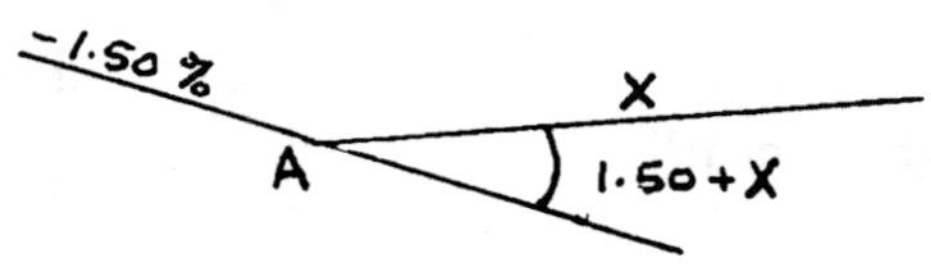

$1.50 + X \leq 2.00$. $\therefore X = 0.5\%$

$A + B$:

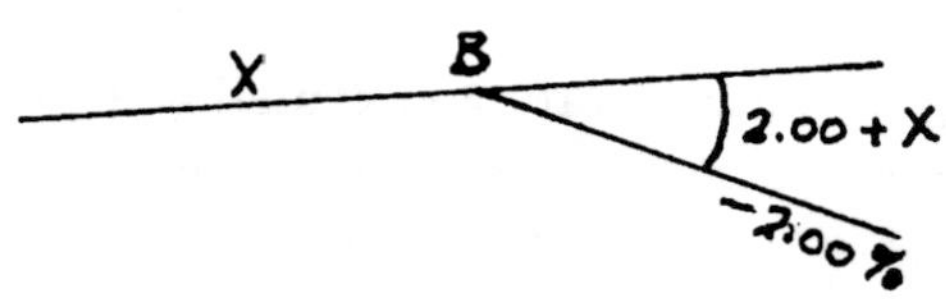

$X + 2.00 \leq 2.00$. $\therefore X = 0.00\%$

Example 4.3

The historical pattern of wind at an airport location has been plotted in a windrose given below. The wind speeds plotted are in the ranges between 0-4, 4-15, 15-31 and 31-47 miles per hour. The airport runway is to be designed for aircraft weighing more than 12,500 pounds. The specifications for wind analysis are given on Page 90 of the NCEES Handbook .

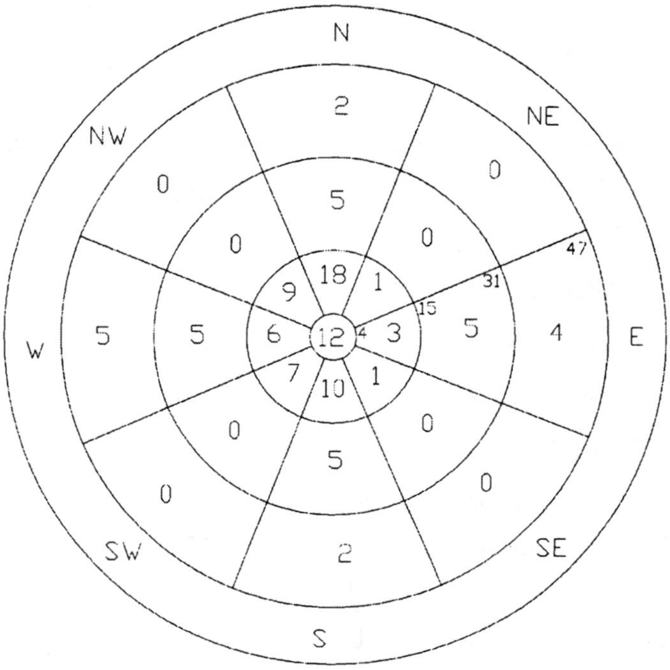

1. The optimal runway orientation at this airport is most nearly:
   (A) N-S
   (B) NW-SE
   (C) E-W
   (D) NE- SW

2. The percentage of time a runway in the N-S direction may be used by aircraft weighing more than 12,500 pounds at this airport without exceeding cross-wind specifications is most nearly:
   (A) 67%
   (B) 81%
   (C) 87%
   (D) 95%

3. The percentage of time the winds come from the NW and have a wind speed of between 4 and 15 miles per hour is:
   (A) 1%
   (B) 9%
   (C) 13%
   (D) 21%

4. The percentage of time a runway in the E-W direction may be used by aircraft taking off and landing towards the East without exceeding cross-wind specifications is most nearly:
   (A) 28%
   (B) 40%
   (C) 56%
   (D) 68%

Solutions:

1. **C** Based on wind data given on the next four pages, the runway orientation and percent uses are:

   N-S: 81%

   NW-SE: 77%

   E-W: 86%

   NE-SW: 77%

   Therefore, E-W maximized wind coverage.

1. **Continued** ---- Runway in N-S direction ----

   p.90 infers the maximum cross wind to be 15 mph.

   % = 12+18+1+3+1+10+7+6+9+5+5+2+2 = 81%

   The N-S runway may be used 81% of the time with cross winds less than 15 mph.

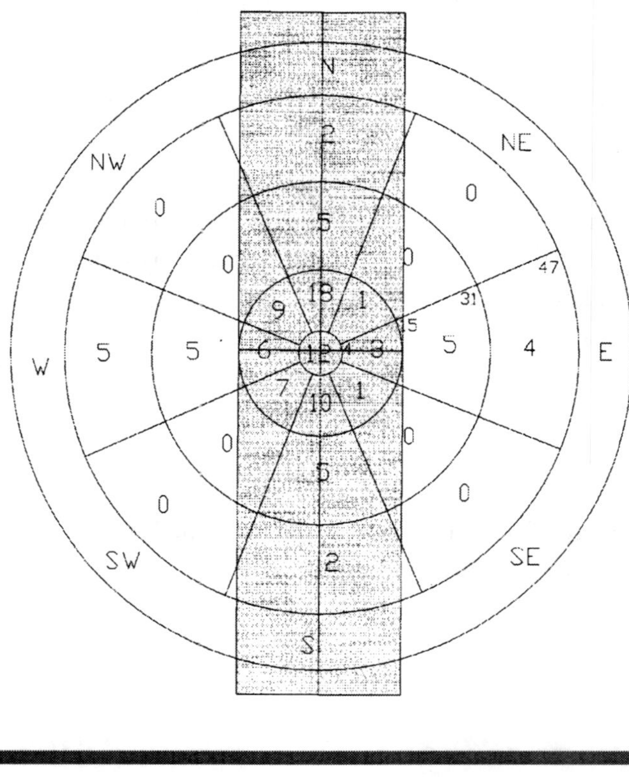

1. **Continued** ---- Runway in the NW-SE direction ----

p.90 infers the maximum cross wind to be 15 mph.

% = 12+18+1+3+1+10+7+6+9+2.5+2.5+2.5+2.5 = 77%

The NW-SE runway may be used 77% of the time with cross winds less than 15 mph.

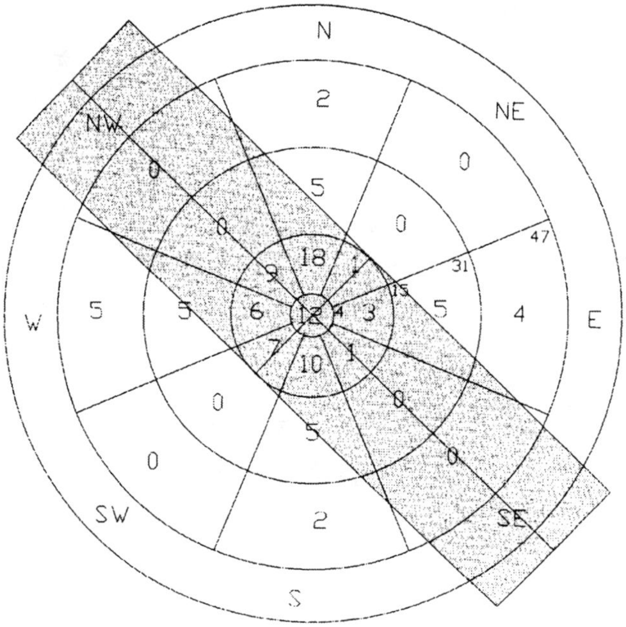

1. **Continued**---- Runway in E-W direction ----

p.90 infers the maximum cross wind to be 15 mph.

% = 12+18+1+3+1+10+7+6+9+5+5+5+4 = 86%

The E-W runway may be used 86% of the time with cross winds less than 15 mph.

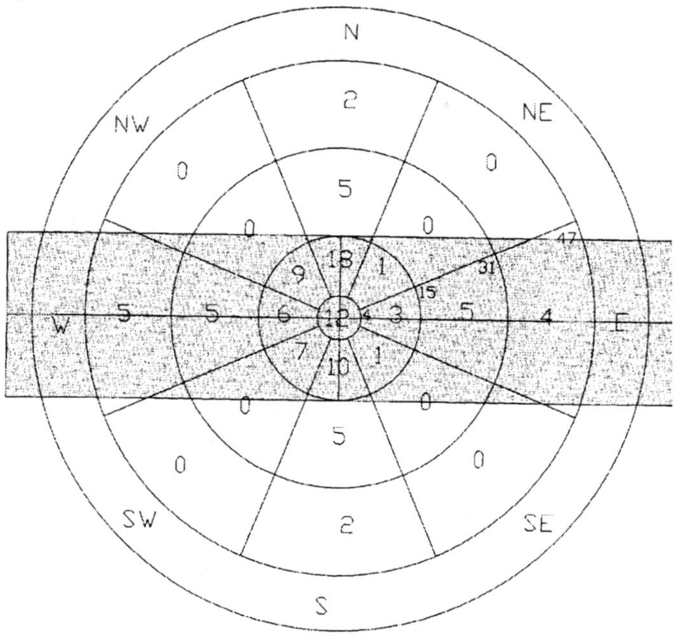

1. **Continued** ---- Runway in NE-SW direction:

p.90 infers the maximum cross wind to be 15 mph.

% = 12+18+1+3+1+10+7+6+9+2.5+2.5+2.5+2.5 = 77%

The NE-SW runway may be used 77% of the time with cross winds less than 15 mph.

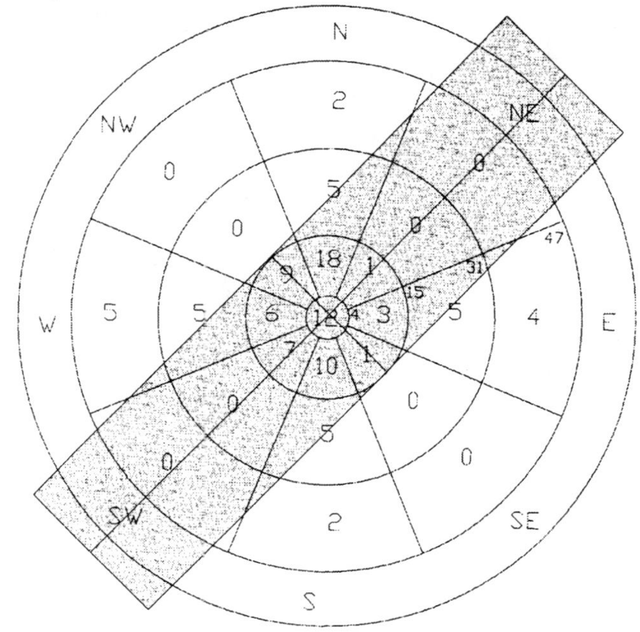

2. **B** For a runway in the N-S direction:

% = 12+18+1+3+1+10+7+6+9+5+5+2+2 = 81%

The N-S runway may be used 81% of the time with cross winds less than 15 mph.

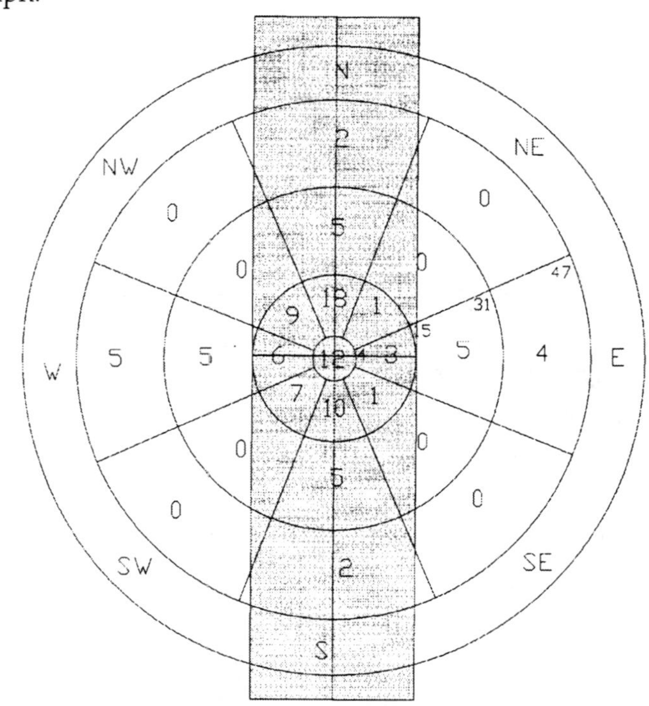

3. **B** The shaded area represents the percentage of time the wind comes from the NW and is between 4 and 15 mph. Therefore, 9% of the time.

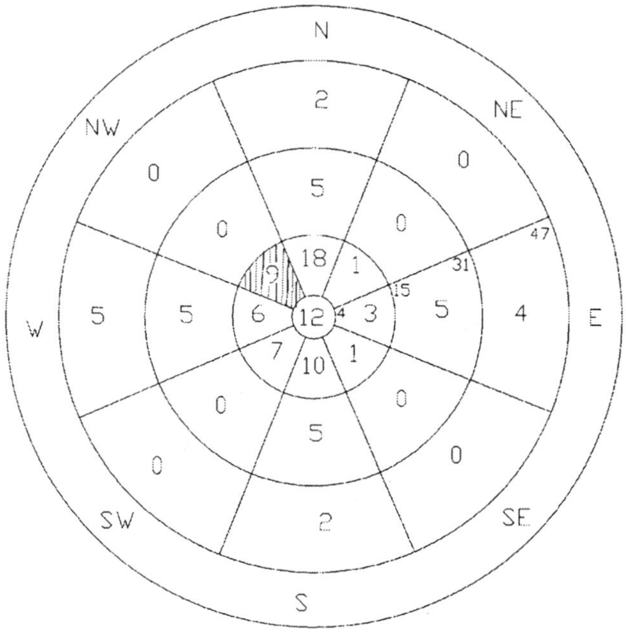

4. **B** The shaded area represents the percentage of time a runway may be used in an East direction so that cross winds are less than 15 mph:

% = 12+9+1+3+1+5+5+4 = 40% of the time.

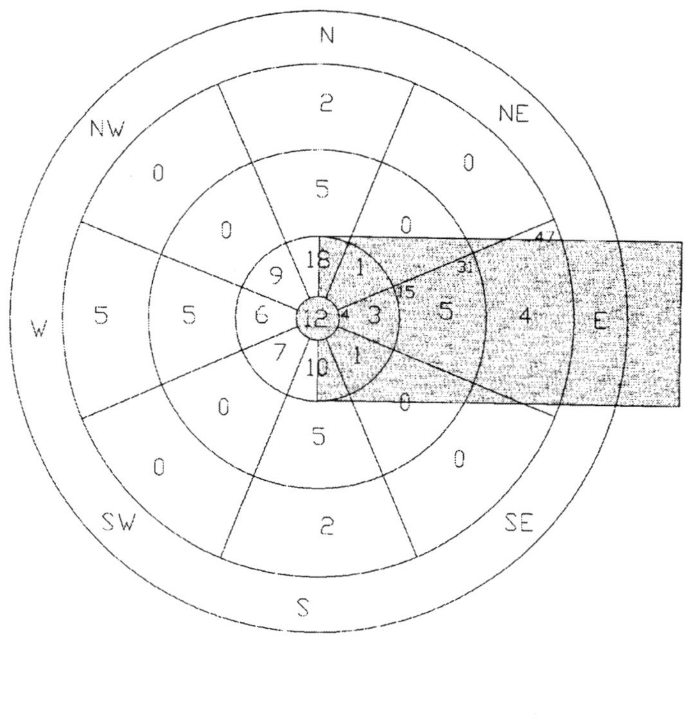

5. Pavement Design

by Gilbert Baladi

Problems involving pavement design can be worked referring to the equations and table on page 91 of the NCEES Handbook. The following examples illustrate the solution technique.

Example 5.1

A coal fired power plant is served by 25 trucks. When loaded, each truck consists of one single-axle of 14,000 pounds (the front axle), two tandem-axles of 32,000 pounds each (the middle axles), and one single-axle of 22,000 pounds (the back axle). Each truck makes 25 daily trips. The weights on the same axles when the truck is empty and leaving the power plant are 10,000, 12,000, and 8,000. A two-lane road is to be designed to serve the power plant and to last for a period of 10 years.

1. What is the load equivalency factor for a 14,000-pound single-axle?
   (A) 0.027
   (B) 0.36
   (C) 0.189
   (D) 1.0

2. What is the load equivalency factor for a 32,000-pound tandem-axle?
   (A) 8.88
   (B) 6.97
   (C) 0.36
   (D) 0.857

3. What is the design ESAL for the incoming traffic?
   (A) 9,704,438
   (B) 388,176
   (C) 970,444
   (D) 26,588

4. What is the design ESAL for the outgoing traffic?
   (A) 344,012
   (B) 943
   (C) 34,402
   (D) 1,376

Solutions:

1. **B** From the table on page 91, the load equivalency factor for a 14,000-pound single-axle load is 0.36.

2. **D** From the table on page 91, the load equivalency factor for a 32,000-pound tandem-axle load is 0.857.

3. **A** To calculate the design ESAL for the incoming traffic, first calculate the ESAL per truck. This implies that the load equivalency factor (LEF) for each axle on the truck must be obtained:

   For the 14,000-pound single-axle, LEF = 0.36.
   For the 32,000-pound tandem-axle, LEF = 0.857.
   For the 220,000-pound single-axle, LEF = 2.18.

   The total ESAL per truck = 1(0.36) + 2(0.857) + 1(2.18) = 4.254 ESAL/truck.

   For 25 trucks, with each truck making 25 trips per day, the total ESAL per day is

   $$25(25)(4.254) = 2658.75 \text{ ESAL/day}$$

   For a 10-year design life, the total design ESAL is

   $$10(365)(2658.75) = 9{,}704{,}438 \text{ ESAL}$$

4. **A** To calculate the design ESAL for the outgoing traffic, follow the same steps as in the solution to question 3, except use the axle weights of the empty truck:

   For the 10,000-pound single-axle, LEF = 0.0877.
   For the 12,000-pound tandem-axle, LEF = 0.0144.
   For the 8,000-pound single-axle, LEF = 0.0343.

   The total ESAL per truck = 1(0.0877) + 2(0.0144) + 1(0.0343) = 0.1508 ESAL/truck.

   For 25 trucks, with each truck making 25 trips per day, the total ESAL per day is

   $$25(25)(0.1508) = 94.25 \text{ ESAL/day}$$

   For a 10-year design life, the total design ESAL is

   $$10(365)(94.25) = 344{,}013 \text{ ESAL}$$

Example 5.2

For the power plant of Example 5.1, the design structural number for the incoming traffic (loaded trucks) is 5.2 and for the outgoing traffic (empty trucks) is 3.0. The design engineer used the following material for road construction:

A 12-inch-thick sand sub-base with a layer coefficient of 0.10
A 6-inch-gravel base with a layer coefficient of 0.13
An asphalt-concrete surface layer with a layer coefficient of 0.40

1. What is the required asphalt-concrete thickness for the incoming traffic (loaded trucks)?
   (A) 8 in.
   (B) 6 in.
   (C) 10 in.
   (D) 4 in.

2. What is the required asphalt-concrete thickness for the outgoing traffic (empty truck)?
   (A) 3 in.
   (B) 3.6 in.
   (C) 2.6 in.
   (D) 4 in.

3. For the material data given above, what is the required gravel thickness that will produce the same structural number as a one-inch-thick asphalt-concrete layer?
   (A) 3.1 in.
   (B) 0.33 in.
   (C) 6 in.
   (D) 1 in.

Solutions:

1. **A** Use the equation on page 91:

$$SN = a_1D_1 + a_2D_2 + a_3D_3$$

Solving for the thickness of the asphalt-concrete yields

$$D_1 = \frac{SN - a_2D_2 - a_3D_3}{a_1}$$

Therefore,

$$D_1 = \frac{5.2 - 0.1(12) - 0.13(6)}{.4} = 8 \text{ in.}$$

2. **C** Solving for D_1 by following the same steps as in the solution to Question 1 yields

$$D_1 = \frac{3.0 - 0.1(12) - 0.13(6)}{.4} = 2.6 \text{ in.}$$

3. **A** The structural number (SN) of a one-inch asphalt-concrete layer is 1(0.4) = 0.4. The required thickness D_2 of the gravel that will yield a structural number of 0.4 can be calculated from the following equation:

$$SN_2 = a_2D_2$$

Solving for D_2 gives

$$D_2 = SN_2/a_2$$
$$= 0.4/0.13 = 3.1 \text{ in.}$$

Example 5.3

Two single-axle trucks travel one road in one direction twice each day. The first truck consists of a 10,000-pound front single-axle and an 18,000-pound rear single-axle. The second truck consists of a 12,000-pound front single-axle and a 24,000-pound rear single-axle. Assume that , for each trip, the first truck delivers a unit damage to the road.

1. For each trip, how many units of damage is delivered to the road by the second truck?
   (A) 3.22
   (B) 2.96
   (C) 1.09
   (D) 0.34

2. For each day, how many units of damage is delivered to the road by the second truck?
   (A) 6.44
   (B) 2.18
   (C) 4.62
   (D) 5.92

Solutions:

1. **B** The relative damage delivered by each truck is proportional to the ESAL of each truck. Thus, the ESAL of each truck must be calculated first.

   Truck 1:

   The LEF for a 10,000-pound single-axle = 0.0877
   The LEF for an 18,000-pound single-axle = 1.0
   The total ESAL of truck 1 = 1.0 + 0.0877 = 1.0877

   Truck 2:

   The LEF for a 12,000-pound single-axle = 0.189
   The LEF for a 24,000-pound single-axle = 3.03
   The total ESAL of truck 2 = 3.03 + 0.189 = 3.219

   The number of units of damage delivered to the pavement by truck 2 is

   $$3.03/1.0877 = 2.96$$

2. **D** The second truck makes two trips per day. The number of units of damage is

   $$2(2.96) = 5.92$$

Example 5.4

The required structural number SN of an asphalt pavement to protect the roadbed soil is 3.5. The required structural number to protect the sub-base material is 2.84. The required structural number to protect the base material is 2.0. The pavement is constructed of the following materials:

Sand sub-base with a layer coefficient of 0.11.
Gravel base with a layer coefficient of 0.14.
Asphalt-concrete surface with a layer coefficient of 0.4.

1. What is the required thickness of the asphalt-concrete layer?
   (A) 6.0 in.
   (B) 5.0 in.
   (C) 4.0 in.
   (D) 3.0 in.

2. What is the required thickness of the base layer?
   (A) 6.0 in.
   (B) 5.0 in.
   (C) 4.0 in.
   (D) 3.0 in.

3. What is the required thickness of the sub-base layer?
   (A) 6.0 in.
   (B) 5.0 in.
   (C) 4.0 in.
   (D) 3.0 in.

Solutions:

1. **B** The AASHTO structural number SN equation is given as

$$SN = a_1D_1 + a_2D_2 + a_3D_3$$

where the a_1 and D_1 are the layer coefficient and layer thickness. The above equation can be rewritten as follows:

$$SN = SN_1 + SN_2 + SN_3$$

where SN is the structural number of the pavement, and SN_i is the SN of layer i. Since the required SN to protect the base material is 2.0, it is inferred that the SN of the asphalt-concrete is 2.0. Therefore:

$$SN_1 = a_1D_1 = 2.0$$

We solve for D_1:

$$D_1 = SN_1/a_1 = 2.0 / 0.4 = 5 \text{ in.}$$

2. **A** Given that the required structural number to protect the sub-base is 2.84, we have

$$SN = a_1D_1 + a_2D_2 = 2.84$$

Since $a_1 = 0.4$, $D_1 = 6$ in., and $a_2 = 0.14$, we solve for D_2 as follows:

$$D_2 = \frac{SN - a_1D_1}{a_2}$$

Therefore,

$$\mathrm{D}_2 = \frac{2.84 - 0.4 \times 5}{0.14} = 6 \text{ in.}$$

3. **A** The required thickness of the sub-base layer can be obtained by solving the following structural number equation:

$$SN = a_1D_1 + a_2D_2 + a_3D_3$$

All variables are known except the thickness D_3 of the sub-base. Solving for D_3:

$$D_3 = \frac{SN - a_1D_1 - a_2D_2}{a_3}$$

Therefore,

$$D_3 = \frac{3.5 - 0.4 \times 5 - 0.14 \times 6}{0.11} = 6 \text{ in.}$$

6. Environmental Engineering

by Susan Masten

To work problems relating to Environmental Engineering, it will be necessary to refer to the equations on page 89 and the figures on page 94 of the NCEES Handbook. It may also be necessary to refer to the information contained in the Chemistry section on pages 59 and 60.

6.1 Hardness Removal

Many groundwaters (and a few surface waters) contain excessive minerals which must be removed prior to distribution of the treated water. Hardness, which is caused by the presence of multivalent cations, usually calcium and magnesium, is problematic because of (1) the formation of slimy, sticky deposits as a result of the interactions between the hardness ions and the soap, (2) the tendency of soaps not to produce a good lather, (3) the formation of hard, rock-like deposits in pipes and boilers.

Hardness may be present as either carbonate or noncarbonate hardness. Carbonate hardness is that portion of the (total) hardness that is associated with carbonate and bicarbonate anions. It is easily removed by precipitation (as calcium carbonate) upon heating of the water. As such, carbonate hardness causes scaling problems in water heaters and hot water pipes. Noncarbonate hardness is that portion of the hardness minerals that is associated with anions other than carbonate and bicarbonate. This form of hardness is more difficult to remove and does not precipitate as the temperature of the water is increased.

The removal of hardness minerals can be accomplished by softening or by ion exchange. Softening is usually accomplished with lime. If insufficient alkalinity is present then soda ash is also used, Following softening, the pH of the treated water must be reduced to approximately 9.0. This is accomplished using recarbonation, in which carbon dioxide is bubbled into the water. The carbon dioxide dissolves and is hydrolyzed to form carbonic acid which lowers the pH. Ion exchange results in the exchange of Ca^{2+} and Mg^{2+} ions with equivalent number of sodium (Na^{2+}) ions. For example, calcium is removed by the reaction:

$$Ca(HCO_3)_2 + Na_2R \rightarrow CaR + 2NaHCO_3$$

where R is the resin support. Refer to the lime-soda softening equations (p. 89) along with the table of equivalent weights, molecular weights and equivalents weights (p.89) in the NCEES Handbook to solve these problems.

Example 6.1

A large University serving 44,000 students and 5,000 faculty and staff obtains its drinking water from an aquifer. Of the 44,000 students half live on campus. The water usage is:

Resident students (those living on campus): 100 gal/capita/day
Nonresident students, faculty and staff: 50 gal/capita/day

A partial analysis of the water has shown that it has the following characteristics (given as mg/L as $CaCO_3$, except for pH):

pH	8.1	Alkalinity	280.0
Ca^{2+}	365.0	Mg^{2}	74.0
Fe^{3+}	2.3	Na^{+}	50.0
K^{+}	26.0	SO_4^{2-}	236.0
CO_2	47.0		

A water has a calcium concentration of 20.0 mg/L (as the ion) after softening. The pH of the softened water is 10.8.

1. Calculate the theoretical amount of lime (in kg/day) that would be required to soften this water. Note that to raise the pH to precipitate magnesium hydroxide, an excess of 1.25 meq/L of lime must be provided.
   A) 3,987
   B) 3,251
   C) 4,608
   D) 2,784

2. Calculate the amount of calcium carbonate ($CaCO_3$) sludge (in kg) produced each day by the reaction of carbon dioxide with lime.
   A) 167
   B) 334
   C) 668
   D) 513

Solutions

1. **C** We must first calculate the concentrations of each of the reactive species using the data provide in the table of equivalent weights given on page 89 of the NCEES Handbook:

$$[CO_2] = \frac{47.0}{50.04} = 0.94 \text{ meq/L}$$

$$[Ca^{2+}] = \frac{365.0}{50.04} = 7.29 \text{ meq/L}$$

$$[Mg^{2+}] = \frac{74.0}{50.04} = 1.48 \text{ meq/L}$$

$$[HCO_3^-] = \frac{280.0}{50.04} = 5.60 \text{ meq/L}$$

(Note: As the pH = 8.1, Alk ≅ $[HCO_3^-]$)

We must also calculate the flow rate for the treatment plant:

$$Q = (22{,}000 \times 100 \text{ gal/cap/day}) + [(22{,}000 + 5000) \times 50 \text{ gal/cap/day}] = 3.55 \text{ MGD}$$

The amount of lime required can be calculated using equations 1,2, and 4 (Lime-Soda Softening Equations) on page 89 of the NCEES Handbook.

	meq/L	**meq/L lime required**	**meq/L soda ash required**
CO_2	0.94	0.94	
$Ca(HCO_3)$	5.60	5.60	
$CaSO_4$	1.69	—	1.69
$MgSO_4$	1.48	1.48	1.48
excess		1.25	
		9.27	

Mass of lime per day =

$$(9.27 \text{ meq } Ca(OH)_2/L) \times (37.046 \text{ mg } Ca(OH)_2/\text{meq}) \times (3.785 \text{ L/gal}) \times (3.55\times10^6 \text{ gal/day}) \times (\text{kg}/10^6 \text{ mg}) = 4608 \text{ kg/day}$$

2. **A** We can calculate the amount of calcium carbonate sludge produced using the stoichiometric equation given on page 89 of the NCEES Handbook:

$$CO_2 + Ca(OH)_2 \rightarrow CaCO_{3\,(s)} + H_2O$$

The concentration of carbon dioxide in the water is 0.47 mmol/L (= 47.0 mg/L as $CaCO_3$/100.086 mg/mmol). Since 1 mole of carbon dioxide reacts to form one mole of calcium carbonate, 0.47 mmol/L of calcium carbonate is produced.

Mass of sludge produced =

$$(0.47 \text{ mmol/L}) \times (100.08 \text{ mg/mmol}) \times (3.55 \times 10^6 \text{ gal/d}) \times (1 \text{ kg}/10^6 \text{ mg}) = 167.0 \text{ kg/d}$$

6.2 Grit Chamber Design

The initial steps in municipal wastewater treatment involve pretreatment using screens, comminutors, grit chambers and/or equalization basins.

Bar screens or racks are used to remove large objects like tree limbs and small animals that may get into the wastewater plant influent through storm sewer entrances or displaced sewer access (manhole) covers. These materials can seriously affect pumps and piping within the plant. Bar screens or racks may be mechanically or manually cleaned. Generally only smaller facilities have manually screened racks. Larger plants may use communitors in place of bar screens. Communitors grind up or chop large objects so that they do not clog pipes or pumps but can be removed in downstream processes.

Grit chambers follow bar screens and are used to remove grit, which consists primarily of sand, cinders, coffee grinds and pea gravel. Grit can originate from storm runoff or from material disposed of in home or commercial garbage disposal units.

Equilization basins are used to dampen diurnal variations in flow, BOD and suspended solids loading rates. It also enhances the degree of treatment in downstream processes. These basins are usually constructed of concrete, although earth or steel may be used. The wastewater in the basin is aerated to provide mixing, to keep the solids in suspension and to prevent the wastewater from becoming septic (anoxic).

The approach velocity is used in designing the first three of these unit operations. Use the equations given on p. 89 of the NCEES Handbook to design a grit chamber.

Example 6.2

A municipal wastewater is pretreated using a horizontal flow grit chamber having the following dimensions:

Length = 60′
Width = 4′
Depth = 6′

The design flow rate is 10.83 cfs. Calculate the approach velocity in the chamber.

(A) 0.05 ft/s
(B) 0.03 ft/s
(C) 0.68 ft/s
(D) 0.45 ft/s

Solution:

A_x = Cross-sectional area in grit chamber = 4′× 6′ = 24 ft^2. Using the equation for approach velocity given on page 89 of the NCEES Handbook, the approach velocity can be calculated as follows:

$$\text{approach velocity} = \frac{Q}{A_x} = \frac{10.83\ \text{ft}^3/\text{s}}{24\ \text{ft}^2} = 0.45\ \text{ft/s}$$

The answer is **D**.

6.3 Primary Settling

Primary treatment of municipal wastewater involves gravity settling of discrete particles. In some cases, coagulant aids may be added to enhance settling. The main uses of primary settling in waste water treatment are the removal of settleable solids and floating material such as oil and grease. Primary settling basins can achieve from 50 to 70% removal of suspended solids and 25 to 40% removal of BOD_5. These basins usually have a detention time from 1.5 to 3 h. Primary sedimentation basins may be rectangular, square, or circular in shape.

The detention time and weir loading are important design parameters for sedimentation basins. Detention time is also used in designing any type of process which involves chemical addition, including rapid mix, flocculation, chlorination and ozonation. Overflow (hydraulic loading) rates are commonly used for designing both sedimentation basins and filtration units.

Example 6.3

A municipal wastewater having a design flow rate of 2.5 MGD is to be treated using 2 primary clarifiers. The clarifiers each have the following characteristics:

Diameter = 43′
Depth = 10′
Weir length = 270′

1. Calculate the detention time (hydraulic residence time) in each of the clarifiers.
   (A) 4.16 hr
   (B) 2.08 hr
   (C) 0.52 hr
   (D) 1.04 hr

2. Calculate hydraulic loading (overflow) rate in units of ft^3/ft^2-hr.
   (A) 2.40
   (B) 4.79
   (C) 9.59
   (D) 19.18

3. What is the weir loading rate in units of $ft^3/ft\text{-}hr$?
   (A) 25.8
   (B) 12.9
   (C) 6.45
   (D) 51.6

Solutions:

1. **B** Use the equation for detention time (t_d) given on page 89 of the NCEES Handbook. Since the total flow is divided between two clarifiers, each clarifier is to be designed based upon a flow rate of 1.25 MGD:

$$Q = (1.25 \times 10^6 \text{ gal/day})(0.1337 \text{ ft}^3/\text{gal})(0.417 \text{ day/hr}) = 6962 \text{ ft}^3/\text{hr}$$

$$t_d = V/Q = \frac{H\pi d^2/4}{6962 \text{ ft}^3/\text{hr}} = \frac{10 \text{ ft } \pi(43 \text{ ft})^2/4}{6962 \text{ ft}^3/\text{hr}} = 2.08 \text{ hr}$$

where d = diameter and H = depth of the clarifier.

2. **B** The equation for the overflow rate is

$$\text{Overflow rate} = Q/A = \frac{6962 \text{ ft}^3/\text{hr}}{\pi d^2/4} = 4.79 \text{ ft}^3/\text{ft}^2\text{-hr}$$

3. **A** The equation for the weir loading rate provides

$$\text{Weir loading rate} = Q/L = \frac{6962 \text{ ft}^3/\text{hr}}{270 \text{ ft}} = 25.8 \text{ ft}^3/\text{ft-hr}$$

6.4 Activated Sludge

Activated sludge treatment uses a consortium of microorganisms in a well-mixed tank to oxidize organic matter under aerobic conditions. The main purpose of activated sludge treatment is to remove soluble BOD from the wastewater. The main components of the system include a biological reaction basin (aeration basin), a clarifier for sedimentation of the biomass and recycle sludge pumps to recycle a portion of the biomass (return activated sludge, RAS) back to the head of the aeration basin. The activated sludge-wastewater mixture is referred to as mixed liquor suspended solids (MLSS). Typically, the concentration of MLSS ranges from 2,000 to 4,000 mg/L. The concentration of microorganisms is often quantified as mixed liquor volatile suspended solids (MLVSS). Because these systems use aerobic organisms to consume the organic matter, either air or pure oxygen must be supplied in the aeration basins. This is usually accomplished by either diffused aerators or by mechanical aerators, the former being more efficient.

Among the major parameters used in designing activated sludge systems are: F:M (organic loading rate), the volumetric organic loading rate, the solids retention time and the solids loading rate to the secondary clarifier.

Example 6.4

A municipal wastewater treatment plant is to be designed. The design flow rate is 2.5 MGD. The wastewater has the following characteristics after primary treatment:

Influent BOD = 200 mg/L
Influent Suspended Solids = 200 mg/L

The wastewater must have the following characteristics after secondary treatment.

Effluent BOD = 15 mg/L
Effluent SS = 20 mg/L

A preliminary design for a complete-mix activated sludge system has yielded the following design parameters:

Sludge volume index (SVI) = 125
Return activated sludge concentration = 8000 mg/L
MLSS = 3500 mg/L
Recycle = 50 %
Depth = 15 ft
Width = 50 ft
Length = 100 ft

Sedimentation of the waste activated sludge is accomplished using secondary clarification. Two secondary clarifiers (each 28 ft. in diameter) are proposed.

1. Calculate *F:M* in units of mg BOD applied/mg MLSS-day.
   (A) 0.255
   (B) 0.128
   (C) 0.111
   (D) 0.510

2. Calculate the volumetric organic loading rate in units of kg BOD/m^3-day.
   (A) 0.45
   (B) 15.6
   (C) 35.6
   (D) 0.89

3. Calculate the solids residence time.
   (A) 10.7 hr
   (B) 4.7 hr
   (C) 0.44 hr
   (D) 1.01 hr

4. Calculate the solids loading rate in lb/ft^2-day to the secondary clarifier.
   (A) 135
   (B) 118.4
   (C) 59.2
   (D) 270

Solutions:

1. **A** $V_A = (15 \text{ ft})(50 \text{ ft})(100 \text{ ft}) = 75{,}000 \text{ ft}^3$

   The equation for the *F:M* ratio can be obtained from page 89 of the NCEES Handbook.

$$F{:}M = \frac{QS_0}{V_A X_A} = \frac{2.5\times10^6 \text{ gal / day}(0.1337 \text{ ft}^3 \text{ / gal})(200 \text{ mg / L})}{75{,}000 \text{ ft}^3(3{,}500 \text{ mg / L})}$$

$$= 0.255 \text{ mg BOD applied/mg MLSS-day}$$

2. **D** The equation for the volumetric organic loading rate is:

   Volumetric organic loading rate $= QS_0 / V =$

$$\frac{2.5\times10^6 \text{ gal / day}(200 \text{ mg / L})(0.1337 \text{ ft}^3 \text{ / gal})(10^3 \text{ L / m}^3)(1 \text{ kg / }10^6 \text{ mg})}{75{,}000 \text{ ft}^3}$$

$$= 0.89 \text{ kg BOD/m}^3\text{-day}$$

3. **B** The equation for the solids residence time is:

$$\text{Solids residence time} = \frac{V_A X_A}{(Q_w X_w + Q_e X_e)} =$$

$$\frac{75{,}000 \text{ ft}^3(3500 \text{ mg / L})}{\{0.5(2.5\times10^6 \text{gal / day})(8000 \text{ mg / L}) + 2.5\times10^6 \text{ gal / day}(20 \text{ mg / L})\}0.1337 \text{ ft}^3 \text{ / gal}}$$

$$= 0.19 \text{ day} = 4.7 \text{ hr}$$

4. **C** The equation for the solids residence time can be obtained from page 89 of the NCEES Handbook.

   Solids loading rate $= QX/A =$

$$\frac{1.25\times10^6 \text{ gal / day}(3.785 \text{ L / gal})(3500 \text{ mg / L})(1 \text{ g / }10^3 \text{ mg})(1 \text{ lb / 454 g})}{\pi(28 \text{ ft})^2 / 4}$$

$$= 59.2 \text{ lb/ft}^2\text{-day}$$

6.5 Fixed-film Processes

Tricking filters and rotating biological contactors (RBCs), i.e., attached growth processes, use a consortium of microorganisms that are attached to the surface of a medium to oxidize BOD in wastewater. For this reason, they are referred to as fixed-film or attached-growth processes. The medium used in trickling filters is stationary, while the disks used in RBCs rotate around a central axis in a basin containing the wastewater. As with activated sludge treatment, the purpose of these processes is to reduce the concentration of soluble BOD in the wastewater. As the wastewater contacts the microorganisms in the biofilm on the trickling filter medium (usually rock or plastic) or on the RBC plates, the microorganisms use the oxygen demanding material in the wastewater as a growth substrate. As with activated sludge, oxygen must be provided for these processes to be effective.

Example 6.5

The wastewater flow rate from the City of Oz is 1.20 MGD. The influent BOD_5 and suspended solids (SS) are 240 mg/L and 220 mg/L, respectively. The wastewater is treated by primary sedimentation which removes 33% of the BOD_5 and 35% of the SS followed by a 4-stage RBC unit. The effluent BOD_5 must be less than 20 mg/L. The total surface area of the discs is 700,000 ft^2. Calculate the organic loading rate (based on the surface area) in units of lb BOD/10^3 ft^2-day to the RBC unit.

(A) 4.2

(B) 1.41

(C) 2.30

(D) 0.53

Solution:

The equation for the organic loading rate can be obtained from page 89 of the NCEES Handbook:

$$\text{Organic loading rate} = \frac{QS_o}{A_m} =$$

$$\frac{1.20 \times 10^6 \text{ gal / day}(240 \text{ mg / L})(0.67)(3.785 \text{ L / gal})(1 \text{ mg / } 10^3 \text{ g})(\text{lb / } 454 \text{ g})}{700{,}000 \text{ ft}^3}$$

$$= 2.30 \text{ lb BOD/}10^3 \text{ ft}^2\text{-day}$$

The answer is **C**.

6.6 Sludge Production

Sludge is produced by a number of processes. For example, in water treatment chemical sludges are produced by the coagulation of naturally occurring organic matter and turbidity in surface waters and by the softening of groundwaters. Sludges are also produced during primary and secondary clarification of municipal wastewater. Since sludge contains predominately water, sludge processing units must be designed based upon the wet weight of the sludge, rather than the dry weight.

Example 6.6

The primary sludge from the City of Oz (see Example 6.5) is 3.5% solids. The density of the sludge is 1020 kg/m^3.

1. Calculate the sludge production rate on a dry weight basis.
   (A) 1,000 kg/day
   (B) 382 kg/day
   (C) 350 kg/day
   (D) 1,100 kg/day

2. Calculate the sludge production rate on a wet weight basis.
   (A) 28.0 m^3/day
   (B) 9.8 m^3/day
   (C) 30.8 m^3/day
   (D) 10.7 m^3/day

Solutions:

1. **C** The sludge production rate can be calculated based using the flow rate, the influent SS concentration and the percent removal:

   Sludge production rate =

   $$(1.20 \times 10^6 \text{ gal/day})(220 \text{ mg/L})(0.35)(3.785 \text{ L/gal})(\text{kg}/10^6 \text{ mg}) = 350 \text{ kg/day}.$$

2. **B** The equation for the sludge production rate on a wet weight basis can be obtained from page 89 of the NCEES Handbook. Note the parameter, density, in the numerator is the density of the sludge, not the density of water.

   Sludge production rate on a wet weight basis =

   $$\frac{M(100)}{\rho(\%\ solids)} = \frac{350 \text{ kg / day} \times 100}{1020 \text{ kg / m}^3 \times 3.5} = 9.8 \text{ m}^3\text{/day}$$

6.7 Chemical Dosing

Numerous chemicals must be added during treatment of water and wastewater. The mass of chemicals that are to be added on a daily basis must be calculated to ensure that a sufficient supply of the required chemicals are maintained on site.

Example 6.7

A water treatment plant having an average daily flow rate of 17.5 MGD adds lime during softening at a dosage of 245.0 mg/L. Calculate the mass of lime that must be supplied each day.

(A) 40,000 lb

(B) 24,000 lb

(C) 35,800 lb

(D) 61,000 lb

Solution:

The mass of a chemical that is to be added can be calculated using the flow rate and the dosing rate. The equation is provided on page 89 of the NCEES Handbook:

$$\frac{\text{mass}}{\text{day}} = \frac{Q\ \text{MGD} \times \text{Concentration mg / L} \times 8.34\ \text{lb / MGal}}{\text{mg / L}}$$

$$= 17.5\ \text{MGD} \times 245\ \text{mg / L} \times 8.34 = 35{,}800\ \text{lb / day}$$

The answer is **C**.

6.8 Biochemical Oxygen Demand (BOD)

Biochemical oxygen demand (BOD) is the amount of oxygen required by microorganisms, predominately bacteria, to oxidize organic matter under aerobic conditions. The BOD test is conducted in the dark (to avoid algae growth) at 20 °C and in the presence of an excess of nutrients. This assay (BOD_5) is conventionally run for 5 days. If a wastewater is analyzed for BOD on a daily basis, one would observe that the concentration of BOD increases exponentially and asymptotes towards a constant value, known as the ultimate BOD. The ultimate BOD is usually reached after about 20 days.

Example 6.8

A 5 day BOD was determined to be 250.0 mg/L, while the 20 day BOD (assume that this is the ultimate BOD) was 300.0 mg/L. Determine the BOD rate constant (base e).

(A) 0.36 day^{-1}
(B) 0.037 day^{-1}
(C) 0.16 day^{-1}
(D) 0.016 day^{-1}

Solution:

Use the equation for BOD exertion given on page 89 of the NCEES Handbook to solve this problem:

$$y_t = L(1 - e^{-kt})$$

Solving for k yields:

$$k = -\frac{\ln\left(1 - \frac{y_t}{L}\right)}{t} = -\frac{\ln\left(1 - \frac{250}{300}\right)}{5} = 0.36\ \text{day}^{-1}$$

The answer is **A**.

6.9 Sewage Flow Ratio Curves

Wastewater flows vary depending upon a variety of factors including the season of the year, time of day, affluence of the community, and weather conditions. Wastewater treatment plants are designed based upon low, average and peak dry weather flows. The sewage flow ratio curves (p. 94 of the NCEES Handbook) can be used to estimate the peak and minimum dry weather flows based upon the size of the population served by the treatment plant. These curves are based upon equations and graphical relationships developed from case studies.

Example 6.9

A city of 100,000 people has an average wastewater flow of 10 MGD. The maximum and minimum flows can be estimated by the sewage flow ratio curves. Using the equations $\frac{14}{4+\sqrt{P}}$ (Curve B) and $\frac{5}{P^{0.107}}$ (Curve A_2) estimate the minimum flow.

(A) 20 MGD
(B) 5 MGD
(C) 4.3 MGD
(D) 3 MGD

Solution:

Using the equation $\frac{5}{P^{0.107}}$ (Curve A_2), we can estimate the minimum flow. The population is 100 thousand, so read the ratio for the point where the line for 100 thousand intersects with curve A_2. The ratio is 0.5. Therefore, the minimum flow is 0.5×10 MGD = 5 MGD

The answer is **B**.

6.10 Hydraulic-Elements Graph

Wastewater and storm run-off are commonly collected in circular pipes. In many cases, these sewers do not flow full and it is time-consuming to calculate the hydraulic radii and the cross-sectional area for pipes that are partially full. The Hydraulic-Elements Graph for Circular Sewers (presented on p. 94 of the NCEES Handbook) can be used to calculate partial-flow values from full-flow conditions.

Example 6.10

A circular sewer when flowing full flows at a flow rate of 10 m^3/s and a velocity of 6 m/s.

1. What is the velocity when $Q = 6\ m^3/s$?
   (A) 4.2 m/s
   (B) 5.6 m/s
   (C) 6.3 m/s
   (D) 7.2 m/s

2. What is the full-flow depth if the depth of flow when $Q = 6\ m^3/s$ is 80 cm?
   (A) 2.15 m
   (B) 1.95 m
   (C) 1.65 m
   (D) 1.45 m

Solutions:

Use the chart provided on page 94 of the NCEES Handbook. The ratio of the flows ($Q/Q_f = 6/10$) is 0.60. Enter this value on the chart and read off the corresponding values for the velocity and depth of flow. Use the dashed lines (for constant Darcy-Weisbach friction factor *f* and Manning's constant *n*).

1. **C** Read 1.05 on the horizontal axis using the dashed *V*-line:

$$V = 1.05 \times 6 = 6.3\ m/s$$

2. **D** Read 0.55 on the vertical axis. Then

$$\frac{d}{D} = 0.55 \quad \text{or} \quad D = \frac{0.8}{0.55} = 1.45\ m$$

7. Hydrology and Fluid Flow

by Merle C. Potter

The equation used to solve problems related to hydrology is found on page 90 of the NCEES Handbook. Other equations on fluid mechanics are also relavent to this subject, particularly those relating to flow rate (discharge or volumetric flow rate) ans mass flux(mass flow rate). Problems involving fluid flows will use the equations found on pages 34 to 42 of the Handbook. Also, review the problems in the chapter on Fluid Mechanics in the Mechanical Engineering part of this book, especially those in the sections on the energy equation and fans, pumps and turbines.

Example 7.1

The runoff from a 50-hectare parcel, in the Chicago area, flows through a 180-meter-long storm sewer pipe. The pipe is designed for a 10-year, 30-minute duration storm. The ground elevation of the pipe inlet is 201 m and it is 200 m for the pipe outlet. The dimensionless runoff coefficient is 0.3 and the Manning n is 0.015. Two intensity-duration curves for the Chicago area follow.

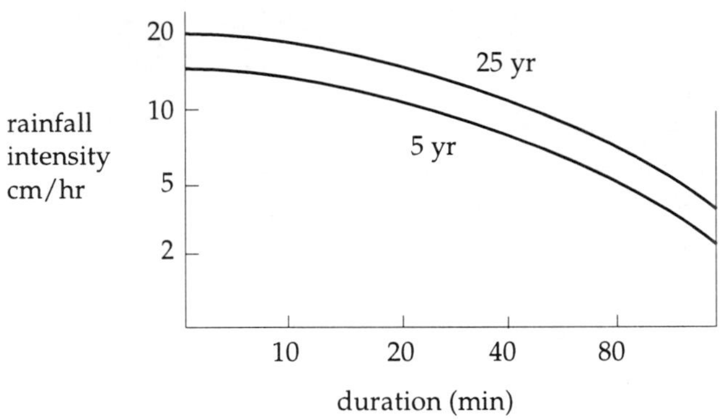

1. Estimate the flow rate.
   - (A) 4 m^3/s
   - (B) 3 m^3/s
   - (C) 2 m^3/s
   - (D) 1 m^3/s

2. The slope of the sewer pipe is:
   (A) 0.0078
   (B) 0.0067
   (C) 0.0056
   (D) 0.0045

3. The diameter of the sewer pipe should be sized at:
   (A) 4 m
   (B) 3 m
   (C) 2 m
   (D) 1.5 m

Solutions:

1. **A** Use the formula at the top of p.90:

$$Q = ciA = 0.3 \times 10 \times 50 \times (10^4 \times \frac{1}{100} \times \frac{1}{3600}) = 4.16 \quad \text{m}^3/\text{s}$$

The 10^4 changes hectares to m^2; the 100 changes cm to m; the 3600 changes hours to seconds. Find i on the given figure.

2. **C** The slope is:

$$S = \Delta z / \Delta L = (201 - 200) / 180 = 0.005556$$

3. **D** Use Manning's equation on p. 38 to find the diameter assuming the pipe flows full:

$$V = Q / \pi D^2 / 4 = (1/n)(D/4)^{2/3} S^{1/2}$$

$$4.16 / \pi D^2 / 4 = (1/0.015)(D^{2/3} / 4^{2/3})0.00556^{1/2} \qquad \therefore D = 1.45\text{m}$$

We used the hydraulic radius $R = D/4$.

Example 7.2

A 2-m-dameter, 200-m-long, cast-iron pipe transports water from a reservoir with surface elevation 726 m to an 89%-efficient turbine which has its outlet at 696 m. The turbine operates such that the flow rate is 6 m^3/s. Use kinematic viscosity $\nu = 10^{-6}$ m^2/s.

1. A 15-m-wide, 1.2-m deep river feeds the reservoir from above. Estimate the river's slope if the Manning n is 0.035.
   (A) 0.013
   (B) 0.0013
   (C) 0.00013
   (D) 0.000013

2. Approximate the losses up to the inlet of the turbine.
   (A) 20 m
   (B) 10 m
   (C) 2.5 m
   (D) 0.25 m

3. What is the expected output of the turbine?
   (A) 1560 kW
   (B) 960 kW
   (C) 270 kW
   (D) 43.2 kW

4. Estimate the maximum possible velocity entering a blade on the turbine.
   (A) 15 m/s
   (B) 28 m/s
   (C) 37 m/s
   (D) 43 m/s

Solutions:

1. **C** Use the Manning equation on p. 38:

$$Q=\frac{1}{n}R^{2/3}S^{1/2} \quad \text{or} \quad 6=\frac{1}{0.035}(15\times 1.2)^{2/3}\sqrt{S}. \qquad \therefore S=0.00013$$

2. **D** To find the head loss we must know the velocity, the Reynolds number and the *e/D* ratio:

$$V=Q/A=6/\pi=1.91 \ \text{m/s}. \quad e/D=0.25/2000=0.000125.$$

$$\text{Re}=\frac{VD}{\nu}=\frac{1.91\times 2}{10^{-6}}=3.8\times 10^{-6}.$$

$\therefore$ from Moody diagram: $f=0.013$

Finally,

$$h_f=f\frac{L}{D}\frac{V^2}{2g}=0.013\frac{200}{2}\frac{1.91^2}{2\times 9.8}=0.24 \text{ m}$$

3. **A** Use the energy equation:

$$\frac{p_1}{\gamma}+z_1+\frac{V_1^2}{2g}=\frac{p_2}{\gamma}+z_2+\frac{V_2^2}{2g}+h_f+H_T$$

$$0+726+0=0+696+0+0.013\frac{200}{2}\frac{1.91^2}{2\times 9.8}+H_T. \quad \therefore H_T=30\,\text{m}$$

$$\dot{W}=Q\gamma H_T\eta=6\times 9800\times 30\times 0.89=1560000 \text{ W} \quad \text{or} \quad 1560 \text{ kW}$$

We have used an expression similar to the pump power equation on p.37, except we must multiply when efficiency is used for a turbine.

4. **B** The maximum velocity is limited by the absolute pressure going to zero (actually, it cannot be lower than the vapor pressure, but the vapor pressure is nearly absolute zero for river temperatures). The energy equation from the surface to the blade inlet provides us with the maximum velocity:

$$\frac{p_1}{\gamma}+z_1+\frac{V_1^2}{2g}=\frac{p_2}{\gamma}+z_2+\frac{V_2^2}{2g}+h_f$$

$$0+726+0=\frac{-100\,000}{9800}+696+\frac{V_2^2}{2\times 9.8}+0.013\frac{200}{2}\frac{1.91^2}{2\times 9.8}$$

$$\therefore V_2=28.1\,\text{m/s}$$

Civil Engineering Practice Test

This practice exam has been developed to help test-takers prepare for the afternoon Discipline-Specific test in Civil Engineering. In addition, if test-takers are undecided, taking this test may also help them determine whether they should take the afternoon DS test or the afternoon General test.

The subjects tested are those subjects that the NCEES states will make up the exam. We have placed the same number of problems from each area of Civil Engineering that can be expected to be on the actual exam. The difficulty level can, of course, be quite different on the actual exam; we have, however, attempted to create an exam with essentially the same level of difficulty as one would expect on the actual exam.

If you find the problems on this exam to be more difficult than the problems on the afternoon General test, you may wish to take the General test in the afternoon. About one half of all Civil Engineering examinees take the afternoon General test. This practice exam should help you make that decision before the day of the test.

cut here!

FUNDAMENTALS OF ENGINEERING EXAM

Afternoon Session—Civil Exam

(Simulated answer form with topical breakout and scoring grid.)

BE SURE EACH MARK IS DARK AND COMPLETELY FILLS THE INTENDED SPACE AS ILLUSTRATED HERE: ●.

CONCRETE	SURVEYING	WATER TREATMENT	SOILS	CONTRACTS
1 Ⓐ Ⓑ Ⓒ Ⓓ	13 Ⓐ Ⓑ Ⓒ Ⓓ	25 Ⓐ Ⓑ Ⓒ Ⓓ	37 Ⓐ Ⓑ Ⓒ Ⓓ	49 Ⓐ Ⓑ Ⓒ Ⓓ
2 Ⓐ Ⓑ Ⓒ Ⓓ	14 Ⓐ Ⓑ Ⓒ Ⓓ	26 Ⓐ Ⓑ Ⓒ Ⓓ	38 Ⓐ Ⓑ Ⓒ Ⓓ	50 Ⓐ Ⓑ Ⓒ Ⓓ
3 Ⓐ Ⓑ Ⓒ Ⓓ	15 Ⓐ Ⓑ Ⓒ Ⓓ	27 Ⓐ Ⓑ Ⓒ Ⓓ	39 Ⓐ Ⓑ Ⓒ Ⓓ	51 Ⓐ Ⓑ Ⓒ Ⓓ
4 Ⓐ Ⓑ Ⓒ Ⓓ	16 Ⓐ Ⓑ Ⓒ Ⓓ	28 Ⓐ Ⓑ Ⓒ Ⓓ	40 Ⓐ Ⓑ Ⓒ Ⓓ	52 Ⓐ Ⓑ Ⓒ Ⓓ
5 Ⓐ Ⓑ Ⓒ Ⓓ	17 Ⓐ Ⓑ Ⓒ Ⓓ		41 Ⓐ Ⓑ Ⓒ Ⓓ	53 Ⓐ Ⓑ Ⓒ Ⓓ
6 Ⓐ Ⓑ Ⓒ Ⓓ	18 Ⓐ Ⓑ Ⓒ Ⓓ		42 Ⓐ Ⓑ Ⓒ Ⓓ	54 Ⓐ Ⓑ Ⓒ Ⓓ
Score: ______	Score: ______	Score: ______	Score: ______	Score: ______

STEEL	HYDRAULICS/ HYDROLOGY	ENVIRONMENTAL	TRANSPORTATION	COMPUTERS
7 Ⓐ Ⓑ Ⓒ Ⓓ	19 Ⓐ Ⓑ Ⓒ Ⓓ	29 Ⓐ Ⓑ Ⓒ Ⓓ	43 Ⓐ Ⓑ Ⓒ Ⓓ	55 Ⓐ Ⓑ Ⓒ Ⓓ
8 Ⓐ Ⓑ Ⓒ Ⓓ	20 Ⓐ Ⓑ Ⓒ Ⓓ	30 Ⓐ Ⓑ Ⓒ Ⓓ	44 Ⓐ Ⓑ Ⓒ Ⓓ	56 Ⓐ Ⓑ Ⓒ Ⓓ
9 Ⓐ Ⓑ Ⓒ Ⓓ	21 Ⓐ Ⓑ Ⓒ Ⓓ	31 Ⓐ Ⓑ Ⓒ Ⓓ	45 Ⓐ Ⓑ Ⓒ Ⓓ	57 Ⓐ Ⓑ Ⓒ Ⓓ
10 Ⓐ Ⓑ Ⓒ Ⓓ	22 Ⓐ Ⓑ Ⓒ Ⓓ	32 Ⓐ Ⓑ Ⓒ Ⓓ	46 Ⓐ Ⓑ Ⓒ Ⓓ	58 Ⓐ Ⓑ Ⓒ Ⓓ
11 Ⓐ Ⓑ Ⓒ Ⓓ	23 Ⓐ Ⓑ Ⓒ Ⓓ	33 Ⓐ Ⓑ Ⓒ Ⓓ	47 Ⓐ Ⓑ Ⓒ Ⓓ	59 Ⓐ Ⓑ Ⓒ Ⓓ
12 Ⓐ Ⓑ Ⓒ Ⓓ	24 Ⓐ Ⓑ Ⓒ Ⓓ	34 Ⓐ Ⓑ Ⓒ Ⓓ	48 Ⓐ Ⓑ Ⓒ Ⓓ	60 Ⓐ Ⓑ Ⓒ Ⓓ
Score: ______	Score: ______	35 Ⓐ Ⓑ Ⓒ Ⓓ	Score: ______	Score: ______
		36 Ⓐ Ⓑ Ⓒ Ⓓ		
		Score: ______		

Civil Discipline Exam

Questions 1 – 3

Questions 1-3 relate to the reinforced concrete beam shown below.

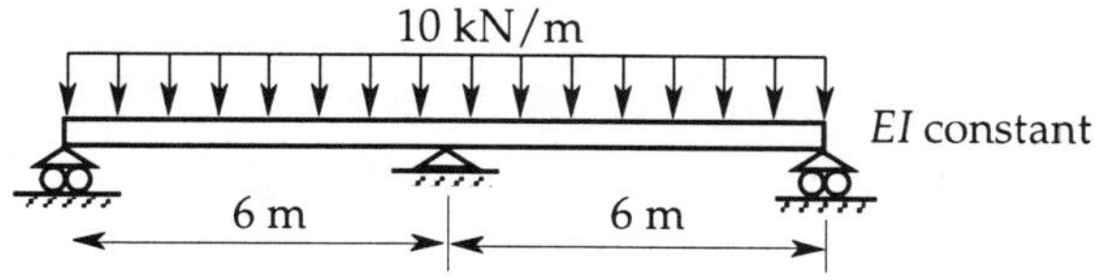

1. The beam is
   (A) determinate
   (B) indeterminate
   (C) improperly restrained
   (D) none of the above

2. The value of the reaction force at the center support is most nearly:
   (A) 20 kN
   (B) 25 KN
   (C) 30 kN
   (D) 75 kN

3. Tension reinforcement at the center support should be located:
   (A) near the top surface of the beam
   (B) at mid-depth
   (C) near the bottom surface of the beam
   (D) no tension reinforcement needed

Questions 4 – 6

Questions 4–6 relate to the cross-section of a short reinforced concrete column shown below. The 28-day compression strength is 4000 psi. Reinforcement is Grade 60. Longitudinal bars are #9, and spiral reinforcing is #3 (not shown).

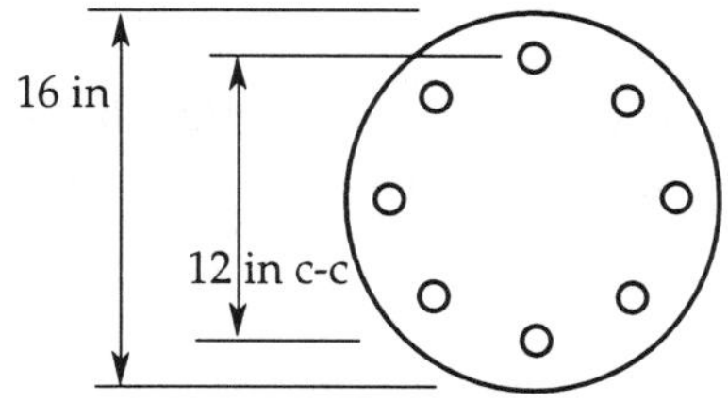

4. The reinforcement ratio is most nearly:
   (A) 0.02
   (B) 0.04
   (C) 0.06
   (D) 0.08

5. Clear cover is most nearly:
   (A) 1 inch
   (B) 1-1/2 inch
   (C) 2 inches
   (D) 3-1/2 inches

6. The load capacity of the column is most nearly:
(A) 560 kips
(B) 680 kips
(C) 800 kips
(D) 970 kips

Questions 7 - 9

Questions 7–9 relate to a flat bar tension member made of steel with tensile yield strength of 50 ksi and ultimate strength of 65 ksi. The figure below illustrates the holes for a bolted connection at the end of the member. The effective hole diameter is 1.00 inch.

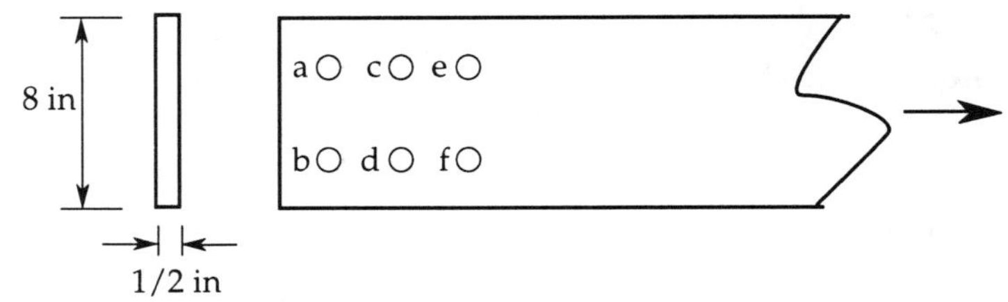

7. The net area is most nearly:
(A) 1.0 in^2
(B) 2.5 in^2
(C) 3.0 in^2
(D) 4.0 in^2

8. A block shear failure involves fracturing along a path through holes:
(A) e-f only
(B) a-c-e only
(C) a-c-e-f-d-b
(D) a-b only

9. Assume that tension yielding is the critical limit state and that dead load is 30 kips. The maximum permissible live load is most nearly:
(A) 90 kips
(B) 100 kips
(C) 120 kips
(D) 150 kips

Questions 10 – 12

Questions 10-12 relate to the cross section of a built-up compression member shown below:

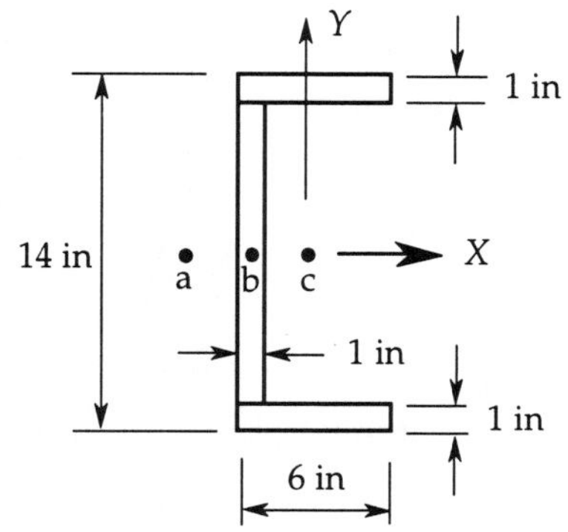

10. The shear center is most nearly located at point:
 (A) a
 (B) b
 (C) c
 (D) This section does not have a shear center

11. The moment of inertia about the X-axis is most nearly:
 (A) 5 cm^4
 (B) 24 cm^4
 (C) 200 cm^4
 (D) 650 cm^4

12. Assume that the moment of inertia about the Y-axis is 74.5 cm^4, that the member is 2.40 meters long, and that the ends are restrained from both rotation and lateral translation. The slenderness ratio for buckling about the Y-axis is most nearly:
 (A) 20
 (B) 90
 (C) 200
 (D) 430

Questions 13 – 18

Questions 13-18 refer to the plat of a suburban lot shown below. The front boundary is a circular curve and the side boundaries are radial to that curve. The table gives plane coordinates of the corners of the lot.

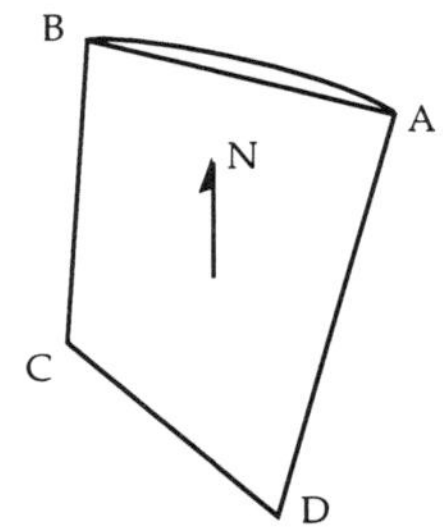

Pt	N	E
A	339.74	310.09
B	391.50	116.90
C	198.29	100.00
D	100.00	198.30

13. The angle of intersection of the curve is most nearly:
    (A) 10°
    (B) 20°
    (C) 70°
    (D) 80°

14. The chord length AB is most nearly
    (A) 193
    (B) 200
    (C) 201
    (D) 207

15. The radius of the curve is most nearly:
    (A) 400
    (B) 567
    (C) 576
    (D) 585

16. The curve length AB is most nearly:
    (A) 193
    (B) 200
    (C) 201
    (D) 207

17. The bearing of line AB is most nearly:
    (A) N 15° W
    (B) W 15° N
    (C) N 60° W
    (D) N 75° W

18. The area bounded by traverse ABCD is most nearly:
    (A) 36400
    (B) 43560
    (C) 50000
    (D) 72800

Questions 19 – 22

A construction project requires that water be delivered to a small reservoir at the site at elevation 400 meters. Water is available in a lake at elevation 390 meters. There is 4-cm-diameter, galvanized iron piping and a pump available. The pump has characteristics shown. It will require 400 meters of pipe. Use $\nu_{water} = 10^{-6}\ \text{m}^2/\text{s}$.

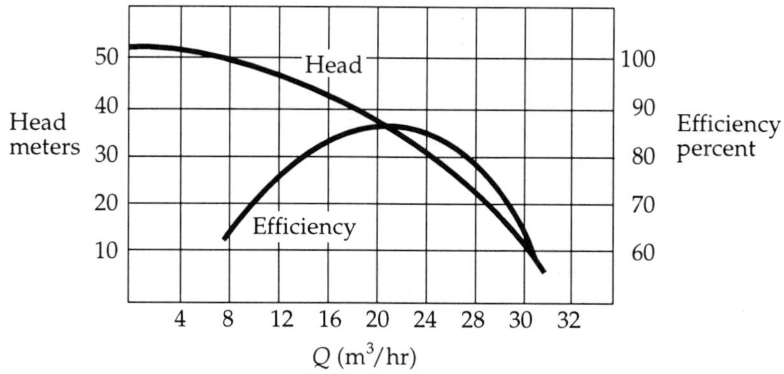

19. In the analysis of the piping system, we:
    (A) must include minor losses
    (B) may neglect minor losses
    (C) must include the pressure drop from inlet to exit
    (D) must include the kinetic energy at the exit

20. If the flow is controlled with a valve to be $6\ \text{m}^3/\text{hr}$, the head loss over 400 meters of pipe is most nearly:
    (A) 1 m
    (B) 2 m
    (C) 4 m
    (D) 6 m

21. Estimate the flow rate assuming the water flows freely through the pipe from the river to the reservoir:
    (A) $8\ \text{m}^3/\text{hr}$
    (B) $12\ \text{m}^3/\text{hr}$
    (C) $16\ \text{m}^3/\text{hr}$
    (D) $20\ \text{m}^3/\text{hr}$

22. The pump can be placed anywhere along the length of the pipe. If it is placed just before the reservoir it may not perform because:
    (A) pump efficiency is too low
    (B) turbulence is too high
    (C) surface tension prevents proper flow
    (D) vapor pressure may be a factor

Questions 23 – 24

The runoff coefficient for a particular 50 acre industrial area is determined to be 0.8. A storm produces rain intensity of 0.5 in./hr. A 3-ft-diameter storm sewer transports the water from the area. The Manning roughness coefficient is 0.012.

23. Estimate the velocity of the water in the storm sewer if it flows full with no surcharge.
    (A) 2 fps
    (B) 3 fps
    (C) 4 fps
    (D) 5 fps

24. What slope should the sewer have if it is to flow full with no surcharge?
    (A) 0.002
    (B) 0.0012
    (C) 0.00029
    (D) 0.00077

Questions 25 – 28

A partial ion analysis has been performed on a groundwater that is to be treated using conventional lime/ soda ash softening. The flow rate of the water to be treated is 15 MGD. The ion concentrations reported below are all given in units of mg/L as $CaCO_3$. The pH of the water is 7.8.

$Ca^{2+} = 240.0$
$Mg^{2+} = 60.0$
$Fe^{3+} = 1.3$
$Mn^{4+} = 0.65$
$Na^{+} = 45.0$
$HCO_3^{-} = 220.0$
$SO_4^{2-} = 90.0$
$CO_2 = 15.0$

25. Calculate the total hardness of this water.
    (A) 302 mg/L as $CaCO_3$
    (B) 347 mg/L as $CaCO_3$
    (C) 180 mg/L as $CaCO_3$
    (D) 547 mg/L as $CaCO_3$

26. Calculate the carbonate hardness.
    (A) 122 mg/L as $CaCO_3$
    (B) 167 mg/L as $CaCO_3$
    (C) 180 mg/L as $CaCO_3$
    (D) 302 mg/L as $CaCO_3$

27. Calculate the amount of lime (in kg) that must be added each day to react with the carbon dioxide present in the water.

(A) 12 618 kg/d
(B) 31 544 kg/d
(C) 631 kg/d
(D) 7565 kg/d

28. Calculate the amount of soda ash that must be required to soften this water.

(A) 3610 kg/d
(B) 7220 kg/d
(C) 903 kg/d
(D) 4815 kg/d

Questions 29 – 32

A municipal wastewater treatment plant is being designed to treat wastewater at a flow rate of 0.75 m^3/s. A horizontal flow grit chamber is to be used to pretreat the wastewater. The approach velocity in the chamber is to be 0.35 m/s. The length of the chamber is to be 15 times the width and the depth is to be 1.8 m.

A circular primary sedimentation basin is to be designed for this facility. The tank is designed to have an overflow rate of 40 m^3/m^2/day and a detention time of 2.0 h. The weir loading rate is to be 250 m^3/m/day.

29. Calculate the width of the grit chamber.

(A) 0.13 m
(B) 0.95 m
(C) 1.2 m
(D) 2.4 m

30. Calculate the volume of the grit chamber.

(A) 0.46 m^3
(B) 24.4 m^3
(C) 78 m^3
(D) 39 m^3

31. Calculate the diameter of the sedimentation basin.

(A) 23 m
(B) 18 m
(C) 9.1 m
(D) 45 m

32. Calculate the length of weir required.

(A) 1620 m
(B) 810 m
(C) 260 m
(D) 130 m

Questions 33 – 36

The Westwood Wastewater Treatment Plant (WWTP) uses primary sedimentation followed by activated sludge. Ferric chloride is added at the end of the activated sludge basin to precipitate phosphate. Following secondary clarification, tertiary filtration is used to remove turbidity. The final effluent is treated with chlorine for disinfection followed by dechlorination using sodium bisulfite. Just before discharge into the Red Oak River the waste water is aerated to increase the dissolved oxygen level.

33. The main sewer entering the Westwood WWTP is 4 ft in diameter. The sewer flows at full capacity at a flow rate of 7.5 ft^3/s and a velocity of 0.6 ft/s. What is the depth of water in the sewer when the wastewater flow is at a flow rate of 5.25 ft^3/s? (Note: Assume that the Manning's constant, n, is constant with depth.)

    (A) 34 in
    (B) 32 in
    (C) 29 in
    (D) 24 in

34. The influent wastewater has BOD_5 and suspended solids (SS) concentrations of 240 mg/L and 260 mg/L, respectively. Primary sedimentation results in a 70% decrease in the suspended solids concentration and a 30% decrease in the BOD_5. Calculate the amount of primary sludge produced each day, assuming a flow rate of 7.5 ft^3/s.

    (A) 4700 kg/d
    (B) 3300 kg/d
    (C) 10,900 kg/d
    (D) 150 kg/d

35. What should be the volume of the aeration basin (for activated sludge treatment) if the organic loading rate (volumetric) is to be 75 lb BOD_5/ 1000 ft^3-d. Use the peak flow rate of 7.5 ft^3/s.

    (A) 12,000 ft^3
    (B) 4330 ft^3
    (C) 28,000 ft^3
    (D) 90,500 ft^3

36. The water flow meter at the Westwood water plant is not operating. The plant superintendent tells you that each of the four dual media filters (each 10.00 ft × 20.0 ft) are loaded at a velocity of 600 ft/d. The depth of the filters is 9.00 ft. What is the total flow rate through the four filters?

    (A) 5.6 ft^3/s
    (B) 1.4 ft^3/s
    (C) 5.0 ft^3/s
    (D) 2.5 ft^3/s

Questions 37 – 39

A saturated, normally consolidated clay has a water content of 30 percent, a liquid limit of 40 percent, and a plastic limit of 20 percent. The specific gravity of solids is 2.70.

37. Estimate the compression index for this soil.
   (A) 0.300
   (B) 0.360
   (C) 0.270
   (D) 0.0027

38. Estimate the void ratio for this soil.
   (A) 1.080
   (B) 0.810
   (C) 0.300
   (D) 0.429

39. The vertical stress increase caused by a newly-constructed building causes the effective stress on this soil in the ground to double. How much will the void ratio of this clay decrease as the building settles?
   (A) 0.187
   (B) 0.081
   (C) 0.013
   (D) 0.500

Questions 40 – 42

A retaining wall is as shown. The sand in the backfill and foundation has a unit weight of 125 lb/ft^3 and a friction angle of 32 degrees.

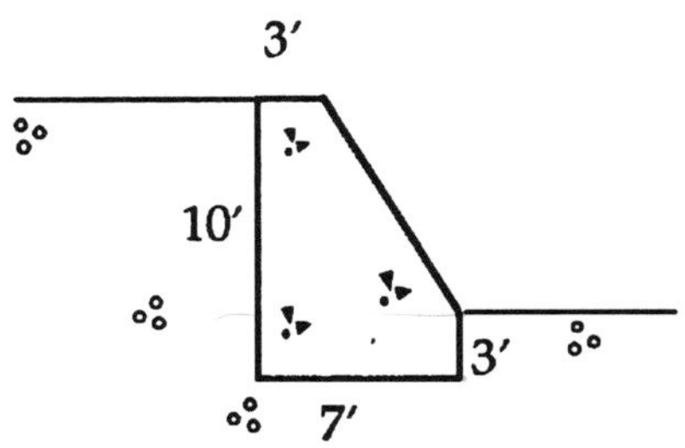

40. What is the total active earth force per one-foot length of wall?
   (A) 417 lb
   (B) 1919 lb
   (C) 1250 lb
   (D) 384 lb

41. What is the overturning moment per one-foot length of wall?
   (A) 6396 ft-lb
   (B) 19,190 ft-lb
   (C) 9594 ft-lb
   (D) 1279 ft-lb

42. If the wall is made of concrete with a unit weight of 150 lb/ft^3, what is the factor of safety against sliding? Take the factor of safety as the ratio of the resisting forces to the driving forces.
(A) 2.00
(B) 1.05
(C) 0.95
(D) 3.69

Questions 43 – 45

A 2-directional 24-foot wide asphalt concrete (AC) surfaced road is to be constructed along with a total of 12-foot asphalt concrete shoulders. The road and shoulders cross section includes 12-inch granular subbase, 6-inch gravel base, and 6-inch asphalt concrete. The asphalt paver places the AC in two courses: 4-inch base course and 2-inch surface course. The paver places the AC in 12-ft wide strips at an average speed of 2 miles per day.

43. If the density of the placed and compacted AC is 145 pcf, then the total weight in pounds of the AC material (both surface and base courses) in a one-square-yard surface area is about:
(A) 752.5
(B) 652.5
(C) 600.0
(D) 700.0

44. The density of the AC material in the trucks is 120 pcf. The daily volume (the number of cubic yards) of the AC surface course in the delivery trucks is about:
(A) 1045
(B) 1245
(C) 945
(D) 2145

45. The new road was designed and constructed to carry the 20-year design period of traffic. The projected two-directional traffic in terms of 18,000 pounds equivalent single axle load (ESAL) over the 20-year design period was 2,900,000 ESAL. The design ESAL (the number of ESAL used in the design of the road) is about:
(A) 2,900,000
(B) 1,450,000
(C) 5,800,000
(D) None of the above

Questions 46 – 48

A freeway in a rural area has the following design criteria:

Design Speed	70 mph
Wet Coefficient of Friction	0.28
Perception of Reaction Time	2.5 secs
Height of Eye	3.5 ft
Height of Object	0.5 ft
Top of Vehicle	4.25 ft

46. The distance traveled by a vehicle during the perception and reaction of the driver before stopping is most nearly:
(A) 260 ft
(B) 290 ft
(C) 320 ft
(D) 350 ft

47. The distance traveled by the vehicle while decelerating to a stop (leaving skid marks) is most nearly:
(A) 500 ft
(B) 550 ft
(C) 580 ft
(D) 620 ft

48. If the maximum allowable superelevation and lateral friction force are 0.08 and 0.10, respectively, the minimum design radius is most nearly:
(A) 1745 ft
(B) 1815 ft
(C) 1910 ft
(D) 1975 ft

49. During the construction of a concrete framed office building, it is discovered that a concrete beam has insufficient steel reinforcement to support the code required floor loads. After review, it is determined that the contractor has installed the reinforcement as detailed in the contract documents and that the consulting engineer inadvertently had placed the wrong reinforcing information on the contract drawings. A general note on the drawings indicated that "the contractor shall install all work in accordance with all codes and ordinances". The contractor claims that he deserves extra compensation to replace the under-reinforced beam. The engineer argues that the "compliance with laws" note places the responsibility to install the work to meet code on the contractor and therefore the contractor is required to comply regardless of the information presented on the contract drawings. The engineer rules that the contractor must replace the beam without additional compensation. Which statement below best reflects the proper allocation of liability in this case?

(A) The contractor shall remove and replace the beam without extra compensation because of the "compliance with laws" note.

(B) The contractor shall remove and replace the beam and shall be compensated by the owner for the extra work.

(C) The contractor shall remove and replace the beam and shall be compensated by the owner for the extra work. The owner may then attempt to obtain reimbursement from the engineer for the extra work required by the engineer's incorrect drawing.

(D) The engineer and contractor are equally responsible and should share in the cost of removing and replacing the beam.

50. During negotiations between an owner and consulting engineer for the provision of engineering design services, the owner insists on including the following language in the contract:

"The engineer shall provide engineering design services in a timely manner and with the highest standard of care of his or her profession."

The engineer should do which of the following?

(A) Recognize that owners have a right to receive quality services and agree to the clause without modification.

(B) Refuse to enter into the agreement because of the "timely manner" provision.

(C) Refuse to enter into the agreement because of the "standard of care" provision.

(D) Attempt to negotiate a new clause with a lower standard of care provision.

51. A "general contract" is let for the construction of an electrical power substation. Which answer below best describes the contractual lines of privity between the parties in this form of construction project management?

(A) The consulting engineer will have a contract with the owner, but will not have a contract with either the general contractor or the subcontractors.

(B) The consulting engineer will have a contract with the owner, and also a separate contract with the general contractor.

(C) The consulting engineer will have separate contracts with the owner, general contractor, and the subcontractors.

(D) The consulting engineer will have a contract with the general contractor but will not have agreements with either the owner or the subcontractors.

52. Which statement is true regarding the legality, enforceability and binding of contracts?

(A) For a contract to be enforceable it must always be in writing.

(B) Contracts written for an illegal purpose will still be enforced by a court.

(C) Consideration must be evident in the contract. Consideration refers to fairness of the agreement.

(D) Mutual agreement of the parties must be evident.

53. Often in legal disputes involving construction and engineering projects the courts will use past precedent setting legal cases to help interpret and decide the case before the court. This general body of law, which has been developed through decisions of courts, is referred to as which of the following?

(A) Tort Law

(B) Statute Law

(C) Common Law

(D) Public Law

54. Which of the duties below would not be considered as a construction contract administration duty of a consulting engineer?

(A) Preparation of initial design drawings

(B) Field observation of the construction project

(C) Review of the contractor's payment requests

(D) Processing of contract change orders

55. A spreadsheet has cell $B2$ set to 4. Then cell $B3$ is set to the formula $\$B\$2 + B2$. This formula is copied into cells $B4$, $B5$, $B6$, and $B7$ in order. The number in $B6$ is:

(A) 20

(B) 24

(C) 32

(D) 16

56. The pseudo-code segment is given the following list of integer values as data: 17, 15, 12, 18, 19, 25. After execution, what values will be displayed?

```
a := 100;  b := 0;  c := 0;
for i := 1 to 5
   read( x );
   if( x  a)
      a := x;
   if( x  b)
      b := x;
   c := c + x;
endfor
print( a, b, c/5 );
```

(A) 12, 25, 21

(B) 12, 17, 16

(C) 12, 19, 16

(D) 12, 25, 21

57. A file contains the integers: 3, 2, -4, 5, 1, -2, ... which are processed by the pseudo-code program:

```
a := 0;
i := 0;
while( i  3)
  read(b);
  if( b  0 )
    a = a + b*b;
  i := i + 1;
endwhile;
print( a/i);
```

What will be the approximate value printed?

(A) 11

(B) 11.25

(C) 7.8

(D) 9.5

Questions 58 – 59

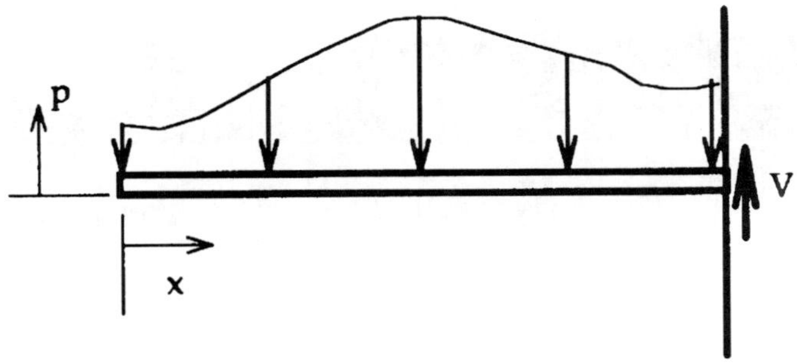

A 48-inch-long beam is loaded with a varying distributed load p which is a function of distance x along the beam. The shear V is the integral of p with respect to x, or the area under the $p(x)$ curve. Values of p at points along the beam are as follows:

x, inches	0	12	24	36	48
p, lb/in	5	11	18	14	10

58. Determine V using the forward trapezoidal rule.
(A) 606 lb
(B) 604 lb
(C) 576 lb
(D) 557 lb

59. Determine V using the Simpson's rule.
(A) 606 lb
(B) 604 lb
(C) 576 lb
(D) 557 lb

60. A root of $P(x) = x^3 - 4x + 4$ near $x = -1$ is most nearly:
(A) –0.949
(B) –1.043
(C) –1.099
(D) –0.903

Answers to the Civil Examination

Concrete

1. B
2. D
3. A
4. B
5. A
6. D

Steel

7. C
8. C
9. A
10. A
11. D
12. B

Surveying13.

13. B
14. B
15. C
16. C
17. D
18. A

Hydraulics/ Hydrology

19. B
20. A
21. A
22. D
23. B
24. D

Water Treatment

25. A
26. C
27. C
28. D

Environmental

29. C
30. D
31. D
32. C
33. A
34. B
35. D
36. A

Soils

37. C
38. B
39. B
40. B
41. A
42. D

Transportation

43. B
44. C
45. B
46. A
47. C
48. B

Contracts

49. C
50. D
51. A
52. D
53. C
54. A

Computers

55. A
56. C
57. D
58. A
59. B
60. D

tear out answer sheet

Solutions to the Civil Exam

1. **B** Number of reactions > number of independent equations of static equilibrium.

2. **D** By flexibility method:

$$(5wL^4)/(384EI) = (PL^3)/(48EI)$$

Therefore,

$$P = 0.625wL$$
$$= 0.625 \times 10 \text{ kN/m} \times 12 \text{ m} = 75 \text{ kN}$$

3. **A** Moment is negative at center, so the top of the beam is in tension.

4. **B**

$$A_g = \pi d^2 / 4 = 201 \text{ in}^2$$
$$A_s = 8 \times 1.00 = 8.0 \text{ in}^2$$
$$\rho_g = A_s / A_g = 8.00 / 201 = 0.04$$

5. **A** 16/2 - 12/2 - 1.128/2 - 0.375 = 1.06 inch

6. **D**

$$A_{concrete} = A_g - A_s = 201 - 8 = 193 \text{ in}^2$$
$$P_o = 0.85 f_c' A_{concrete} + f_y A_s$$
$$= 0.85 \times 4 \text{ ksi} \times 193 \text{ in}^2 + 60 \text{ ksi} \times 8 \text{ in}^2 = 1136 \text{ kips}$$
$$P_n = \phi P_o = 0.85 \times 1136 = 966 \text{ kips}$$

7. **C**

$$A_n = b_n t = (b - \Sigma d)t$$
$$= (8.0 - 2 \times 1.0) \times 0.5 = 3.0 \text{ in}^2$$

8. **C** By definition of block shear.

9. **A** By ASD:

$F_t = 0.6F_y = 0.6 \times 50 = 30$ ksi

$P_n = F_t A_g = 30 \text{ ksi} \times 4.0 \text{ in}^2 = 120$ kips. $\therefore$ 120 - 30 = 90 kips

By LRFD:

$$\phi P_n = 0.9 F_y A_g = 0.9 \times 50 \text{ ksi} \times 4.0 \text{ in}^2 = 180 \text{ kips}$$
$$1.2\text{D} + 1.6\text{L} \le \phi \text{P}_n$$

Therefore,

$$L \le (\phi P_n - 1.2D)/(1.6 = 180 - 1.2 \times 30)/1.6 = 90 \text{ kips}$$

10. **A**

11. **D** $(6 \times 14^3)/12 - (5 \times 12^3)/12 = 652 \text{ cm}^4$

12. **B**

$$r = \sqrt{I/A} = \sqrt{74.5/24} = 1.76 \text{ cm}$$
$$KL/r = 0.65 \times 240/1.76 = 89$$

13. **B** $AzBC = \tan^{-1}[(X_C - X_B)/(Y_C - Y_B)] =$

$$\tan^{-1}[(100.00 - 116.90)/(198.29 - 391.50)] = 185°$$

$AzAD = \tan{-}^{1}[(198.30 - 310.09)/(100.00 - 339.74)] = 205° \quad \therefore 205 - 185 = 20°$

14. **B**

$$\sqrt{(X_B - X_A)^2 + (Y_B - Y_A)^2}$$
$$= \sqrt{(116.90 - 310.09)^2 + (391.50 - 339.74)^2} = 200$$

15. **C** $(L.C.)/2\sin(I/2) = 200.00/2\sin 10° = 575.88$

16. **C** $L = RI = 575.88 \times 20\pi/180 = 201.02$

17. **D** $\tan^{-1}[(116.90 - 310.09)/(391.50 - 339.74)] = -75°. \quad \therefore N\ 75°\ W$

18. **A** $[X_A(Y_B - Y_D) + X_B(Y_C - Y_A) + \ldots]/2 =$

$[310.09(391.50 - 100.00) + 116.90(198.29 - 339.74) + 100.00(100.00 - 391.50)$

$+198.30(339.74 - 198.29)]/2 = 36\,378$

19. **B** The relative roughness is $e/D = 0.15/40 = 0.00375$. Assuming $\text{Re} > 10^5$, or so, $f \cong 0.026$. If fL/D is quite large we can neglect the minor losses:

$$f\frac{L}{D} = 0.026\frac{400}{0.04} = 260$$

Therefore, minor losses are negligible; we know that $C_{\text{minor}} \leq 1$ (see p.37 of the Handbook). Compare C_{minor} with fL/D.

20. **A** Calculate the friction factor using the Moody diagram:

$$V = \frac{Q}{A} = \frac{6/3600}{\pi \times 0.02^2} = 1.33 \text{ m/s}, \quad \text{Re} = \frac{VD}{\nu} = \frac{1.33 \times 0.04}{10^{-6}} = 5.8 \times 10^4.$$

$$\frac{e}{D} = 0.00375. \qquad \therefore f = 0.03.$$

Then, the head loss is

$$h_f = f\frac{L}{D}\frac{V^2}{2g} = 0.03\frac{400}{0.04}\frac{1.33^2}{2 \times 9.8} = 1.0 \text{ m}.$$

21. **A** Apply the energy equation (we use Q as m^3/hr):

$$\frac{p_1}{\gamma}+z_1+\frac{V_1^2}{2g}+h_P=\frac{p_2}{\gamma}+z_2+\frac{V_2^2}{2g}+h_f$$

$$0+390+0+h_P=0+400+0+0.028\frac{400}{0.04}\frac{Q^2/(\pi\times 0.02^2)}{2\times 9.8\times 3600^2}$$

$$h_P=10+0.7Q^2$$

Solve this system equation and the characteristic curve simultaneously by trial-and-error:

try $h_P = 40$: $Q = 6.56$ (from equation)

$Q = 19$ (from curve)

try $h_P = 50$: $Q = 7.56$ (from equation)

$Q = 8$ (from curve)

The approximate flow rate is $Q = 8\ \text{m}^3/\text{hr}$.

22. **D** The pressure decreases along the pipe up to the pipe inlet. Cavitation (vaporization of the water) might result if the inlet pressure falls below the vapor pressure. This condition must be checked to avoid possible cavitation.

23. **B** The average velocity of water in the sewer is

$$V=\frac{Q}{A}=\frac{ciA_{surface}}{A_{pipe}}=\frac{0.8\times 0.5\times 50}{\pi\times 1.5^2}=2.83\text{ fps}$$

Note: The coefficient C requires the surface area to be in acres.

24. **D** Use Manning's equation:

$$V=\frac{1.486}{n}R^{2/3}S_0^{1/2}$$

$$2.83=\frac{1.486}{0.012}\left(\frac{\pi\times 1.5^2}{\pi\times 3}\right)^{2/3}S_0^{1/2}. \qquad \therefore\ S_0=0.000766$$

25. **A** The total hardness is the sum of the concentrations of the polyvalent cations:

$[Ca^{2+}] + [Mg^{2+}] + [Fe^{3+}] + [Mn^{4+}] = 240 + 60 + 1.3 + 0.65 =$ 302 mg/L as $CaCO_3$

26. **C** The carbonate hardness is equal to the total hardness or the alkalinity, whichever is less. At a pH of 7.8, the alkalinity is approximately equal to the concentration of bicarbonate. As such, the alkalinity in this water is 220 mg/L. In this situation, since the alkalinity is less than the total hardness and the carbonate hardness cannot exceed the alkalinity, the carbonate hardness equals 220 mg/L.

27. C We must first calculate the concentrations of carbon dioxide in units of meq/L using the data provided in the table of equivalent weights given on p. 89 of the NCEES Handbook:

$$[CO_2] = \frac{15.0}{50.04} = 0.3 \text{ meq/L}$$

The amount of lime required can be calculated using equations 1 (Lime-Soda Softening Equations) on p. 89 of the NCEES Handbook:

	meq/L	meq/L lime required
CO_2:	0.30	0.30

Mass of lime per day =

$(0.30 \text{ meq } Ca(OH)_2/L) \times (37.046 \text{ mg } Ca(OH)_2/\text{meq}) \times (3.785 \text{ L/gal})$
$\times (15 \times 10^6 \text{ gal/day})(\text{kg}/10^6 \text{ mg}) = 631 \text{ kg/day}$

28. D We must first calculate how much calcium and magnesium is associated with sulfate. Since the concentration of calcium is 240 mg/L as $CaCO_3$ and the alkalinity is 220 mg/L as $CaCO_3$, we can assume that the portion of the calcium that is not associated with carbonate or bicarbonate is complexed with sulfate:

$$[Ca^{2+}] = \frac{20.0}{50.04} = 0.4 \text{ meq/L}$$

The concentration of magnesium is 60 mg/L. Since the sulfate concentration is greater than the sum of the calcium that is associate with sulfate plus the magnesium, we can assume that all of the magnesium is associated with sulfate:

$$[Mg^{2+}] = \frac{60.0}{50.04} = 1.2 \text{ meq/L}$$

The amount of soda ash required is calculated from the stoichiometry of equation 3 and 5 in the Section on Lime-Soda Ash Softening Equations on p. 89 of the NCEES Handbook:

	meq/L	meq/L soda ash required
$CaSO_4$	0.40	0.40
$MgSO_4$	1.20	1.20
		1.60

Mass of soda ash per day =

$(1.60 \text{ meq } Na_2CO_3/L) \times (53 \text{ mg } Na_2CO_3/\text{meq}) \times (3.785 \text{ L/gal})$
$\times (15 \times 10^6 \text{ gal/day}) \times (\text{kg}/10^6 \text{ mg}) = 4815 \text{ kg/day}$

29. C The width of the grit chamber is calculated using the equation for the approach velocity (p. 89 of the NCEES Handbook):

$$\text{Approach velocity} = Q / A_x$$

The approach velocity was given as 0.35 m/s. The flow rate was stipulated to be 0.75 m^3/s. Therefore, the cross-sectional area A_x can be calculated by dividing the flow rate by the approach velocity:

$$A_x = \frac{Q}{V} = \frac{0.75}{0.35} = 2.14 \text{ m}^2$$

The cross-sectional area, i.e., the area of the basin that is perpendicular to the direction of flow, is also equal to the depth D times the width W:

$$A_x = D \times W = 1.8W = 2.14 \text{ m}^2$$

Therefore, $W = 1.2$ m.

30. **D** The volume of the grit chamber is

$$V = LA_x = LDW = (15 \times 1.2) \times 1.8 \times 1.2 = 39 \text{ m}^3$$

where V = volume of the basin and L = the length of the grit chamber.

31. **D** The diameter of the sedimentation is calculated from the cross-sectional area of the basin. The cross-sectional area is determined using the equation for the hydraulic loading rate given on p. 89 of the NCEES Handbook. The hydraulic loading rate or overflow rate was given to be 40 $m^3/m^2 \cdot$ day.

$$A = \frac{Q}{\text{hydraulic loading rate}} = \frac{0.75 \times 3600 \times 24}{40} = 1620 \text{ m}^2$$

$$D^2 = 4A/\pi \quad \text{so that} \quad D = \sqrt{4 \times 1620/\pi} = 45 \text{ m}$$

32. **C** The length of the weir is calculated using the weir loading rate. The appropriate equation is given on p. 89 of the NCEES Handbook:

$$\text{Weir loading rate} = Q/L$$

where L = length of the weir:

$$L = (0.75)(3600 \times 24)/250 = 260 \text{ m}$$

33. **A** Use the chart provided on p. 94 of the NCEES Handbook. The ratio of the flows (Q/Q_f = 5.25/7.5) is 0.70. Enter this value on the chart and read off the corresponding values for the depth of flow. Use the dotted lines (for constant Darcy-Weisbach friction factor f and Manning's constant n):

Depth = 0.7 x full-flow depth = (0.7)(48 in) = 34 in

34. **B** The sludge production rate can be calculated using the flow rate, the influent SS concentration, and the percent removal:

Sludge production rate =

$$(7.5 \text{ ft}^3/\text{s})(86{,}400 \text{ s/d})(260 \text{ mg/L})(0.70) \times (0.028317 \text{ m}^3/\text{ft}^3)$$

$$\times (1000 \text{ L/m}^3)/(10^{-6} \text{ mg/kg}) = 3300 \text{ kg/day}$$

35. **D** The volume of the aeration basin can be calculated from the volumetric organic loading rate given on p. 89 of the NCEES Handbook:

Volume =

$$\frac{(240\,\text{mg/L})(0.70)(7.5\,\text{ft}^3/\text{s})(86{,}400\,\text{s/d})(10^{-3}\,\text{g/mg})(1.0\,\text{lb}/454\text{g})(28.32\,\text{L}/\text{ft}^3)}{(0.075\,\text{lb BOD}/\text{ft}^3\cdot\text{d})}$$

$= 90{,}500\ \text{ft}^3$

36. **A** The flow rate through the four filters can be calculated from either the equation for the overflow rate or the equation for the approach velocity (both are given on p. 89 of the NCEES Handbook):

Overflow rate = Q/A Approach velocity = Q/A_x

The velocity given is both the overflow rate and the approach velocity. Therefore, the total flow rate through all of the four filters is

$$\frac{4(10\ \text{ft})(20\ \text{ft})(600\ \text{ft/d})}{86{,}400\ \text{s/d}} = 5.6\ \text{ft}^3/\text{s}$$

37. **C** $C_c = 0.009(LL - 10) = 0.009(40 - 10) = 0.27$

38. **B** $e = \dfrac{wG}{S} = \dfrac{0.3 \times 2.70}{1.00} = 0.81$

39. **B** Using the relation for the compression index:

$$\Delta e = C_c \Delta \log p = C_c(\log p_2 - \log p_1) = C_c(\log 2p_1 - \log p_1) = 0.27 \log 2 = 0.081$$

40. **B** $P_a = \dfrac{1}{2}\gamma K_a H^2 = \dfrac{1}{2} \times 125 \times 0.307 \times 10^2 = 1919\ \text{lb/ft}$

41. **A** $M_o = P_a \bar{y} = 1919(10/3) = 6396\ \text{ft-lb/ft}$

42 **D** We use the relationship $FS = (W \tan\phi + P_p)/P_a$. The various variables are

$$W = 3 \times 7 \times 150 + 0.5(3 + 7) \times 7 \times 150 = 8400\ \text{lb/ft}$$

$$P_p = \frac{1}{2}\gamma K_p H^2 = 0.5 \times 125 \times 3.255 \times 3^2 = 1831\ \text{lb/ft}$$

The factor of safety is then

$$FS = \frac{8400 \tan 32^\circ + 1831}{1919} = 3.69$$

43. **B**

44. **C**

45. **B**

46. **A** $x = 1.47vt = 1.47 \times 70 \times 2.5 = 257\ \text{ft}$

47. **C** $d = \dfrac{V^2}{30f} = \dfrac{70^2}{30 \times 0.28} = 583\ \text{ft}$

48. **B** $eff = \frac{V^2}{15R}$. $0.08 + 0.10 = \frac{70^2}{15R}$. $\therefore R = 1814$ ft

49. **C** The contractor is entitled to rely on the accuracy of the engineer's design. The engineer is the party licensed to perform engineering design and has the duty to design his or her work in accordance with applicable codes and ordinances. Generally, construction contracts will have clauses that will require the contractor to notify the engineer if the contractor becomes aware of a design error, but this duty does not extend to a requirement to investigate or evaluate the structural design produced by the engineer for adequacy. The "compliance with laws" clause in the scenario is a general responsibility which requires the contractor to follow laws in his or her construction operations, but this clause does not make the contractor responsible for the engineer's work. Because the contract for construction is held by the owner, the owner would have to pay for the extra charges. The owner would be likely to attempt to collect reimbursement from the engineer for work made necessary by the engineer's mistake.

50. **D** In assessing negligence an engineer is generally held by the courts and society to practice at a reasonable level equal to the average standard of care within the profession and location of the service. In other words, for the purpose of evaluating negligence the court would look to what the generally accepted engineering practice would be. Engineers are not expected to provide perfect service. By agreeing to the "highest standard of care language" required by the owner, the engineer would be agreeing to provide a service quality far in excess of what may be possible and he or she would be contractually obligated to provide such a service. Owners frequently will include "highest standard of care" clauses in their contracts, and while they occasionally will truly be seeking such a service, many times they do not really intend to impose such an extreme standard of care on the engineer. When the engineer encounters such risky clauses it is best to find out what the true intent of the owner is rather than simply rejecting the contract or agreeing to the offensive language.

Providing the engineering service in a timely manner does not seem like an unreasonable request by the owner and should not be a basis for rejecting the contract. However, the engineer may want to include language which provides protection from service delays caused by conditions outside of the engineer's control.

51. **A** The three principle contractual organization systems used in construction are 1) general contract, 2) construction management, and 3) design build. In the "general contract" the consulting engineer will be hired by the owner to develop the design and contract documents, as well as to assist in the preparation of the bid documents and provide contract administrative services during the construction phase. The contract documents produced by the engineer will form the basis of the owner's agreement with the contractor. Although the engineer will work closely with the contractor during the construction phase, the engineer

will not have a contractual line of privity with the contractor under the "general contract" approach. Design/build project delivery is gaining in popularity and involves a single contract with the owner to provide both design and construction services. In design/build agreements, the engineer would be likely to have a contractual agreement with the construction company.

52. **D** There are five essential elements for making a contract binding.

1. The contract must be established for a legal purpose.
2. The contract must be in proper legal form.
3. There must be consideration (an exchange of something of value, i.e., service in exchange for a fee).
4. The parties must have capacity (the parties must be mentally com petent, of legal age, and have the power to enter into the contract).
5. There must be an agreement by both the parties.

Oral contracts may be legally binding in some instances, depending on the circumstances and purpose of the contract. Oral contracts may be difficult to enforce, however, and should not be used for engineering and construction agreements.

53. **C** Past precedent setting legal cases decided at the appellate and supreme court levels will frequently be used in settling legal disputes. The body of law developed from these legal decisions is referred to as "common law".

Statute law is the body of law which is developed by governments creating statutes.

Public law refers to the body of law that governs the actions of governments.

Contract law refers to the body of law that governs contracts between parties.

Tort Law is the body of law that governs actions for wrongful acts against another party not stemming from a contract breech.

54. **A** There are typically five distinct service phases that an engineer will provide for an owner on a construction project. These phases are as follows:

- Schematic or preliminary design phase (preparation of initial design alternatives for the owner)
- Design development phase (refinement of the design with more detailed design decisions)
- Contract documents phase (preparation of the detailed construction drawings and specifications as well as preparation of project general conditions)

- Bidding and negotiation phase (aiding the owner in soliciting, receiving and evaluating contractor's bids and proposals)
- Contract Administration (services conducted during the construction phase)

Contract administrative services are those services provided by the engineer during the construction phase. In this project phase, the engineer is aiding the owner and contractor as they pursue the completion of their contract obligations. The specific contract administrative services provided by the owner during the construction phase of the project should be defined by the owner/engineer agreement and could entail processing applications for payment and change orders, field observation, project reporting, attending project meetings, clarifying the contract documents, rendering decisions for owner/contractor disputes, and other administrative duties.

55. **A**

56. **C**

57. **D**

58. **A** The required integral is

$$V = \int_0^{48} p(x)dx$$

There are four intervals within the range (i.e., 1 to 12, 12 to 24, etc.), so

$$n = 4 \quad \text{and} \quad \Delta x = \frac{b-a}{n} = 12$$

The trapezoidal rule yields

$$V = \frac{\Delta x}{2}\left[f(0) + 2f(1) + 2f(2) + 2f(3) + f(4)\right]$$
$$= \frac{12}{2}(5 + 2\times 11 + 2\times 18 + 2\times 14 + 10) = 606 \text{ lb}$$

59. **B** Simpson's rule provides a better estimate:

$$V = \frac{\Delta x}{3}\left[f(0) + 4f(1) + 2f(2) + 4f(3) + f(4)\right]$$
$$= \frac{12}{3}(5 + 4\times 11 + 2\times 18 + 4\times 14 + 10) = 604 \text{ lb}$$

60. **D** The derivative of $P(x)$ is $P'(x) = 3x^2 - 8x$. The formula for providing a next estimate, given an initial estimate a_1^1, is

$$a_1^2 = a_1^1 - \frac{P(a_1^1)}{P'(a_1^1)} = -1 - \frac{-1^3 - 4(-1)^2 + 4}{3(-1)^2 - 8(-1)} = -0.909$$

$$a_1^3 = a_1^2 - \frac{P(a_1^2)}{P'(a_1^2)} = -0.909 - \frac{-0.909^3 - 4(-0.909)^2 + 4}{3(-0.909)^2 - 8(-0.909)} = -0.903$$

Note: The exponent on a_i^j represents the iteration number.

The D/S Review for Mechanical Engineering

edited by M. C. Potter

Overview of the ME/DS Topics

Chapters

Special Concerns about the NCEES Handbook

Most of the problems on the Mechanical Engineering test of the afternoon session can be worked using pages 108 to 116 of the NCEES Handbook. Many of the problems, however, on the afternoon Mechanical Engineering test will come from the material in pages 1 to 81 of the NCEES Handbook, material used for the General tests of the morning and afternoon sessions. Since all engineers must take the General morning exam, it is assumed that you are familiar with that material. You must realize, however, that you will have to make reference to equations, tables, and charts located in those first 81 pages of the NCEES Handbook when you are working problems in the afternoon Mechanical Engineering session. The general engineering sections that you should be particularly familiar with are the sections on Mathematics, Statics, Dynamics, Mechanics of Materials, Thermodynamics, Heat Transfer, Control Systems, and Fluid Mechanics. A review of general engineering material is found in our *Fundamentals of Engineering*, a review book written specifically for the FE/EIT exam.

Do I Have to Take the DS Test?

If, after you study this Mechanical Engineering review section, and take the Mechanical Engineering Discipline Practice Exam, you feel that you could more easily pass the General exam of the afternoon session, you should register to take the afternoon General test. If you have already registered for the Mechanical Engineering afternoon test, you may contact your state board and request a change. Make sure your state board recognizes a *pass* on the General afternoon test for your particular discipline before you make that decision; the vast majority of the state boards will recognize a pass on any of the afternoon tests by any registrant.

This Review is Focused on the NCEES Handbook

In order to provide a very efficient, yet effective review, this DS Review has been designed to review only the material associated with the formulas and figures presented in the NCEES Handbook. The Handbook is the only material allowed in the exam. Thus it makes an efficient base to study around. It is assumed that you have a copy of the NCEES Handbook dated 1996 or later. If you do not, you can obtain one from Great Lakes Press by calling 1-800-837-0201.

We Offer Other Aides for the A.M. Session & Prep in General

We also have an effective review for the General tests, both morning and afternoon, entitled *Fundamentals of Engineering*. It is used at hundreds of engineering schools across the nation. We also distribute the HP-48G and GX calculators, which are pre-programmed with many of the equations you'll need when working problems on the FE/EIT exam. Also, we've developed *Jump Start your HP-48G/X*, a 150-page book designed to efficiently help engineers get the most out of the HP48G/X. You can order these items by calling 1-800-837-0201.

1. Thermodynamics

by Merle C. Potter

Many of the questions relating to Thermodynamics can be answered using the equations and material found in pages 43-47 of the NCEES Handbook and the tables and figures on pages 48-52. So, make sure that you're familiar with that material. Additional material is reviewed in the following sections. Thermodynamics occurs on pages 108, 109, 111, 112, and 113 of the Handbook. This section reviews those parts that are included on Thermodynamics. It should be emphasized, however, that you should be familiar with the material for the morning session of the FE/EIT exam. Our *Fundamentals of Engineering Review*, covers these subjects succinctly.

1.1 The Two-Stage Refrigeration Cycle

Refer to the *T-s* diagram and the component diagram on page 108 of the NCEES Handbook. The temperature of state 7 must be slightly higher than that of state 4 to allow for condensation of the lower cycle refrigerant and evaporation of the upper cycle refrigerant; in the limit the temperature of state 7 could equal that of state 4.

Example 1.1

The upper cycle of a two-stage refrigeration cycle operates between pressures of 300 kPa and 1000 kPa while the lower cycle operates between 30 kPa and 200 kPa. The R-134a, the refrigerant in both cycles, produces 300 MJ/hr of refrigeration.

1. What is the highest temperature in the upper cycle?
   (A) 50°C
   (B) 46°C
   (C) 42°C
   (D) 38°C

2. What is the lowest temperature accomplished with this two-stage cycle?
   (A) -30°C
   (B) -35°C
   (C) -40°C
   (D) -50°C

3. What is the quality of state B?
   (A) 0.22
   (B) 0.20
   (C) 0.18
   (D) 0.16

4. What mass flux is required in the lower cycle?
   (A) 27 kg/s
   (B) 12.2 kg/s
   (C) 8.6 kg/s
   (D) 0.46 kg/s

5. What is the total power requirement?
   (A) 20 kW
   (B) 30 kW
   (C) 40 kW
   (D) 50 kW

6. Calculate the COP of the two-stage cycle.
   (A) 5
   (B) 4
   (C) 3
   (D) 2

Solutions:

1. **B** Using the diagram on page 50 of the NCEES Handbook, the entropy of state 1 must equal that of state 2. Following a $s = C$ line we find $T_2 = 46°C$. The solid lines are $T = C$ lines.

2. **D** The temperature of state 7 is found from the diagram to be $T_y = -50°C$ using $P_y = 0.03$ MPa.

3. **A** Across a valve $h_y = h_g = 187$ kJ/kg. Use $h_g = h_f + xh_{fg}$ or $187 = 137 + x_8(369 - 137)$. $\therefore x_8 = 0.22$.

4. **D** The first law: $\dot{Q} = \dot{m}(h_5 - h_8)$ or $\frac{300\,000}{3600} = \dot{m}(369 - 187)$. $\therefore \dot{m} = 0.46$ kg/s.

5. **C** Find $\dot{m}$ of the upper cycle: $\dot{m}_u(h_1 - h_4) = \dot{m}_\ell(h_6 - h_7)$. Using $\dot{m}_\ell = 0.46$ kg/s. $\dot{w} = \dot{w}_u + \dot{w}_\ell = \dot{m}_u(h_2 - h_1) + \dot{m}_\ell(h_6 - h_5) = .72(428 - 400) + 0.46(408 - 369) = 38\,\text{kW}$

6. **D** $\text{COP} = \frac{\text{output}}{\text{input}} = \frac{\dot{Q}_\ell}{\dot{W}} = \frac{300\,000/3600}{38} = 2.2$.

1.2 The Air Refrigeration Cycle

Refer to the *T-s* diagram and the component diagram on page 108 of the NCEES Handbook. Such a cycle is often used to cool an aircraft with air as the refrigerant. $\dot{Q}_{\text{in}}$ enters the conditioned space and is the desired effect in a refrigerator, whereas $\dot{Q}_{\text{out}}$ is the desired effect in a heat pump.

Example 1.2

A heat pump operates with air as the working fluid between pressures of 100 kPa and 600 kPa. The lowest temperature in the cycle is –10°C and 100 MJ/hr of heat is produced by moving 6 kg/min of air.

1. What is the highest temperature in the cycle?
   (A) 440°C
   (B) 380°C
   (C) 320°C
   (D) 260°C

2. What is the ratio of $\dot{W}_T/\dot{W}_C$?
   (A) 2.4
   (B) 1.2
   (C) 0.6
   (D) 0.4

3. What is $\dot{Q}_{\text{in}}$?
   (A) 33 kW
   (B) 29 kW
   (C) 24 kW
   (D) 17 kW

4. What is the COP of the cycle?
   (A) 4.0
   (B) 3.5
   (C) 3.0
   (D) 2.5

Solutions:

1. **A** The temperature of state 3 is

$$T_3 = T_4\left(P_3/P_4\right)^{k-1/k} = 263\left(600/100\right)^{.4/1.4} = 439\ \text{K}.$$

The first law: $\dot{Q}_{\text{out}} = \dot{m}\left(h_2 - h_3\right)$ or $\dfrac{100\ 000}{3600} = \dfrac{6 \times 1.0}{60}\left(T_2 - 439\right)$

where $c_p = 1.0\ \text{kJ/kg}\cdot\text{K}$. $\therefore\ T_2 = 717\ \text{K}$ or $444°\text{C}$.

2. **C** $\dot{W}_T = \dot{m}(h_3 - h_4) = \frac{6}{60} \times 1.0(439 - 263) = 17.6 \text{ kW}$.

   $\dot{W}_C = \dot{m}(h_2 - h_1) = \frac{6}{60} \times 1.0(717 - 430) = 28.7 \text{ kW}$,

   where $T_1 = 717 \times (100/600)^{.4/1.4} = 430 \text{ K}$.

   $\therefore \ \dot{W}_T / \dot{W}_C = 17.6/28.7 = 0.613$.

3. **D** $\dot{Q}_{\text{in}} = \dot{m}(h_1 - h_4) = \frac{6}{60} \times 1.0(430 - 263) = 16.7 \text{ kJ/s}$.

4. **D** For a heat pump $\text{COP} = \frac{\text{output}}{\text{input}} = \frac{Q_{\text{out}}}{W_{\text{net}}} = \frac{100\ 000/3600}{28.7 - 17.6} = 2.5$

1.3 HVAC

Heating, ventilating and air-conditioning problems are worked using the psychrometric chart on page 51. Dry-bulb temperature is on the horizontal axis, specific humidity ω is on the right-hand vertical axis, enthalpy h is on the left-hand slanted scale, dew-point temperature T_{dp} is along the left curved scale and is associated with a horizontal line, and wet-bulb temperature T_{wb} is also along the left curved line but is associated with a slanted constant-h line. Constant specific volume lines are the steeper-sloped lines. It's important to note that h and ω are given per kilogram of dry air.

Example 1.3

Air at a temperature of 30°C and a wet-bulb temperature of 24°C is conditioned to 20°C and 40% humidity. The mass flux of air is 50 kg/min.

1. What is the dew-point temperature entering the unit?
   (A) 30°C
   (B) 24°C
   (C) 21.6°C
   (D) 16.8°C

2. What is the partial pressure of the water vapor entering the refrigeration unit?
   (A) 1.86 kPa
   (B) 2.63 kPa
   (C) 3.47 kPa
   (D) 4.25 kPa

3. Estimate the temperature of the exiting condensate.
   (A) 6°C
   (B) 10°C
   (C) 14°C
   (D) 18°C

4. Calculate the required cooling load.
   (A) 63 kJ/s
   (B) 59 kJ/s
   (C) 52 kJ/s
   (D) 37 kJ/s

5. How many liters of condensate exit the unit in 10 hours?
   (A) 320 L
   (B) 290 L
   (C) 260 L
   (D) 210 L

Solutions:

1. **C** Refer to the figure on the top left of page 109 of the NCEES Handbook. Locate state 1 on the psychrometric chart on page 51. Use 30°C on the horizontal axis and locate the 24°C dashed line (sloping down to the right). Move left along a constant ω line (horizontal) to find $T_{\text{dp}} = 21.6°\text{ C}$.

2. **B** From the table on NCEES p. 48 we find $P_g = 4.246\,\text{kPa}$. The relative humidity from the psychrometric chart is 62%. Then

$$P_v = \phi P_g = 0.62 \times 4.246 = 2.63\,\text{kPa (see p. 45).}$$

3. **C** The high temperature of the condensate is 22°C (move horizontally from state 1). The low temperature of the condensate (move horizontally from state 2) is 6°C. The average is (22 + 6)/2 = 14°C.

4. **D** The cooling load is, using the psychrometric chart and the equation on p. 109:

$$\dot{Q}_{\text{out}} = \dot{m}_a\left[h_2 - h_1 + h_{f3}(\omega_1 - \omega_2)\right]$$
$$= \frac{50}{60}\left[35 - 72.2 + 59(0.0164 - 0.0058)\right] = -36.6\,\text{kJ/s}.$$

We found h_{f3} on p. 48 as h_f at $T = 14°\text{C}$. The negative sign means heat leaves the system.

5. **A** $\dot{m}_w = \dot{m}_a(\omega_1 - \omega_2) = 50(0.0164 - 0.0058) = 0.53\text{ kg/min}$ or 31.8 kg/hr

$31.8\text{ kg/hr} = 0.0318\text{ m}^3\text{/hr}$ or 31.8 L/hr. $\therefore$ 318 L leave in 10 hours.

We used 1000 kg/m^3 as the density of water. This standard value is quite good up to about 100°C (it's 96 kg/m^3 at 100°C).

Example 1.4

One hundred m^3/min of atmospheric air at 5°C and 80% relative humidity is to be heated and humidified to 22°C and 50% relative humidity.

1. If no moisture is added, what would be the relative humidity at 22°C?
   (A) 62%
   (B) 51%
   (C) 43%
   (D) 37%

2. To end up at 22° C, what temperature will the air have just before the moisture is added?
   (A) 35°C
   (B) 32°C
   (C) 29°C
   (D) 27°C

3. How many liters of water are needed in 4 hours to humidify the air?
   (A) 120 L
   (B) 92 L
   (C) 80 L
   (D) 68 L

4. What is the required heating load?
   (A) 30 kJ/s
   (B) 40 kJ/s
   (C) 50 kJ/s
   (D) 60 kJ/s

Solutions:

1. **D** Move horizontally (ω remains constant) on the psychrometric chart on p. 51 to 22°C where $\phi = 37\%$.

2. **B** The air must reach a state like state 1 on the figure at the bottom left of p. 109, with state 2 at 22°C and 50% humidity. From the psychrometric chart we find 31.5°C, assuming a constant h-line from state 2 to the state of interest. (Since $h_3(\omega_2 - \omega_1) << h_2 - h_1$ we can assume that $h_2 = h_1$.)

3. **A** $\dot{m}_w = \dot{m}_a(\omega_2 - \omega_1) = \frac{100}{0.793} - (0.0083 - 0.0043) = 0.504\ \text{kg/min}$. We have used $v_{\text{air}} = 0.793\ \text{m}^3/\text{kg}$ at state 1 from the psychrometric chart. There is 1 L per kg of water, hence $\dot{m}_w = 0.504\ \text{L/min}$. In 4 hours, we need $0.504 \times 4 \times 60 = 121\ \text{L}$ of water.

4. **C** $\dot{Q}_{\text{in}} = m_a\left[(h_2 - h_1) + h_3(\omega_2 - \omega_1)\right]$

$$= \frac{100}{0.793 \times 60}\left[41.8 - 18 + 63(0.0083 - 0.0043)\right] = 50\ \text{kJ/s}$$

Note: $h_3(\omega_2 - \omega_1)$ is negligible regardless of the temperature of the water used for humidification.

Example 1.5

The hot, humid air from a covered pool is at 30°C and 80% humidity. It is mixed with 40 m^3/min of dry, cool atmospheric air at 20°C and 30% humidity. The mixed air stream has a humidity of 50%. Both streams of air are at 100 kPa.

1. What is the mass flux of the dry, cool air stream?
   (A) 0.24 kg/s
   (B) 0.83 kg/s
   (C) 2.6 kg/s
   (D) 32 kg/s

2. What is the partial pressure of the dry air in the hot, humid air stream?
   (A) 94.6 kPa
   (B) 95.3 kPa
   (C) 95.8 kPa
   (D) 96.6 kPa

3. Calculate the specific humidity of the hot, humid air stream.
   (A) 0.0217
   (B) 0.0218
   (C) 0.0219
   (D) 0.0220

4. Estimate the flow rate of the hot, humid air stream.
   (A) 12 m^3/min
   (B) 16 m^3/min
   (C) 22 m^3/min
   (D) 30 m^3/min

5. What is the temperature of the mixed air stream?
   (A) 25.5°C
   (B) 24.5°C
   (C) 23.5°C
   (D) 22.5°C

Solutions:

1. **B** Find v_1 on the psychrometric chart on p. 51: $v_1 = 0.802\ m^3/kg$. Then,

$$\dot{m}_{a1} = 40/(0.802 \times 60) = 0.831\ kg/s\,.$$

2. **D** $P_g = 4.246$ from p. 48. Then $P_v = \phi P_g = 0.8 \times 4.246 = 3.397\ kPa$. Finally, $P_a = P_{atm} - P_v = 100 - 3.397 = 96.6\ kPa$. We used $P_{atm} = 100\ kPa$. (See p. 45.)

3. **C** Because the value for ω is quite accurate we must use an equation (see p. 45): $\omega = 0.622\, P_v/P_a = 0.622 \times 3.397/96.6 = 0.0219$ kg water/kg air.

4. **A** Locate state 3 at 50% humidity and $T_3 = 22.5°\text{C}$ by assuming a straight line between states 1 and 2. Then, using $h_3\dot{m}_{a3} = h_1\dot{m}_{a1} + h_2\dot{m}_{a2}$, $42(\dot{m}_{a2} + .831) = 30.5 \times .831 + 82.8\dot{m}_{a2}$. $\therefore \dot{m}_{a2} = 0.234$ kg/s. Flow rate = $\dot{m}v = .234 \times .848 = 0.199\ \text{m}^3/\text{s}$ or $11.9\ \text{m}^3/\text{min}$

5. **D** From the psychrometric chart we find $T_3 = 22.5°\text{C}$.

1.4 The Otto Cycle

The Otto cycle is the cycle that models the cycle used by a gasoline engine that powers cars and trucks. The process form state 4 to state 1 on page 47 of the NCEES Handbook simulates the exhaust process during which no work is done and the process returns to the initial state of the cycle. Combustion occurs instantaneously while the piston is at the top-dead-center (TDC). The compression ratio r is the maximum volume in the cylinder divided by the volume at TDC. Since air is compressed and expands in an enclosed cylinder, we use closed system equations in all processes.

Example 1.6

An Otto cycle operates between a maximum temperature of 1400°C and a minimum temperature of 20°C. It accepts the air at 100 kPa and has a compression ratio of 8. Each cylinder is 8 cm in diameter and the maximum volume is 0.8 liters.

1. Calculate the mass of air in each cylinder.
   (A) 2.63 grams
   (B) 2.34 grams
   (C) 1.68 grams
   (D) 0.951 grams

2. What is the temperature just before combustion?
   (A) 400°C
   (B) 412°C
   (C) 493°C
   (D) 567°C

3. What is the maximum pressure?
   (A) 8840 kPa
   (B) 7260 kPa
   (C) 6490 kPa
   (D) 4570 kPa

4. Estimate the heat transfer needed each cycle.
   (A) 870 J
   (B) 790 J
   (C) 730 J
   (D) 680 J

5. Estimate the net work output each cycle.
   (A) 320 J
   (B) 390 J
   (C) 450 J
   (D) 520 J

6. What is the cycle efficiency?
   (A) 42%
   (B) 46%
   (C) 52%
   (D) 56%

7. What would be the maximum possible efficiency of any cycle operation between the same two temperatures?
   (A) 61%
   (B) 76%
   (C) 82%
   (D) 87%

Solutions:

1. **D** $m = \dfrac{P_1V_1}{RT_1} = \dfrac{100 \times 0.0008}{0.287 \times 293} = 9.51 \times 10^{-4}$ kg or 0.951 grams.
   $\left(R = 8.314/29 = 0.287 \dfrac{\text{kJ}}{\text{kg} \cdot \text{K}}.\right.$ We use $P_1 = 100\,\text{kPa}$ since R has kJ as a unit.$\left.\right)$

2. **A** Process 1-2 is an isentropic process: $T_2 = T_1(v_1/v_2)^{k-1} = 293(8)^{.4} = 673$ K or 400°C. See p. 43 and the Otto cycle on p. 47.

3. **D** For the isentropic process, $P_2 = P_1(v_1/v_2)^k = 100 \times 8^{1.4} = 1840$ kPa. The next process is at constant volume: $P_3 = P_2(T_3/T_2) = 1840(1673/673)$ $= 4570\,\text{kPa}$. See p. 43.

4. **D** The needed heat transfer occurs at constant volume between state 2 and 3: $Q_{\text{in}} = \Delta U = mc_v\Delta T = 9.51 \times 10^{-4} \times 0.718(1673 - 673) = 0.683\,\text{kJ}$. c_v is found on p. 52. Note: $W_{2\text{-}3} = 0$ since volume is constant.

5. **B** Since $Q_{\text{net}} = W_{\text{net}}$ for a cycle, let's find Q_{out}:
   $$T_4 = T_3(V_3/V_4)^{k-1} = 1673(1/8)^{.4} = 728\text{ K}.$$
   Then, for the constant volume process from 4 to 1:

$$Q_{out} = \Delta U = mc_v\Delta T = 9.51\times10^{-4}\times0.718(728-293) = 0.297\,\text{kJ}.$$
$$\therefore Q_{net} = 0.683-0.297 = 0.386\,\text{kJ}.$$

6. **D** The cycle efficiency is defined as output divided by input. Hence,

$$\eta = \frac{W_{net}}{Q_{in}} = \frac{.386}{.683} = 0.565 \quad \text{or} \quad 56.5\%$$

7. C The Carnot efficiency is the maximum possible efficiency (see p. 47):

$$\eta_{max} = 1-\frac{T_L}{T_H} = 1-\frac{293}{1673} = 0.825 \quad \text{or} \quad 82.5\%$$

1.5 The Diesel Cycle

The Diesel cycle differs from the Otto cycle in that combustion occurs relatively slowly during a constant pressure process while the fuel is being injected, as shown on p. 11 of the NCEES Handbook. During this combustion process the volume changes from V_2 to V_3; the cut-off ratio is defined as $r_c = V_3/V_2$. The compression ratio remains $r = V_1/V_2$. The exhaust process is again modeled as the constant volume process from state 4 to state 1, a zero-work process.

Example 1.7

A Diesel cycle operates with a compression ratio of 20 and a cutoff ratio of 2. Air at 90 kPa and 20°C inlets each 12-cm-dia cylinder. The piston has a 16 cm stroke.

1. Calculate the volume at bottom-dead-center.
   (A) 2.4 L
   (B) 1.9 L
   (C) 1.5 L
   (D) 1.2 L

2. What is the maximum temperature in the cycle?
   (A) 1940°C
   (B) 1850°C
   (C) 1760°C
   (D) 1670°C

3. What is the pressure at top-dead-center?
   (A) 6.0 MPa
   (B) 5.5 MPa
   (C) 5.0 MPa
   (D) 4.5 MPa

4. Calculate the necessary heat addition each cycle.
   (A) 2900 J
   (B) 2300 J
   (C) 1980 J
   (D) 1760 J

5. What is the cycle efficiency?
   (A) 65%
   (B) 60%
   (C) 55%
   (D) 50%

Solutions:

1. **B** The stroke moves the piston from TDC to BDC. Hence, $V_1 - V_2 = 0.16\pi D^2/4$. The compression ratio gives $V_1/V_2 = 20$. This allows

$$V_1 - \frac{V_1}{20} = 0.16\pi \times 0.12^2/4. \quad \therefore V_1 = 0.001905 \text{ m}^3, \text{ or } 1.905 \text{ L}.$$

2. **D** The isentropic process gives $T_2 = T_1 r^{k-1} = 293 \times 20^{.4} = 971 \text{ K}$. The constant pressure process from state 2 to state 3 (see the P-v diagram on p. 111) provides

$$T_3 = T_2 \frac{V_3}{V_2} = 971 \times 2 = 1942 \text{ K} \quad \text{or} \quad 1669°\text{C}.$$

3. **A** State 2 is at top-dead-center. Therefore,

$$P_2 = P_1 (V_1/V_2)^k = 90 \times 20^{1.4} = 5970 \text{ kPa}.$$

4. **C** Heat addition occurs from state 2 to state 3, a constant pressure process:

$$Q_{in} = m\Delta h = \frac{P_1 V_1}{RT_1} c_p (T_3 - T_2)$$
$$= \frac{90 \times 0.001905}{0.287 \times 293} \times 1.00(1942 - 971) = 1.98 \text{ kJ}.$$

5. **A** Recognizing that $W_{net} = Q_{net}$ for a cycle, we first find Q_{out}. We use

$$T_4 = T_3 (V_3/V_4)^{k-1} = T_3 (V_3/V_1)^{k-1} = T_3 (2V_2/V_1)^{k-1}$$
$$= 1942 (2 \times 1/20)^{.4} = 773 \text{ K}.$$

The heat rejected is then

$$Q_{out} = mc_v \Delta T = \frac{90 \times 0.001905}{0.287 \times 293} \times 0.718(773 - 293) = 0.703 \text{ kJ}$$

Finally, the efficiency is

$$\eta = \frac{W_{net}}{Q_{in}} = \frac{Q_{in} - Q_{out}}{Q_{in}} = 1 - \frac{Q_{out}}{Q_{in}}$$
$$= 1 - \frac{0.703}{1.98} = 0.645 \quad \text{or} \quad 64.5\%$$

1.6 The Brayton Cycle

The Brayton cycle, which models the cycle used by a gas turbine engine, involves control volume (open system) components, so that the first law used in all processes shown on p. 112 and p. 113 of the NCEES Handbook takes the form $\dot{Q} - \dot{W}_{\text{shaft}} = \dot{m}\Delta h$. We always neglect kinetic and potential energy changes unless obvious information is given indicating that they should be retained; such information includes diameters of inlet and exit areas and elevations. The simple Brayton cycle is shown on p. 112, and the Brayton cycle with regeneration is shown on p. 113. A regenerator is used if $T_4 > T_2$ (which it usually is) in the simple Brayton cycle; it simply preheats the air entering the combustor. In an actual engine, the turbine and compressor are connected with a common shaft so that the power output is $\dot{W}_{\text{net}} = \dot{W}_T - \dot{W}_C$. An ideal regenerator has an efficiency (usually called the effectiveness) of unity so that $T_2 = T_6$ and $T_3 = T_5$ (refer to the figures on p. 113). The cycles shown are open, but they are often assumed to be closed by adding a heat exchanger between the exit and the inlet, as indicated on the *T-s* diagrams.

Example 1.8

A simple Brayton cycle accepts air at 80 kPa and 30°C. The pressure ratio across the compressor is 6 and the high temperature is 1800°C. The power output is 2 MW.

1. What is the temperature of the air exiting the compressor?
   (A) 500°C
   (B) 450°C
   (C) 340°C
   (D) 230°C

2. Estimate the mass flux of air required.
   (A) 0.76 kg/s
   (B) 0.94 kg/s
   (C) 1.2 kg/s
   (D) 3.2 kg/s

3. What is $\dot{W}_C/\dot{W}_T$, the back-work ratio?
   (A) 0.19
   (B) 0.21
   (C) 0.24
   (D) 0.29

4. A maximum velocity of 200 m/s is allowed. What is the minimum diameter at the compressor inlet?
   (A) 18 cm
   (B) 15 cm
   (C) 12 cm
   (D) 9 cm

5. What is the cycle efficiency?
   (A) 55%
   (B) 50%
   (C) 45%
   (D) 40%

Solutions:

1. **D** For the isentropic process across the compressor,

$$T_2 = T_1(P_2/P_1)^{k-1/k} = 303 \times 6^{0.4/1.4} = 506\,\text{K} \quad \text{or} \quad 233°\text{C}.$$

2. **D** We must find w_C and w_T . The isentropic turbine provides

$$T_4 = T_3(P_4/P_3)^{k-1/k} = 2073^{.4/1.4} = 1242\,\text{K}$$

We used $P_4 = P_1$ and $P_3 = P_2$ (see the P-v diagram on p. 112). We have

$$w_C = h_2 - h_1 = c_p(T_2 - T_1) = 1.0(506 - 303) = 203\,\text{kJ/kg}.$$

$$w_T = h_3 - h_4 = c_p(T_3 - T_4) = 1.0(2073 - 1242) = 831\,\text{kJ/kg}.$$

Finally,

$$\dot{W}_{\text{net}} = \dot{m}(w_T - w_C)$$

$$2000 = \dot{m}(831 - 203). \quad \therefore \dot{m} = 3.18\,\text{kg/s}.$$

3. **C** The back-work ratio is

$$\dot{W}_C/\dot{W}_T = w_C/w_T = 203/831 = 0.244.$$

4. **B** We use $\dot{m} = \rho AV$ where $\rho = P/RT$. At the inlet (using V as velocity):

$$\frac{\pi D^2}{4} = \frac{\dot{m}}{\rho V} = \frac{\dot{m}RT}{PV}$$

$$= \frac{3.18 \times 0.287 \times 303}{80 \times 200}. \quad \therefore D = 0.148 \text{ m} \quad \text{or} \quad 14.8 \text{ cm}$$

5. **D** The efficiency is

$$\eta = \frac{w_{\text{net}}}{q_{\text{in}}} = \frac{w_T - w_C}{q_{\text{in}}} = \frac{c_p(T_3 - T_4) - c_p(T_2 - T_1)}{c_p(T_3 - T_2)}$$

$$= \frac{2073 - 1242 - 506 + 303}{2073 - 506} = 0.401 \quad \text{or} \quad 40.1\%$$

Example 1.9

A 90% effective regenerator is added to the cycle of Example 3.8.

1. Calculate the temperature exiting the regenerator at state 6.
   (A) 310°C
   (B) 370°C
   (C) 420°C
   (D) 490°C

2. Calculate the percentage increase in cycle efficiency.
   (A) 40%
   (B) 52%
   (C) 61%
   (D) 73%

Solutions:

1. **A** The regenerator efficiency provides, using $T_5 = 1242\text{ K}$ and $T_2 = 506\text{ K}$ from Ex. 1.8,

$$\eta = \frac{T_3 - T_2}{T_5 - T_2}. \quad 0.9 = \frac{T_3 - 506}{1242 - 506}. \quad \therefore\ T_3 = 1168\text{ K}.$$

The heat leaving the high-temperature stream equals the heat entering the low-temperature stream:

$$c_p(T_3 - T_2) = c_p(T_5 - T_6). \quad 1168 - 506 = 1242 - T_6. \quad \therefore\ T_6 = 580\,\text{K} \text{ or } 307°\text{C}.$$

2. **A** The efficiency of the cycle is

$$\eta = \frac{w_T - w_C}{q_{in}} = \frac{T_4 - T_5 - (T_2 - T_1)}{T_4 - T_3} = \frac{2073 - 1242 - (506 - 303)}{2073 - 1168} = 0.694$$

The percentage increase is

$$\% = \frac{0.694 - 0.401}{0.401} \times 100 = 73.1\%$$

1.7 The Rankine Cycle

The Rankine cycle models the cycle of a fossil fueled power plant. The simple cycle shown on p. 47 of the NCEES Handbook is modified in most power plants by utilizing a reheater—the steam is intercepted in the superheat region and heated at constant pressure back to a temperature as high as T_3—and a feedwater heater, in which some of the steam in between state 3 and state 4 is used to preheat the water entering the boiler; either an open or closed (or both) feedwater heater is used. Reheating moves state 4 on p. 47 to the right thereby reducing the condensation in the turbine; it does not significantly increase the cycle efficiency if the condenser pressure remains unchanged. Preheating with a heater does increase the cycle efficiency.

Example 1.10

A simple Rankine cycle operates between 45°C and 600°C with a maximum pressure of 1000 kPa. The mass flux is 4 kg/s and the inlet velocity to the pump is not to exceed 20 m/s.

1. If the pump is 75% efficient, what power is needed by the pump?
   (A) 5.3 kW
   (B) 4.3 kW
   (C) 3.8 kW
   (D) 2.7 kW

2. What is the maximum power output of the turbine?
   (A) 3.4 MW
   (B) 3.9 MW
   (C) 4.3 MW
   (D) 4.6 MW

3. If the turbine were 87% efficient, estimate the temperature of the steam at the outlet of the turbine.
   (A) 122°C
   (B) 115°C
   (C) 107°C
   (D) 100°C

4. Calculate the minimum diameter of the pump inlet.
   (A) 128 mm
   (B) 86 mm
   (C) 45 mm
   (D) 16 mm

5. Calculate the efficiency of this cycle, assuming an 87% efficient turbine.
   (A) 26%
   (B) 29%
   (C) 32%
   (D) 35%

6. Select the primary reason why the Rankine cycle is substantially less than the efficiency of a Carnot cycle operating between the same temperature limits.
   (A) The turbine is inefficient
   (B) There is substantial frictional losses in the piping systems.
   (C) Heat transfer occurs across a large temperature difference.
   (D) Too much heat is rejected in the condenser.

Solutions:

1. **A** The first law for a pump results in (see p. 113)

$$w_P = h_2 - h_1 = \frac{v(P_2 - P_1)}{\eta_P} = \frac{0.001(1000 - 9.593)}{0.75} = 1.321\,\text{kJ/kg}$$

$$\therefore \dot{W}_P = \dot{m} w_P = 4 \times 1.321 = 5.28 \text{ kW}.$$

We have reversed the signs so that w_P and $\dot{W}_P$ are positive. Actually, in technical terms they are negative since they are energy input to the system. Also, we found $P_1 = 9.593$ kPa right next to $T_1 = 45°\text{C}$ on p. 48 .

2. **D** An isentropic process through the ideal turbine is assumed for maximum output (use p. 48 and p. 49):

$$s_4 = s_3 = 8.0290 = 0.6387 + x_4(7.5261). \quad \therefore x_4 = 0.982 .$$

$$\therefore h_4 = 188.45 + 0.982 \times 2394.8 = 2540 \text{ kJ/kg}.$$

The first law gives :

$$\dot{W}_T = \dot{m}(h_3 - h_4) = 4(3697.9 - 2540) = 4632 \text{ kW}.$$

3. **C** The efficiency of a turbine is (see p. 44)

$$\eta_T = \frac{h_3 - h_4}{h_3 - h_{4s}}. \quad 0.87 = \frac{3697.9 - h_4}{3697.9 - 2540}. \quad \therefore h_4 = 2691 \text{ kJ/kg}.$$

Using $h = 2691$ kJ/kg (which is greater that h_g), p. 49 allows us to interpolate at

$$P_4 = 0.0096\,\text{MPa}$$
$$\cong 0.01 \text{ MPa}$$
$$\therefore T_4 = 107°\text{C}$$

Note that the steam exits the real turbine in the superheat region. There is no condensation in the turbine.

4. **D** Use $\dot{m} = \rho A V$ where V is velocity. Then, with $\rho = 1/v$ (or use $\rho_{\text{water}} = 1000\,\text{kg/m}^3$)

$$m = AV/v . \quad 4 = \frac{\pi D^2}{4} \times 20/0.001. \quad \therefore D = 0.016 \text{ m} \quad \text{or} \quad 16 \text{ mm}.$$

5. **B** We need the heat input. It is

$$\dot{Q}_{\text{in}} = \dot{m}(h_3 - h_2) = 4[3697.9 - (188.45 + 1.321)] = 14\ 030\,\text{kJ/s}.$$

where $h_2 = h_1 + w_p$. The efficiency is

$$\eta = \frac{\dot{W}_{\text{net}}}{\dot{Q}_{\text{in}}} = \frac{\dot{W}_T - \dot{W}_P}{\dot{Q}_{\text{in}}}$$

$$= \frac{4632 \times 0.87 - 5.28}{14\ 030} = 0.287 \quad \text{or} \quad 29\%$$

6. **C** Heat transfer across a finite temperature difference is more and more irreversible as the temperature difference is increased. In a power plant the temperature difference in the boiler is very large.

Example 1.11

An open feedwater heater is added to the cycle of Example 1.10. Steam is extracted from the ideal turbine at 400 kPa. Two pumps are needed: one to increase the pressure exiting the condenser to 400 kPa, and a second pump to increase the pressure of the mixed stream to 1000 kPa.

1. What should the mass flux of the extracted steam be?
   (A) 0.41 kg/s
   (B) 0.52 kg/s
   (C) 0.61 kg/s
   (D) 0.79 kg/s

2. Estimate the power needed by a second ideal pump?
   (A) 4.6 kW
   (B) 4.2 kW
   (C) 3.8 kW
   (D) 3.2 kW

3. What is the percentage decrease in the power output of the ideal turbine?
   (A) 5%
   (B) 10%
   (C) 15%
   (D) 20%

4. Estimate the percent increase in the ideal cycle efficiency.
   (A) 11
   (B) 9
   (C) 6
   (D) 3

Solutions:

1. **B** An energy balance on the feedwater heater is (see p. 113)

$$\dot{m}_1 h_1 + \dot{m}_2 h_5 = h_6(\dot{m}_1 + \dot{m}_2)$$

where state 5 is the extraction state, and state 6 is the heater exit (shown as state 3 on p. 113). Refer to the Rankine cycle on p. 47 for the basic cycle; state 5 is between 3 and 4 and state 6 is on the saturated liquid line above state 2 at $P_6 = 400$ kPa. Using the steam tables on p. 48 and p. 49,

$$(4 - \dot{m}_2) \times 188.45 + \dot{m}_2 \times 3368 = 605 \times 4. \quad \therefore \dot{m}_2 = 0.524 \text{ kg/s}.$$

We used $s_3 = s_5 = 8.0290$ and $P_5 = 0.4$ MPa so that, by interpolation, $h_5 =$ 3368 kJ/kg. Also, $h_6 = 605$ kJ/kg is the value of h_f at 0.4 MPa by interpolation on p. 48.

2. **D** $w_{P2} = v(P_3 - P_6) = 0.001(1000 - 400) = 0.6 \text{ kJ/kg}$.

$\dot{W}_{P2} = \dot{m} w_{P2} = 4(0.6) = 2.4 \text{ kW}$.

3. **B** After the steam is extracted, there is less to pass through the last part of the turbine. If 4 kg/s enters and 0.524 kg/s is extracted, then 3.476 kg/s passes through the last part. Hence,

$$\begin{aligned}\dot{W}_T &= 4(h_3 - h_5) + 3.476(h_5 - h_4) \\ &= 4(3697.9 - 3368) + 3.476(3368 - 2540) = 4198 \text{ kW}\end{aligned}$$

Using the result of Example 1.10,

$$\% \text{ decrease} = \frac{4632 - 4198}{4632} \times 100 = 9.37\%.$$

4. **D** The heat input required by the boiler is

$$\dot{Q}_{in} = \dot{m}(h_3 - h_6) = 4(3697.9 - 605) = 12\ 370 \text{ kJ/s}.$$

The cycle efficiency is, neglecting the negligible pump power,

$$\eta = \frac{\dot{W}_T}{\dot{Q}_{in}} = \frac{4198}{12\ 370} = 0.339.$$

The efficiency of the ideal cycle of Example 1.10 is

$$\eta_{cycle} = \frac{\dot{W}_T}{\dot{Q}_{in}} = \frac{4632}{14\ 030} = 0.330$$

The percent increase is

$$\% \text{ increase} = \frac{0.339 - 0.330}{0.330} \times 100 = 2.7\%$$

Even though a feedwater heater reduces the turbine power output, it does increase the cycle efficiency; consequently, they are used extensively in the power industry.

2. Heat Transfer

by Craig W. Somerton

2.1 Conduction

Refer to the sketches and equations on pages 53, 55, and 56 of the NCEES Handbook. Examples 2.4, 2.5, and 2.6 are applicable to the afternoon examination for Mechanical and Chemical Engineers.

Example 2.1

A refractory brick wall with thermal conductivity 1.3 W/m·°C is used to insulate a combustion chamber. The inside wall temperature is found to be at 2500°C.

1. For a 0.3 meter thick wall, the outside wall temperature is found to be 600°C. What is the heat transfer rate per wall surface area in W/m^2?
   (A) 200
   (B) 800
   (C) 8200
   (D) 1600

2. If the rate of heat transfer per wall surface area is found to be 6000 W/m^2, what thickness of wall is required to achieve an outside wall temperature of 600°C?
   (A) 0.3 m
   (B) 2.4 m
   (C) 0.6 m
   (D) 0.4 m

Solutions:

1. **C** The heat transfer rate is given by $\dot{Q} = -kA(T_2 - T_1)/L$. Solving for the heat transfer rate per wall surface area yields,

$$\dot{Q}/A = -(1.3)(600 - 2500)/(0.3) \approx 8200 \text{ W/m}^2$$

2. **D** Solving for the wall thickness gives, $L = -k(T_2 - T_1)/(\dot{Q}/A)$.

   Substituting numerical values,

$$L = -(1.3)(600 - 2500)/(6000) = 0.41 \text{ m}$$

Example 2.2

The wall of an office building is composed of 0.005 meters of dry wall (thermal conductivity of 0.17 W/m·°C), 0.05 meters of insulation (thermal conductivity of 0.027 W/m·°C), and 0.03 meters of face brick (thermal conductivity of 1.3 W/m·°C). The outside wall temperature (face brick side) is 32°C and the inside wall temperature (dry wall side) is at 25°C. The wall surface area is 27 m^2.

1. What is the rate of heat transfer through the wall?
   (A) 200 W
   (B) 100 W
   (C) 0.5 W
   (D) 10 W

2. What is the temperature between the face brick and the insulation?
   (A) 31.9°C
   (B) 35°C
   (C) 27.3°C
   (D) 24.9°C

Solutions:

1. **B** The heat transfer rate is given as $\dot{Q} = (T_{out} - T_{in})/R_{tot}$. The total thermal resistance is

$$R_{tot} = R_{dry\ wall} + R_{insulation} + R_{face\ brick}$$

where:

$$R_{dry\ wall} = L_{dry\ wall} / (k_{dry\ wall} A) = 0.005 / (0.17 \times 27) = 1.09 \times 10^{-3}$$

$$R_{insulation} = L_{insulation} / (k_{insulation} A) = 0.05 / (0.027 \times 27) = 6.86 \times 10^{-2}$$

$$R_{face\ brick} = L_{face\ brick} / (k_{face\ brick} A) = 0.03 / (1.3 \times 27) = 8.55 \times 10^{-4}$$

Then the total thermal resistance is 7.05×10^{-2} °C/W which gives a heat transfer rate of approximately 100 W.

2. **A** The temperature between the face brick and the insulation, $T_{FB\text{-}I}$, is given by $T_{FB\text{-}I} = T_{out} - \dot{Q} \cdot R_{face\ brick}$. Using the heat transfer rate and face brick thermal resistance from Problem 1 gives a temperature of 31.9°C.

Example 2.3

Steam at 250°C condenses inside a stainless steel (thermal conductivity of 50 W/m·°C) pipe of inside diameter 0.07 meters, wall thickness 0.006 meters, and length 5 meters.

1. If the outside pipe temperature is determined to be 230°C, what is the heat transfer rate?
   (A) 200 W
   (B) 800 W
   (C) 20 W
   (D) 1600 W

2. What is the radial thermal resistance of the pipe in °C/W?
   (A) 10^{-7}
   (B) 10^{-5}
   (C) 10^{4}
   (D) 10^{-4}

Solutions:

1. **A** The heat transfer rate is given as $\dot{Q} = \dfrac{2\pi kL(T_{\text{in}} - T_{\text{out}})}{\ln(D_{\text{out}} / D_{\text{in}})}$. Substituting numerical values gives 200 kW.

2. **D** The thermal resistance in cylindrical coordinates is given by $R = \dfrac{\ln(D_{\text{out}} / D_{\text{in}})}{2\pi kL}$. Substituting numerical values gives 10^{-4} °C/W.

Example 2.4

Fuel elements for nuclear power systems can be in cylindrical geometry or flat plate geometry. The cylindrical element is of diameter 0.015 m and length 5 m. The flat plate elements are 0.010 m thick, 0.12 m wide and 4.3 m long. The internal heat generation rate per unit volume for both elements is 2000 MW/m^3 and the fuel has a thermal conductivity of 12 W/m·°C.

1. If one side of the flat plate element is at 200°C and the other side is at 180°C, what is the temperature at the center of the element?
   (A) 190°C
   (B) 2300°C
   (C) 8500°C
   (D) 1600°C

2. What is the heat transfer rate per unit length from the cylindrical element in kW/m?
   (A) 200
   (B) 350
   (C) 8200
   (D) 1500

3. For a surface temperature of 550°C, what is the center temperature for the cylindrical element?
   (A) 800°C
   (B) 2900°C
   (C) 1200°C
   (D) 400°C

Solutions:

1. **D** The temperature at the center (x=0) is given by

$$T_c = \frac{\dot{Q}_{gen}(t/2)^2}{2k} + \frac{T_{s1}+T_{s2}}{2}$$

where t is the thickness. Substituting gives 2300°C.

2. **B** The heat transfer rate per unit length may be calculated from $\dot{Q}' = \pi r_0^2 \dot{Q}_{gen}$, which gives a value of 350 kW/m.

3. **B** The temperature at the center (r = 0) is given by

$$h = C(k/H)\text{Ra}_H^n$$

where t is the thickness. Substituting gives 2900°C.

Example 2.5

Bar stock of bronze initially at 300°C (with thermal conductivity 52 W/m·°C, density 8800 kg/m^3, and specific heat 460 J/kg·°C) is cooled in a water bath at 28°C with a heat transfer coefficient of 300 W/m^2·°C. The bar stock has diameter 0.12 m and is 4 meters long.

1. What is the Biot number for the process?
   (A) 5.8
   (B) 0.09
   (C) 0.17
   (D) 1.0

2. Using the lumped capacitance model, at what time will the bar stock reach 100°C?
   (A) 540 s
   (B) 800 s
   (C) 8200 s
   (D) 1600 s

Solutions:

1. **C** The Biot number is defined as

$$\text{Bi} = hV / kA_s = \frac{h\pi r_o^2 L}{k2\pi r_o L}.$$

Calculating gives a value of 0.17.

2. **A** The temperature variation with time is

$$T - T_\infty = (T_i - T_\infty)e^{-(hA_s/\rho c_P V)t}.$$

Solving for t gives

$$t = -\frac{\rho c_P V}{hA_s}\ln\{(T - T_\infty)/(T_i - T_\infty)\}$$

Calculating with $T = 100°C$ yields 540 seconds.

2.2 Convection

Refer to the sketches and equations on pages 53, 55, and 56 of the NCEES Handbook. Question #4 of Example 2.6, and Example 2.7 are applicable to the afternoon tests for Mechanical and Chemical Engineers.

Example 2.6

The outside surface of a steam pipe is in contact with air at 23°C. The pipe has diameter 0.07 m and length 20 m.

1. For a convective heat transfer coefficient of 3 W/m^2·°C and a surface temperature of 175°C, what is the heat transfer rate from the pipe into the air?
   (A) 2.0 kW
   (B) 2.7 kW
   (C) 5.0 kW
   (D) 0.5 kW

2. If the rate of heat transfer is found to be 2.5 kW, what would be the heat transfer coefficient required to maintain a surface temperature of 90°C in W/m^2·°C?
   (A) 5.8
   (B) 16
   (C) 2.5
   (D) 8.5

3. It is proposed to add four fins to the outside surface of the steam pipe, each of which have a cross-sectional area of 0.10 m^2, a length of 0.02 m, perimeter of 40 m, and a thermal conductivity of 175 W/m·°C. For a convective heat transfer coefficient of 4 W/m^2·°C and a heat transfer rate of 3.2 kW, what will be the surface temperature of the steam pipe?
   (A) 151°C
   (B) 186°C
   (C) 212°C
   (D) 103°C

4. If air (with Prandtl number 0.7 and thermal conductivity of 0.026 W/m·°C) is blown over the pipe at a Reynolds number of 10 000, what will be the heat transfer coefficient in W/m^2·°C?
   (A) 19
   (B) 41
   (C) 5
   (D) 0.05

Solutions:

1. **A** The heat transfer rate for convection is given by $\dot{Q} = hA(T_s - T_\infty)$. Calculating gives a heat transfer rate of 20 kW.

2. **D** The heat transfer coefficient can be solved for as $h = \dot{Q}/A(T_s - T_\infty)$ which gives 8.5 W/m^2·°C.

3. **B** The heat transfer rate will be the sum of the convective heat transfer from the bare tube, $\dot{Q}_{bt} = hA_{bt}(T_s - T_\infty)$, plus four times the heat transfer for a fin,

$$\dot{Q}_{\text{fin}} = \sqrt{hpkA_c}\,(T_s - T_\infty)\tanh\{\sqrt{hp/(kA_C)}(L + A_c/p)\}$$

The surface temperature can then be solved for as

$$T_s = T_\infty + \frac{\dot{Q}}{\sqrt{hpkA_c}\,\tanh\{\sqrt{hp/(kA_C)}(L + A_c/p)\} + hA_{bt}}$$

Calculating gives a surface temperature of 186°C.

4. **A** The heat transfer coefficient may be found from the Nusselt number correlation $\text{Nu} = c\text{Re}^n\text{Pr}^{1/3}$. For Re = 10 000, the table on page 55 of NCEES Handbook gives $n = 0.618$ and $c = 0.193$. Calculating gives a Nusselt number of 50.8. With the Nusselt number defined as $\text{Nu} = \frac{hD}{k}$, we obtain a heat transfer coefficient of 19 W/m^2·°C.

Example 2.7

The sides of a travel trailer are found to be at 55°C. Air surrounds the trailer at 29°C with Prandtl number 0.69, thermal conductivity of 0.026 W/m·°C, and kinematic viscosity 1.6×10^{-5} m²/s. A side of particular interest is 2.1 meters high and 4.5 meters long.

1. If the surrounding air is stationary, what is the heat transfer from the side of interest?
   (A) 160 W
   (B) 830 W
   (C) 460 W
   (D) 1200 W

2. What is the heat transfer coefficient (in W/m²·°C) for a wind of 3 m/s in the length direction?
   (A) 50
   (B) 25
   (C) 10
   (D) 5

Solutions:

1. **B** This is a natural convection problem with the heat transfer coefficient given by $h = C(k/H)\text{Ra}_H^n$. The Rayleigh number is defined as $\text{Ra} = \frac{g\beta\,(T_s - T_\infty)H^3}{\nu^2}\text{Pr}$. Calculating gives $\text{Ra} = 2.0 \times 10^{10}$. The table on page 56 of the NCEES Reference Handbook gives n = 1/3 and C = 0.1. Calculating gives a heat transfer coefficient of 3.4 W/m²·°C. The heat transfer is then given by $\dot{Q} = hA(T_s - T_\infty)$, which yields 830 W.

2. **C** For forced convection, begin by calculating the Reynolds number, $\text{Re} = \frac{uL}{\nu}$, which is 844 000 in this case. The appropriate Nusselt number correlation is then $\text{Nu} = 0.0366\text{Re}^{0.8}\text{Pr}^{1/3}$. Calculating gives a Nusselt number of 1780 and a heat transfer coefficient of 10 W/m²·°C.

2.3 Radiation

Refer to the sketches and equations on pages 53, 54, 56 and 57 of the NCEES Reference Handbook. Example 2.9 is applicable to the afternoon examination for Mechanical and Chemical Engineers.

Example 2.8

The floor of an incinerator is at 2200 K and has an emissivity of 0.7. The side walls and the ceiling are found to be at 600 K with an emissivity of 0.85.

1. If the side walls and ceiling are gray, opaque surfaces, what is their reflectivity?
   (A) 0.5
   (B) 1.0
   (C) 0.0
   (D) 0.15

2. What is the radiation heat transfer per unit area emitted by the floor of the incinerator in kW/m^2?
   (A) 930
   (B) 9300
   (C) 1130
   (D) 500

3. What is the net heat transfer rate per unit area from the floor in kW/m^2?
   (A) 775
   (B) 925
   (C) 1050
   (D) 500

Solutions:

1. **D** For a gray, opaque surface $\varepsilon + \rho = 1$, so that with an emissivity of 0.85, the side walls and ceiling must have a reflectivity of 0.15.

2. **A** The radiation emitted by a body is $\dot{Q} = \varepsilon \sigma A T^4$, which gives an emission per unit area for the floor of 930 kW/m^2.

3. **B** Assuming the floor is small compared to its surrounds the net radiation heat transfer rate is given by $\dot{Q} = \varepsilon \sigma A \left(T_{\text{floor}}^4 - T_{\text{c\&w}}^4 \right)$. The net radiation heat transfer rate per unit area is calculated to be 925 kW/m^2.

Example 2.9

A radiation shield is used to protect the passenger compartment from the catalytic converter in an automobile. The catalytic converter is at 900 K with an emissivity of 0.3. The outside surface of the passenger compartment is to be maintained at 350 K and has an emissivity of 0.9. All of the view factors may be taken to be one. For a radiation shield with emissivity of 0.2 on both sides, what is the heat transfer rate per unit area in kW/m^2?

(A) 3.5

(B) 7.0

(C) 10.5

(D) 5.0

Solution:

1. The appropriate equation may be found on page 56 of the NCEES Handbook.

$$\dot{Q}/A = \frac{\sigma\left(T_{cc}^4 - T_{pc}^4\right)}{\frac{1-\varepsilon_{cc}}{\varepsilon_{cc}} + \frac{1}{F_{cc-rs}} + 2\frac{1-\varepsilon_{rs}}{\varepsilon_{rs}} + \frac{1}{F_{rs-pc}} + \frac{1-\varepsilon_{pc}}{\varepsilon_{pc}}}.$$

Calculating gives a value of 3.5 kW/m^2. The answer is **A**.

2.4 Heat Exchangers

Refer to the sketches and equations on pages 54, and 55 of the NCEES Handbook. This section is applicable to the afternoon examination for Mechanical and Chemical Engineers.

Example 2.10

A heat exchanger is composed of a tube bank over which steam condenses at 500°C. Water enters the tubes of diameter 0.015 m at 200°C and leaves at 300°C, flowing at 0.2 m/s. The heat transfer coefficient for the steam condensation is 5000 W/m^2·°C. The properties of the water are:

Thermal Conductivity: 0.6 W/m·°C

Prandtl Number: 0.86

Kinematic Viscosity: 1.31×10^{-7} m^2/s

The configuration correction factor for the heat exchanger may be taken to be 0.85.

1. Neglecting fouling and the thermal resistance of the wall, what is the overall heat transfer coefficient for the heat exchanger in $W/m^2 \cdot °C$?
   (A) 1300
   (B) 5000
   (C) 1900
   (D) 300

2. What is the surface area required for a heat transfer rate of 1 MW?
   (A) 5.0 m^2
   (B) 16 m^2
   (C) 1.0 m^2
   (D) 2.5 m^2

Solutions:

1. **C** The overall heat transfer coefficient is given by $U = \left[1/h_w + 1/h_{stm}\right]^{-1}$. The heat transfer coefficient for the steam has been given, but the heat transfer coefficient for the water must be calculated. Begin by calculating the Reynolds number,

$$\mathrm{Re} = \frac{uD}{\nu} = \frac{(0.2)(0.015)}{1.31 \times 10^{-7}} = 23\ 000$$

This is turbulent flow in a closed conduit. The appropriate Nusselt number correlation is

$$\mathrm{Nu} = 0.026\,\mathrm{Re}^{0.8}\,\mathrm{Pr}^{1/3}$$

Calculating gives a Nusselt number of 76 and a water heat transfer coefficient of 3050 $W/m^2 \cdot °C$. The overall heat transfer coefficient then becomes 1900 $W/m^2 \cdot °C$.

2. **D** The heat transfer rate is given by $\dot{Q} = UAF\Delta T_{lm}$. The log mean temperature difference is

$$\Delta T_{lm} = \frac{(T_{stm} - T_{w,out}) - (T_{stm} - T_{w,in})}{\ln\left\{\frac{(T_{stm} - T_{w,out})}{(T_{stm} - T_{w,in})}\right\}}$$

which calculates out to 247°C. Solving for the area yields a required surface area of 2.5 m^2.

3. Fluid Mechanics

by Merle C. Potter

The subjects relating to fluid mechanics that are included in the Mechanical Engineering Section of the NCEES Handbook are presented on pp. 110 and 111; the material provides for an analysis of fans, pumps and compressors. The other subjects in fluid mechanics are found on pp. 34-42, material that is also used for the a.m. General session. Numerous problems on fluid mechanics in the p.m. exam will be worked using information found in those earlier pages.

We do not review fluid statics in this chapter since it is assumed to be included in the a.m. session only. It is true, however, that many of the problems in the p.m. session will be based on the more elementary material of the a.m. session. So it is imperative that you are prepared to use any material presented on pp. 34-42. We will review here the material that is more likely to be needed to work the problems of the p.m. session. Remember, pressures in fluids are assumed to be presented as gage pressure unless otherwise stated. Pressures in thermodynamics are assumed to be absolute pressure.

3.1 The Energy Equation

The energy equation for pipe flow is presented on p.36. The head loss h_f is calculated using the Darcy equation on p.37 where the friction factor comes from the Moody diagram on p.41. If there are geometry changes, such as fittings, a loss coefficient is used; it is given on p.37 for an entrance or an exit and must be given in the problem for other geometry changes (recall that for an accelerated flow, such as through a nozzle, the loss is usually assumed to be negligible). For multi-path pipelines, see p.38.

Example 3.1

A 4-cm-diameter, cast iron pipe transports water at 0.01 m^3/s. The nozzle contracts to 2-cm diameter at the exit. The total length of the pipe is 50 m. Use $\nu = 10^{-6}\ m^2/s$ and $p_{vapor} = 2$ kPa absolute.

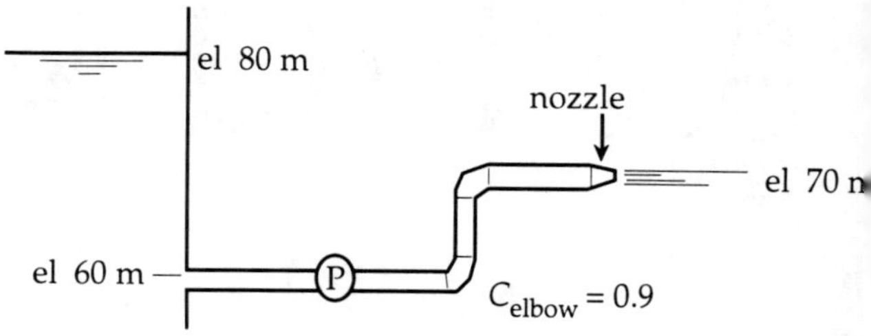

1. Calculate the average velocity in the pipe.
   (A) 2 m/s
   (B) 6 m/s
   (C) 8 m/s
   (D) 10 m/s

2. Estimate the pressure just before the nozzle.
   (A) 500 kPa
   (B) 400 kPa
   (C) 300 kPa
   (D) 200 kPa

3. What is the friction factor in the pipe?
   (A) 0.015
   (B) 0.020
   (C) 0.025
   (D) 0.032

4. What is the maximum distance from the reservoir that the pump can be located?
   (A) 25.7 m
   (B) 20.6 m
   (C) 15.4 m
   (D) 9.7 m

5. The energy grade line is located how far above the exit?
   (A) 68 m
   (B) 52 m
   (C) 43 m
   (D) 28 m

6. Estimate the pump power needed if it is 75% efficient.
   (A) 29 kW
   (B) 23 kW
   (C) 17.5 kW
   (D) 13.1 kW

Solutions:

1. **C** The average velocity is

$$V = \frac{Q}{A} = \frac{.01}{\pi \times .02^2} = 7.96 \text{ m/s}$$

2. **A** There is negligible friction loss across a contraction. Bernoulli's equation provides, with $V_{exit} = A_1 V / A_2 = 4V = 4 \times 7.96 = 31.8$ m/s,

$$\frac{V_1^2}{2g} + \frac{p_1}{\gamma} + z_1 = \frac{V_2^2}{2g} + \frac{p_2}{\gamma} + z_2 \; \frac{V_1^2}{2g} + \frac{p_1}{\gamma} + z_1 = \frac{V_2^2}{2g} + \frac{p_2}{\gamma} + z_2$$

$$\frac{7.96^2}{2 \times 9.8} + \frac{p_1}{9800} = \frac{31.2^2}{2 \times 9.8} + 0. \qquad p_1 = 504\,000 \text{ Pa} \quad \text{or} \quad 504 \text{ kPa}$$

3. **D** We must use the Moody diagram on p.41:

$$\text{Re} = \frac{VD}{\nu} = \frac{7.96 \times .04}{10^{-6}} = 3.2 \times 10^5,\; e/D = .25/40 = .00625.$$

$$\therefore f = .032.$$

4. **D** The inlet pressure must not be less than the vapor pressure, $p_v = -98$ kPa (a vacuum). Apply the energy equation between the surface and the pump inlet:

$$\frac{V_1^2}{2g} + \frac{p_1}{\gamma} + z_1 = \frac{V_2^2}{2g} + \frac{p_2}{\gamma} + z_2 + h_f + h_{f,\text{minor}}$$

$$0 + 0 + 80 = \frac{7.96^2}{2 \times 9.8} + \frac{-98000}{9800} + 60 + .032 \frac{L}{.04} \frac{7.96^2}{2 \times 9.8} + .5 \frac{7.96^2}{2 \times 9.8}$$

$$\therefore L = 9.72 \text{ m}.$$

5. **B** The energy grade line is at the elevation

$$z = z_{exit} + \frac{p_{exit}}{\gamma} + \frac{V_{exit}^2}{2g}$$

$$= 70 + 0 + \frac{31.8^2}{2 \times 9.8} = 121.6 \text{ m}$$

or 51.6 m above the exit.

6. **B** Apply the energy equation between the reservoir surface and the exit (we must add the pump power to the left-hand side):

$$\frac{\dot{W}_P}{\gamma Q} + \frac{V_1^2}{2g} + \frac{p_1}{\gamma} + z_1 = \frac{V_2^2}{2g} + \frac{p_2}{\gamma} + z_2 + h_f + h_{f,\text{min or}}$$

$$\frac{\dot{W}_P}{9800 \times .01} + 80 = 70 + \frac{31.8^2}{2 \times 9.8} + .032 \frac{50}{.04} \frac{7.96^2}{2 \times 9.8} + (.5 + .9 \times 2) \frac{7.96^2}{2 \times 9.8}$$

$$\therefore \dot{W}_P = 17\,500 \text{ W}$$

This is the power the water must receive. The pump is only 75% efficient; hence, it requires

$$\dot{W} = \frac{17.5}{.75} = 23.2 \text{ kW}.$$

3.2 Impulse-Momentum

If a force is desired due to the motion of a fluid, it usually is necessary to use the integral form of the momentum equation that results from Newton's second law applied to a control volume; it is presented on pp. 37 and 38 for several flow situations. When applying the momentum equation, make sure the velocity is treated as a vector quantity. Note that the equations given on p.38 for moving blades are for a series of blades, not for a single blade such as would be encountered when a train dips a deflector into a trough of water. Example 3.3 illustrates the solution for a single blade.

Example 3.2

The pressure at the inlet of the horizontal bend is 800 kPa.

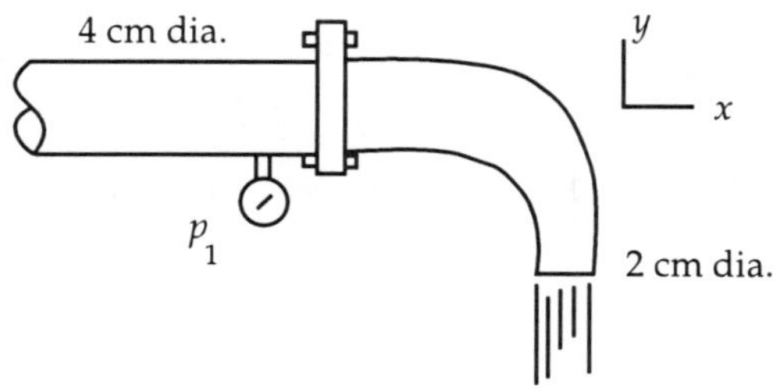

1. What is the velocity at the inlet?
   (A) 41.3 m/s
   (B) 34.2 m/s
   (C) 26.8 m/s
   (D) 10.3 m/s

2. What force is acting on the bend in the y-direction?
   (A) 640 N
   (B) 530 N
   (C) 420 N
   (D) 340 N

3. What force acts on the bend in the x-direction?
   (A) 1140 N
   (B) 990 N
   (C) 830 N
   (D) 530 N

Solutions:

1. **D** The losses over the short bend are negligible. Hence, Bernoulli's equation is used:

$$\frac{V_1^2}{2g}+\frac{p_1}{\gamma}=\frac{V_2^2}{2g}+\frac{p_2}{\gamma}$$

$$\frac{V_1^2}{2 \times 9.8} + \frac{800000}{9800} = \frac{(4V_1)^2}{2 \times 9.8}. \quad \therefore V_1 = 10.3 \text{ m/s}$$

We used continuity to give $V_2 = A_1 V_1 / A_2 = 4V_1$ and $p_2 = 0$ since it exits to the atmosphere. Because the bend is horizontal we ignore the elevation terms.

2. **B** Apply the momentum equation to a control volume containing only water in the y-direction recognizing that $\rho_1 Q_1 = \rho_2 Q_2$:

$$\sum F_y = \rho_1 Q_1 V_{2y} - \rho_2 Q_2 V_{1y}$$

$$F_y = 1000(\pi \times .01^2 \times 4 \times 10.3) \times (-4 \times 10.3) = -533 \text{ N}$$

The negative sign means the force on the water acts in the negative y-direction. An equal and opposite reaction acts on the bend in the positive y-direction.

3. **A** Remember to include the pressure force on the control volume of water:

$$F_y = 1000(\pi \times .01^2 \times 4 \times 10.3) \times (-4 \times 10.3) = -533 \text{ N}$$

$$800\,000 \times \pi \times .02^2 + F_x = (\pi \times .02^2 \times 10.3) \times 1000 \times (-10.3)$$

$$\therefore F_x = -1139 \text{ N}$$

Example 3.3

The velocity in the water jet is 100 m/s.

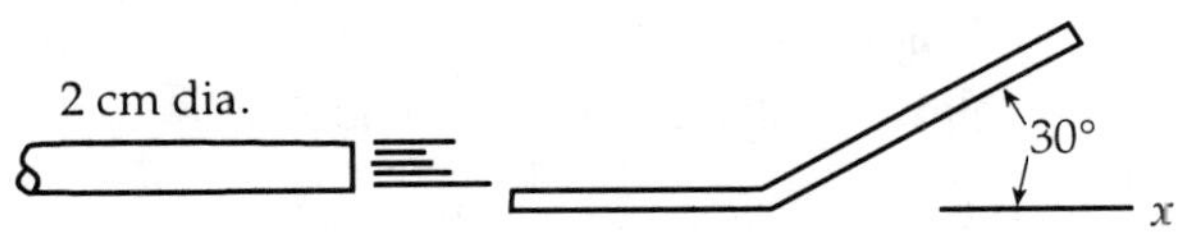

1. The x-component force if this is a fixed blade is:
   (A) 560
   (B) 520 N
   (C) 480 N
   (D) 420 N

2. The x-component force on a single blade moving at 20 m/s in the x-direction is:
   (A) 480 N
   (B) 420 N
   (C) 340 N
   (D) 270 N

3. The x-component force on a series of blades moving at 20 m/s in the x-direction is:
   (A) 520 N
   (B) 480 N
   (C) 420 N
   (D) 340 N

4. The power developed by the series of blades moving at 20 m/s for the jet is:
   (A) 8640 W
   (B) 6740 W
   (C) 5270 W
   (D) 4380 W

Solutions:

1. **D** The flow rate is

$$Q = AV = \pi \times .01^2 \times 100 = .0314 \text{ m/s}$$

Since the water is open to the atmosphere no pressure change occurs, and, since we ignore any slight elevation change, Bernoulli's equation requires the magnitude of the velocity to remain constant, i.e., $V_2 = V_1$. Momentum gives

$$F_x = .0314 \times 1000(100 \cos 30^o - 100) = \text{-}421 \text{ N.}$$

2. **D** A single blade moves away from a stationary jet so that all of the fluid exiting the jet does not have its momentum changed; the flow rate to use is based on the relative speed:

$$Q_r = AV_r = \pi \times .01^2 \times (100 - 20) = .0251 \text{ m/s}$$

The force is then, using relative velocities on the moving blade,

$$F_x = .0251 \times 1000(80 \cos 30^o - 80) = -269 \text{ N.}$$

3. **D** For a series of blades, all of the fluid exiting the jet has its momentum changed (use relative velocities on the blade):

$$F_x = .0314 \times 1000(80 \cos 30^o - 80) = -337 \text{ N.}$$

4. **B** The power is force times velocity:

$$\dot{W} = F \times V = 337 \times 20 = 674 \text{ N.}$$

Example 3.4

There are four nozzles on an impulse turbine. The blades turn the water through a 120-degree angle. The pressure in the 4-cm-diameter source pipe is 1200 kPa. The source pipe is reduced by the nozzle to 1-cm-diameter. The 1.2-m diameter turbine rotates at 400 rpm.

1. Calculate the velocity in the exiting nozzle.
   (A) 93 m/s
   (B) 76 m/s
   (C) 57 m/s
   (D) 49 m/s

2. What is the speed of a blade?
   (A) 48 m/s
   (B) 32 m/s
   (C) 25 m/s
   (D) 19 m/s

3. What average angle does the water exiting the blade make with the incoming water?
   (A) 120°
   (B) 94°
   (C) 81°
   (D) 58°

4. Calculate the x-component force acting on the blades due to one of the nozzles.
   (A) 108 N
   (B) 139 N
   (C) 176 N
   (D) 192 N

5. What is the maximum output of the turbine?
   (A) 13.9 W
   (B) 16.2 W
   (C) 18.7 W
   (D) 20.2 W

Solutions:

1. **D** Refer to the diagram on p. 38. Let V_n be the velocity in the nozzle. Assume no losses in the nozzle and use Bernoulli's equation and continuity:

$$\frac{V_n^2}{2g}+\frac{p_n}{\gamma}=\frac{V_1^2}{2g}+\frac{p_1}{\gamma}$$

$$\frac{V_n^2}{2g}+\frac{1200\,000}{9800}=\frac{(16V_n)^2}{2g}.$$

$$\therefore V_n = 3.068 \text{ m/s} \quad \text{and} \quad V_1 = 49.1 \text{ m/s}$$

2. **C** The velocity is

$$v = r\omega = .6\times(400\times 2\pi/60) = 25.1 \text{ m/s}.$$

3. **D** Refer to the velocity polygon of the figure at the blade exit on p. 38. V_2 makes an angle β with respect to V_1. Use $u = V_1 - v = 49.1 - 25.1 = 24$ m/s:

x-direction: $25.1\text{-}24\cos 60^{\circ} = V_2\cos\beta$

y-direction: $24\sin 60^{\circ} = V_2\sin\beta$

$\therefore \beta = 57.8^{\circ}.$

4. **B** Use the momentum equation on p. 38 (relative velocity u = const on the blade):

$$F_x = 1000 \times \pi \times .02^2 \times 3.068(24 - 24\cos 120^o) = 139 \text{ N}.$$

5. **A** There are four nozzles:

$$\dot{W} = 4F_x v$$
$$= 4 \times 139 \times 25.1 = 13\,900 \text{ W}.$$

3.3 Fans, Pumps and Turbines

The continuity equation on p.35 and the energy equation on p.36 are often necessary when solving certain problems involving fans, pumps, and compressors. The head loss expressions on p.37 along with the pump power equation (h on p. 37 and H on p.110 are the same) may also be of use along with the information presented on pp.110-111 of the NCEES Handbook. Other materials on pp.34-40 may also be required.

Example 3.5

The pump has characteristic curves shown. The velocity in the 10-cm-diameter pipe is 4 m/s.

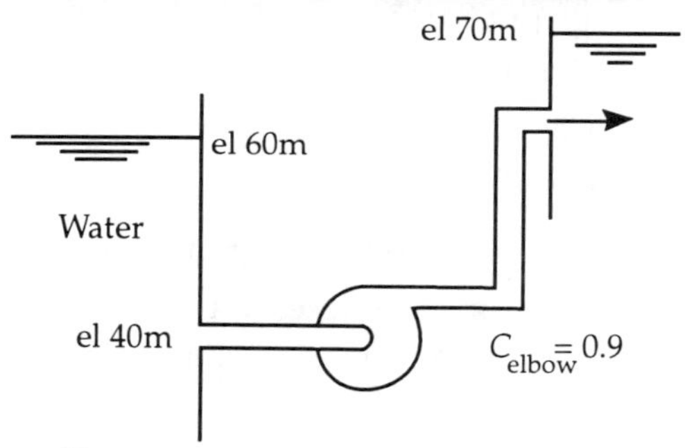

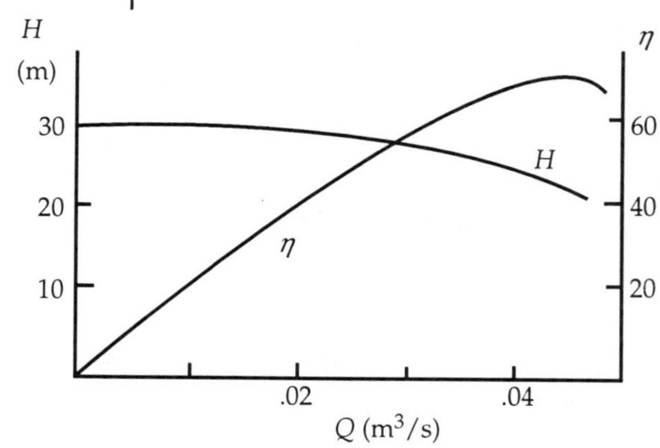

1. The number of liters of water pumped per minute are:
   (A) 0.031
   (B) 31
   (C) 189
   (D) 1890

2. The ideal power required by the pump is:
   (A) 2.4 kW
   (B) 8.3 kW
   (C) 22.9 kW
   (D) 410 kW

3. The power required by the pump is:
   (A) 13.4 kW
   (B) 36.9 kW
   (C) 660 kW
   (D) 862 kW

4. The operating point is:
   (A) the intersection of the head curve $H(Q)$ and the efficiency curve.
   (B) the intersection of the energy equation and the efficiency curve.
   (C) the intersection of the head curve $H(Q)$ and the energy equation.
   (D) the the intersection of the head curve $H(Q)$ and the NPSH curve.

5. Estimate the pressure rise across the pump.
   (A) 120 kPa
   (B) 210 kPa
   (C) 260 kPa
   (D) 360 kPa

6. Estimate the inlet pressure to the pump if one quarter of the losses occur prior to the pump.
   (A) -20 kPa
   (B) 20 kPa
   (C) 80 kPa
   (D) 150 kPa

7. The head loss across one of the elbows is:
   (A) 14 m
   (B) 7 m
   (C) 0.73 m
   (D) 0.31 m

Solutions:

1. **D** Continuity: $Q = VA = 4 \times \pi \times .05^2 = .0314 \text{ m}^3/\text{s}$

$$= 31.4 \text{ L/s}$$
$$= 1885 \text{ L/min.}$$

2. **B** From p.37 with $\eta = 1.0$ for the ideal case:

$$\dot{W} = Q\gamma H = .0314 \times 9800 \times 27 = 8300 \text{ W} \quad \text{or} \quad 8.3 \text{ kW}$$

We found $H = 27$ m from the figure at $Q = .0314 \text{ m}^3/\text{s}$.

3. **A** The power, using the efficiency of .62 from the figure, is

$$\dot{W} = 8.3/\eta = 8.3/.62 = 13.4 \text{ kW}.$$

4. **C**

5. **C** The energy equation across the pump is

$$H_P = \frac{\Delta p}{\gamma}. \qquad 27 = \frac{\Delta p}{9800}. \quad \therefore \quad \Delta p = 9800 \times 27 = 265\,000 \text{kPa}$$

or 265 kPa.

6. **D** The energy equation from the surface to the pump inlet is:

$$0=\frac{V_i^2}{2g}+\frac{p_i}{\gamma}+z_i-\frac{V_1^2}{2g}-\frac{p_1}{\gamma}-z_1+h_f$$

$$0=\frac{4^2}{2\times 9.8}+\frac{p_i}{9800}+40-0-0-60+\frac{h_f}{4}$$

The energy equation from the left surface to the right surface gives

$$H_P=70-60-h_f=27. \;\therefore\; h_f=17 \text{ m}.$$

The first energy equation then gives

$$p_i=146\,000 \text{ Pa or } 146 \text{ kPa}.$$

7. C Across an elbow we write

$$h_f=C\frac{V^2}{2g} \text{ or } h_f=0.9\times\frac{4^2}{2\times 9.8}=0.73 \text{ m}.$$

Example 3.6

A fan with an 80-cm-diameter impeller operates at a speed of 800 rpm moving air at standard conditions at 500 m^3/min with the characteristic curves shown. A larger similar fan with a 1.2-m-diameter impeller is to move a gas with a density of 0.88 kg/m^3 operating at 1200 rpm.

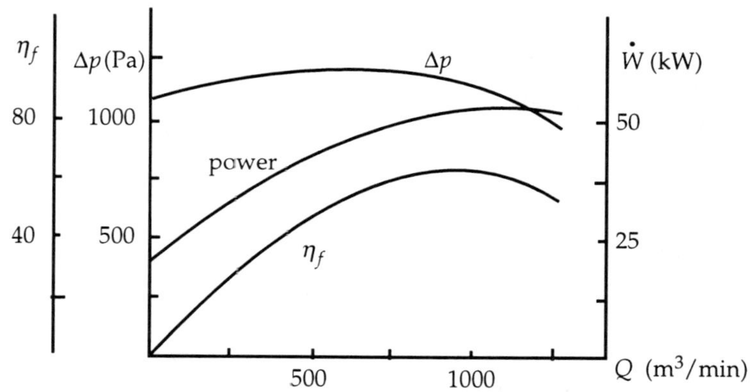

1. What will be the flow rate through the larger fan?

(A) 99 m^3/min

(B) 200 m^3/min

(C) 1200 m^3/min

(D) 2500 m^3/min

2. What pressure rise can be expected across the larger fan?

(A) 4500 Pa

(B) 4000 Pa

(C) 3500 Pa

(D) 3000 Pa

3. The power requirement of the larger fan is about:
   (A) 1000 kW
   (B) 750 kW
   (C) 500 kW
   (D) 250 kW

4. The efficiency of the larger fan is expected to be:
   (A) 64%
   (B) 56%
   (C) 43%
   (D) 32%

Solutions:

1. **D** Use the scaling laws:

$$Q_2 = Q_1\left(\frac{N_2}{N_1}\right)\left(\frac{D_2}{D_1}\right)^3 = 500 \times \frac{1200}{800}\left(\frac{120}{80}\right)^3 = 2530 \text{ m}^3/\text{s}.$$

2. **A** Use the scaling laws:

$$\Delta p_2 = \Delta p_1\left(\frac{\rho_2}{\rho_1}\right)\left(\frac{N_2}{N_1}\right)^2\left(\frac{D_2}{D_1}\right)^2 = 1200 \times \frac{0.88}{1.2}\left(\frac{1200}{800}\right)^2\left(\frac{120}{80}\right)^2 = 4455 \text{ kPa}.$$

Often the scaling laws present the pressure rise as a head H in meters.

3. **A** Use the scaling laws:

$$\dot{W}_2 = \dot{W}_1\left(\frac{\rho_2}{\rho_1}\right)\left(\frac{N_2}{N_1}\right)^3\left(\frac{D_2}{D_1}\right)^5 = 40\frac{.88}{1.2}\left(\frac{1200}{800}\right)^3\left(\frac{120}{80}\right)^5 = 752 \text{ kW}.$$

We obtained the power of 40 kW from the characteristic curve.

4. **C** The efficiency is assumed to be the same on similar machines.

Example 3.7

A compressor operates at 1500 rpm with the characteristic curve shown. It inlets air at standard conditions.

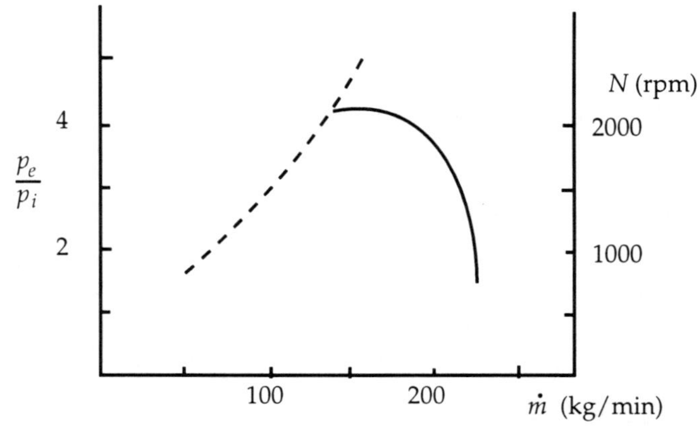

1. The outlet pressure is most nearly:
   (A) 100 kPa
   (B) 200 kPa
   (C) 300 kPa
   (D) 400 kPa

2. Surge limits the compressor at low flow rates. What limits the compressor at the higher flow rates?
   (A) Choking
   (B) Stall
   (C) Cavitation
   (D) Vibrations

3. What is the ideal temperature exiting the compressor?
   (A) 95 °C
   (B) 120 °C
   (C) 150 °C
   (D) 180 °C

4. If the compressor efficiency is 78%, what is the exiting temperature?
   (A) 95 °C
   (B) 120 °C
   (C) 150 °C
   (D) 180 °C

5. If the kinetic energy change is negligible, estimate the minimum power requirement.
   (A) 390 kW
   (B) 240 kW
   (C) 150 kW
   (D) 67 kW

Solutions:

1. **D** From the characteristic curves: $p_e = 3p_i = 3 \times 100 = 300$ kPa where the standard value of 100 kPa is used.

2. **A** Choking occurs due to the Mach number reaching 1.0 somewhere in the blade passages.

3. **B** Assume an isentropic process (see p.43):

$$T_e = T_i \left(\frac{P_e}{P_i} \right)^{(k-1)/k} = 288 \times 3^{.4/1.4} = 394 \text{ K or } 121 \text{ }^{\text{o}}\text{C}.$$

4. **C** Use the efficiency relationship on p.111:

$$\eta = \frac{T_{es} - T_i}{T_e - T_i} = \frac{394 - 288}{T_e - 288} = 0.78. \quad \therefore \quad T_e = 424 \text{ K or } 151 \text{ }^{\text{o}}\text{C}.$$

5. **A** Neglecting kinetic energy change, the energy equation is

$$\dot{W} = \dot{m}\left[c_p \Delta T \right] = \frac{220}{60} \times 1.0 \times (394 - 288) = 389 \text{ kW}.$$

The units are kW because we used 1.0 kJ/kg-°C for the specific heat. We obtained the mass flux from the characteristic curve.

4. Vibrations

by Merle C. Potter

Vibrations is a subject that is not included in the Mechanical Engineering section at the end of the NCEES Handbook. It is, however, one of the subjects listed in the exam guidelines for possible inclusion on the exam. We find the relevant material on p. 16 and p. 24 of the NCEES Handbook. Free vibrations are presented on p. 24 and forced vibrations using the information on p. 16. Torsional vibrations are included on p. 24.

4.1 Free Vibrations

If no external force acts on the mass, the vibrations that result are called free vibrations. Free vibrations result for both damped and undamped systems. If the damping coefficient is zero, undamped motion occurs. This is the situation shown on p. 24. With no damping, the motion occurs indefinitely.

Example 4.1

A 2-kg mass is suspended by a linear undamped spring with spring constant 3.2 kN/m. The mass is given an initial velocity of 10 m/s from the equilibrium position.

1. Calculate the natural frequency of the mass.
   (A) 40 rev/s
   (B) 40 rad/s
   (C) 1.28 rad/s
   (D) 1.28 rev/s

2. How long does it take for the mass to undergo one complete cycle?
   (A) 0.16 s
   (B) 0.64 s
   (C) 3.2 s
   (D) 6.4 s

3. What is the maximum deflection of the mass from the equilibrium position?
   (A) 8 m
   (B) 4 m
   (C) 2 m
   (D) 0.25 m

4. What equation describes the motion?
   (A) 8 sin 1.28t
   (B) 4 cos 40t
   (C) 0.25 sin 40t
   (D) 0.25 cos 40t

Solutions:

1. **B** The frequency is given by

$$\omega_n = \sqrt{k/m} = \sqrt{3200/2} = 40 \text{ rad/s}.$$

2. **A** This is the period of oscillation. It is

$$\tau = 2\pi/\omega_n = 2\pi/40 = 0.156 \text{ seconds}$$

3. **E** Using $x_o = 0$ and $v_o = 10$ m/s,

$$x(t) = \frac{10}{40}\sin 40t.$$

The maximum deflection occurs when sin 40t = 1;

$$x_{max} = 10/40 = 0.25 \text{ m}.$$

4. **C** The equation was found to be $x(t)$ = 0.25 sin 40t.

Example 4.2

A 4-kg mass is suspended by a linear spring with spring constant 6.4 kN/m and is damped so that the damping force is $c\dot{x}$ where $c = 400$ kg/s. The mass is given an initial velocity of 60 m/s from the equilibrium position.

1. The motion that results is termed:
   (A) underdamped
   (B) overdamped
   (C) critically damped
   (D) supercritically damped

2. The motion is given by:
   (A) $2\sinh 40t$
   (B) $\cos 1.28t$
   (C) $e^{-20t} - e^{-80t}$
   (D) $2\sin 40t$

3. The maximum deflection of the mass is:
   (A) 2.46 s
   (B) 0.23 s
   (C) 0.12 s
   (D) 0.046 s

4. What is the time required for the mass to again reach the equilibrium position?
   (A) 0.023 s
   (B) 0.23 s
   (C) 0.46 s
   (D) ∞

Solutions:

1. **B** The differential equation is $4\ddot{x}+400\dot{x}+6400x=0$. In the standard form of p. 16, this is $\ddot{x}+100\dot{x}+1600x=0$. The characteristic equation is $r^2+100r+1600=0$. This can be factored as $(r + 80)(r + 20) = 0$, so that the solution is

$$x(t)=C_1e^{-20t}+C_2e^{-80t}$$

This is overdamped since the solution is exponential. Also, we see that $a^2>b$ or $50^2>1600$.

2. **C** The motion is found by determining C_1 and C_2 above:

$$\left.\begin{aligned}0&=C_1+C_2\\60&=-20c_1-80c_2\end{aligned}\right\}\therefore C_1=1,\ C_2=-1\ ,$$

The solution is $x(t)=e^{-20t}-e^{-80t}$.

3. **D** We find the maximum deflection from $\dot{x}=0$:

$$-20e^{-20t}+80e^{-80t}=0. \quad \therefore e^{-20t}=4e^{-80t}, \quad \text{or } e^{60t}=4.$$

$$\therefore\ 60t=\ln 4 \quad \text{or} \quad t=0.0231 \text{ seconds}.$$

4. **D** We let $x = 0$ and solve for t:

$$0=e^{-20t}-e^{-80t}. \quad \therefore e^{-20t}=e^{-80t}, \quad \text{or } e^{60t}=1.$$

$$\therefore\ 60t=\ln 1 \quad \text{or} \quad t=0.$$

This is the initial state. The next time the mass reaches the equilibrium state is $t=\infty$.

4.2 Forced Vibrations

A forced vibration occurs when a force $f(t)$ acts on the mass resulting in the differential equation $m\ddot{x} + c\dot{x} + kx = f(t)$. This is a nonhomogeneous equation that can be solved using the techniques described on p. 16. The forcing function $f(t)$ usually is exponential or sinusoidal. If the damping coefficient is zero, then resonance occurs whenever ω of the sinusoidal forcing function equals ω_n of the undamped spring-mass system. The situation of resonance is not included as a solution $y_p(x)$ on p. 16, so it will probably not be on the FE/EIT exam. The solution listed there is only valid if $\omega \neq \omega_n$ where $\omega_n = \sqrt{k/m}$. If the forcing function is a product of exponential and sinusoidal functions, Laplace transforms are used.

Example 4.3

An undamped spring-mass system, with $m = 2$ kg and $k = 1.8$ kN/m is acted on by a force $f(t)$. It starts from rest in the equilibrium position.

1. If $f(t) = 200 \cos 20t$, what is $x(t)$?
   (A) $2\cos 20t$
   (B) $2\cos 30t$
   (C) $\cos 20t - \cos 30t$
   (D) $2\cos 20t - 2\cos 30t$

2. If $f(t) = 500\ e^{-10t}$, what is the particular solution $x_p(t)$?
   (A) $2.5e^{-10t}$
   (B) $2.5\cos 10t$
   (C) $0.25e^{-10t}$
   (D) $0.25\cos 10t$

3. What sinusoidal forcing function frequency would lead to the condition of resonance?
   (A) 30 rad/s
   (B) 30 rpm
   (C) 20 rev/s
   (D) 20 rad/s

Solutions:

1. **D** The solution to the homogeneous equation $2\ddot{x} + 1800x = 0$ is

$$y_h(t) = C_1 \cos 30t + C_2 \sin 30t.$$

We assume a particular solution of the form $A\sin 20t + B\cos 20t$. Substitute this into the given equation $2\ddot{x} + 1800x = 200\cos 20t$ and obtain

$$-800A\sin 20t - 800B\cos 20t + 1800A\sin 20t + 1800B\cos 20t = 200\cos 20t.$$

Equating coefficients of sin 20t and cos 20t results in the two equations:

$$100A = 0 \quad \therefore A = 0.$$
$$100B = 200. \quad \therefore B = 2.$$

The solution is $x_p(t) + x_h(t)$:

$$x(t) = 2\cos 20t + C_1 \cos 30t + C_2 \sin 30t$$

Using $x(0) = 0$ and $\dot{x}(0) = 0$ provides

$$0 = 2 + C_1. \quad \therefore C_1 = -2.$$
$$0 = 0 + C_1(0) + 30C_2. \quad \therefore C_2 = 0.$$

Finally,

$$x(t) = 2\cos 2t - 2\cos 30t$$

2. **C** The particular solution is assumed to be Ae^{-10t}. Substitute into

$$2\ddot{x} + 1800x = 500e^{-10t}$$

and we obtain

$$200Ae^{-t} + 1800Ae^{-t} = 500e^{-10t}$$
$$2000A = 500. \quad \therefore A = 0.25.$$

3. **A** Resonance occurs when $\omega = \omega_n = 30$ rad/s.

Example 4.4

An undamped spring-mass system, with $m = 2$ kg and $k = 32$ N/m is acted on by a force $34e^{-2t}$ sin 4t. It starts from rest in the equilibrium position. Determine $x(t)$.

(A) $\cos 4t - 0.25\sin 4t - e^{-2t}\cos 4t - 0.25e^{-2t}\sin 4t$

(B) $-\cos 4t + 0.25\sin 4t + e^{-2t}\cos 4t + 0.25e^{-2t}\sin 4t$

(C) $-e^{-2t}\cos 4t + 0.25e^{-2t}\sin 4t + \cos 4t + 0.25\sin 4t$

(D) $-0.25\sin 4t + \sin 4t + 0.25e^{-2t}\cos 4t + e^{-2t}\sin 4t$

Solution:

The differential equation is $2\ddot{x} + 32x = 34e^{-2t}\sin 4t$. Taking the Laplace transform using the table on p. 17:

$$2s^2X(s) - 2s\,\cancelto{0}{x(0)} - 2\cancelto{0}{\dot{x}(0)} + 32X(s) = \frac{136}{(s+2)^2 + 4^2}.$$

Hence,

$$X(s) = \frac{68}{(s^2+16)[s^2+4s+20]} = \frac{As+B}{s^2+16} + \frac{Cs+D}{s^2+4s+20} = \frac{-s+1}{s^2+16} + \frac{s+3}{s^2+4s+20}$$
$$= -\frac{s}{s^2+16} + \frac{1}{s^2+16} + \frac{s+2}{(s+2)^2+4^2} + \frac{1}{(s+2)^2+4^2}$$

Take the inverse using the table on p. 17:

$$x(t) = -\cos 4t + \frac{1}{4}\sin 4t + e^{-2t}\cos 4t + \frac{1}{4}e^{-2t}\sin 4t.$$

The answer is ***B***.

4.3 Torsional Vibrations

A metal shaft can undergo a torsional vibration when subjected to a torque and then suddenly released. Every rotating shaft has an associated natural frequency, given on p. 24 of the NCEES Handbook. If a periodic torque is applied to such a shaft, resonance could occur if the applied frequency is a *multiple* of the natural frequency, a result from a two-dimensional analysis of the shaft.

Example 4.5

A solid, 10-cm-diameter steel shaft is set into a torsional vibration. The shaft is 4 m long and the density of the steel is 9600 kg/m^3.

1. What is the natural frequency of the shaft?
   (A) 735 Hz
   (B) 532 Hz
   (C) 363 Hz
   (D) 117 Hz

2. How long does it take for one complete vibrational cycle?
   (A) 0.0053 s
   (B) 0.0061 s
   (C) 0.0069 s
   (D) 0.0085 s

3. If the shaft were hollow with an 8-cm-inner diameter, what would be the natural frequency?
   (A) 78 Hz
   (B) 234 Hz
   (C) 633 Hz
   (D) 1130 Hz

Solutions:

1. **D** Using $G = 8.3 \times 10^{10}$ Pa from p. 32, $J = \pi a^4/2$ from p. 26, and $I = MR^2/2$ from p. 28 we use ω_n from p. 24 and find

$$\omega_n = \sqrt{\frac{GJ}{IL}} = \sqrt{\frac{8.3 \times 10^{10} \times \pi \times .05^4/2}{\left[\left(\pi \times .05^2 \times 4 \times 9600\right) \times .05^2/2\right] \times 4}} = 735 \text{ rad/s}$$

Then, referring to p. 68,

$$f = \frac{\omega_n}{2\pi} = \frac{735}{2\pi} = 117 \text{ Hz}$$

2. **D** The time to complete a cycle is

$$\tau = \frac{1}{117} = 0.00855 \text{ seconds}$$

3. **B** Referring to pages 26 and 28 we find

$$\omega_n = \sqrt{\frac{GL}{IL}} = \left[\frac{8.3\times10^{10}\times\pi\left(.05^4-.04^4\right)/2}{\pi\times\left(.05^2-.04^2\right)\times4\times9600\times\left(.05^2+.04^2\right)/2}\right]^{1/2} = 1470 \text{ rad/s}$$

$\therefore f = 1470/2\pi = 234$ Hz

5. Machine Design

by Brian S. Thompson

The equation summaries needed to work problems in the general area of Machine Design are found on pages 113–116 of the NCEES Reference Handbook. It will also be necessary to refer to the material on the Mechanics of Materials located on pages 29–34, so make sure you are very familiar with that material. The following examples illustrate the material presented in the NCEES Reference Handbook.

Example 5.1

A machine element is subjected to an alternating stress of 31.7 MPa and a mean stress of 31.7 MPa. It is manufactured from a metal with an ultimate strength of 1400 MPa and a yield strength of 588 MPa.

1. Using the Modified Goodman Theory, what is the upper bound on the fatigue strength of the machine element?
   (A) 1400 MPa
   (B) 588 MPa
   (C) 2800 MPa
   (D) 32.4 MPa

2. If the Soderberg Theory is employed, what is the upper bound on the fatigue strength of the machine element?
   (A) 1400 MPa
   (B) 32.4 MPa
   (C) 33.5 MPa
   (D) 30.7 MPa

3. Will fatigue failure occur if the magnitude of the alternating stress is increased to 50 MPa and the fatigue strength is 100 MPa, using the Modified Goodman Theory?
   (A) Yes
   (B) No
   (C) Insufficient Data
   (D) Conflicting Data

4. Will fatigue failure be predicted by the Soderberg Theory, if the magnitude of the alternating stress is increased to 90 MPa, and the magnitude of the mean stress increases to 70 MPa but the endurance limit remains at 100 MPa and the yield tensile strength is 588 MPa?
   (A) Yes
   (B) No
   (C) Insufficient Data
   (D) Conflicting Data

5. What is the fatigue strength of the machine element if the surface factor is 0.9, the size factor is 0.6, the load factor is 1.0, the temperature factor is 0.99, the stress concentration factor is 2.4 and the miscellaneous effects factor is 1.0?
   (A) 1400 MPa
   (B) 588 MPa
   (C) 905 MPa
   (D) 1796 MPa

Solutions:

1. **D** The equation formulating the Modified Goodman Theory, given on page 113, can be algebraically manipulated to yield the quotient

$$S_e \leq \frac{\sigma_a}{1-\sigma_m / S_{ut}}$$

Upon substituting the numerical data the limiting value of the fatigue strength is 32.4 MPa.

2. **C** The equation formulating the Soderberg Theory on page 113 has the same format as the Modified Goodman Theory, hence the same algebraic procedure yields the quotient

$$S_e \leq \frac{\sigma_a}{1-\sigma_m / S_y}$$

Substituting the values of the stresses and the strengths yields a limit of 33.5 MPa for the fatigue strength.

3. **B** Substitute the stress and strength data into the inequality describing the Modified Goodman Theory. Hence the left-hand-side of the equation is written

$$\frac{50}{100}+\frac{31.7}{1400}<1$$

Since this is not greater than unity, this theory does not predict failure by fatigue.

4. **A** The left-hand-side of the equation describing the Soderberg inequality becomes

$$\frac{90}{100}+\frac{70}{588}>1$$

upon substituting the values of the strengths and stresses. Upon evaluation these quotients sum to 1.019. Since this is greater than 1.000, fatigue failure is predicted.

5. C The endurance limit for the machine element S_e is determined using the equation on page 113 for a material with an ultimate tensile strength of 1400 MPa. The value is $0.504S_{ut}$. This value, and the values for the various modifying factors, are substituted into the equation at the top left hand column of page 114. Thus

$$S_e = 0.9\times0.6\times1.0\times0.99\times2.4\times1.0\times0.504\times1400 = 905.3 \text{ MPa}$$

Example 5.2

A cantilever beam of circular cross-section 50 mm and length 120 mm is manufactured from aluminum. It is subjected to an axial tensile load of 9.0 kN, a torsional load of 72 N.m and a transverse point load at the tip of the beam of 200 N.

1. What is the axial stress due to the action of the axial load acting alone?
   (A) 400 MPa
   (B) 458 MPa
   (C) 4.58 MPa
   (D) 3.65 MPa

2. What is the axial strain associated with the axial load?
   (A) 6.638×10^{-5}
   (B) 6.664×10^{-4}
   (C) 6.621×10^{-3}
   (D) 5.348×10^{-4}

3. What is the maximum normal axial stress caused by all loads acting simultaneously?
   (A) 1.956 MPa
   (B) 6.536 MPa
   (C) 1.956 MPa
   (D) 6.536 MPa

4. What is the angle of twist of the end of the cantilever caused by the torsional load?
   (A) 0.5347 rad
   (B) 0.0005029 rad
   (C) 0.5029 rad
   (D) 0.0005347 rad

5. Determine the torsional stiffness of the cantilever.
   (A) 143.17 kN·m / rad
   (B) 143.17 kN·m / rad
   (C) 314.67 kN·m / rad
   (D) 0.31457 kN·m / rad

6. Calculate the maximum shear stress.
   (A) 3 GPa
   (B) 4.39 MPa
   (C) 0.0039 GPa
   (D) 6.9825 MPa

7. Calculate the von Mises stress.
   (A) 0.0358 GPa
   (B) 3.7725 GPa
   (C) 0.1216 GPa
   (D) 12.16 MPa

Solutions:

1. **C** The action of the axial load alone, creates a stress given by the equation at the top of the right hand column of page 29. Upon substitution this becomes

$$\frac{P}{A} = \frac{9000 \times 4}{\pi \times 50 \times 50 \times 10^{-6}} = 4\ 580\ 000 \text{ Pa} \quad \text{or} \quad 4.58 \text{ MPa}$$

2. **A** The axial strain associated with this stress is also determined from an equation at the top of the page 29, namely $\delta / L = \sigma / E$. The Young's modulus is obtained from the Table on page 32. Hence the strain

$$\varepsilon = \frac{4.58 \times 10^6}{6.9 \times 10^{10}} = 6.638 \times 10^{-5}$$

3. **B** The maximum normal stress from all loads acting simultaneously is a combination of the bending stress and the axial stress. The governing equation is taken from pages 29 and 31. Thus

$$\sigma = \frac{P}{A} + \frac{My}{I}$$

$$= 4.58 \times 10^6 + \frac{200 \times 0.12 \times 25 \times 10^{-3}}{\pi \times \left(50 \times 10^{-3}\right)^4 / 64}$$

$$= 6\ 536\ 000 \text{ Pa} \quad \text{or} \quad 6.536 \text{ MPa}$$

4. **B** The angle of twist of the cantilever is calculated from page 31. The angle ϕ is given by the quotient TL/GJ. The shear modulus for aluminum, 2.8×10^{10}, is obtained from the table on page 32. Hence

$$\phi = \frac{72 \times 0.12}{2.8 \times 10^{10} \times \pi(50 \times 10^{-3})^4 / 32} = 0.0005029 \text{ rad}$$

5. **A** The torsional stiffness can be determined from the equation on page 31. Thus

$$k = \frac{GJ}{L} = \frac{2.8 \times 10^{10} \times \pi \times (50 \times 10^{-3})^4}{0.12 \times 32} = 143\ 170 \text{ N} \cdot \text{m / rad}$$

6. **B** The maximum shear stress can be calculated from page 114. Namely,

$$\tau_{max} = \frac{2}{\pi d^3}\left[(8M + Fd)^2 + (8T)^2\right]^{0.5}$$

Upon substitution of the numerical data from the question this can be rewritten as:

$$\tau_{max} = \frac{2}{\pi(50 \times 10^{-3})^3}\left[\left\{(8 \times 200 \times 0.12) + \left(9000 \times 50 \times 10^{-3}\right)\right\}^2 + (8 \times 72)^2\right]^{0.5}$$

This quotient reduces to $\tau_{max} = 4.39$ MPa.

7. **A** The von Mises stress can be calculated from the equation on page 114. Namely

$$\sigma' = \frac{4}{\pi \times d^3} \times \left[(8M + Fd)^2 + (48T)^2\right]^{0.5}$$

This becomes

$$\sigma' = \frac{4}{\pi \times (50 \times 10^{-3})^3} \times \left[\left\{(8 \times 200 \times 0.12) + (9000 \times 0.05)\right\}^2 + (48 \times 72)^2\right]^{0.5}$$

$$= 35\ 800\ 000 \text{ Pa} \quad \text{or} \quad 35.8 \text{ MPa}$$

Example 5.3

What is the diameter of a steel shaft subjected to a fatigue environment characterized by an alternating moment of 120 N.m, a mean moment of 50 Nm, an alternating torque of 0 N.m, a constant torque of 100 N.m and a fatigue strength reduction factor of 0.5? The safety factor is assumed to be 2.5, and the steel shaft has an S_e of 300 MPa and S_y of 600 MPa.

(A) 50 mm

(B) 32.56 mm

(C) 32.56 cm

(D) 25.16 mm

Solution:

The fatigue loading section on page 114 presents the governing equation:

$$d^3 = \frac{32n}{\pi} \times \left[\left(\frac{M_m}{S_y} + \frac{K_f M_a}{S_e}\right)^2 + \left(\frac{T_m}{S_y} + \frac{K_{fs} T_a}{S_e}\right)^2\right]^{0.5}$$

This becomes upon substitution:

$$d^3 = \frac{32 \times 2.5}{\pi} \times \left[\left(\frac{50}{600 \times 10^6} + \frac{0.5 \times 120}{300 \times 10^6} \right)^2 + \left(\frac{100}{600 \times 10^6} + 0 \right)^2 \right]^{0.5}$$

$$= 15.92 \times 10^{-6} \quad \text{or } d = 0.02516 \text{ m} \quad \text{or} \quad 25.16 \text{ mm}$$

The answer is **D**.

Example 5.4

A double-start square-thread power screw on a production machine has a load of 7.2 kN imposed upon it. The mean collar diameter is 40 mm, the major thread diameter is 34 mm and the pitch is 4 mm, and the coefficient of friction for the thread and also for the collar are both 0.1.

1. What is the thread depth?
   (A) 4 mm
   (B) 8 mm
   (C) 2 mm
   (D) 1 mm

2. What is the thread width?
   (A) 4 mm
   (B) 6 mm
   (C) 2 mm
   (D) 3 mm

3. What is the mean diameter?
   (A) 30 mm
   (B) 32 mm
   (C) 28 mm
   (D) 24 mm

4. What is the lead?
   (A) 6 mm
   (B) 4 mm
   (C) 2 mm
   (D) 8 mm

5. What is the torque required to turn the screw in a direction aided by the load?
   (A) 32.1 $kN \cdot m$
   (B) 15.67 $N \cdot m$
   (C) 16.7 $N \cdot m$
   (D) 44 $kN \cdot m$

6. What is the torque required to turn the screw in a direction resisted by the load?
   (A) 12.6 $N \cdot m$
   (B) 34.9 $N \cdot m$
   (C) 38.4 $kN \cdot m$
   (D) 56 $N \cdot m$

7. What is the efficiency of the power screw?
   (A) 26.2 %
   (B) 56.3%
   (C) 45.6%
   (D) 67.8%

Solutions:

1. **C** The thread depth is equal to half the pitch: namely 2 mm.

2. **C** The thread width equals the thread depth: 2 mm.

3. **B** The mean thread diameter $d_m = d_{\max} - \text{half the pitch}$. Therefore $d_m = 34 - 2 = 32\,\text{mm}$.

4. **D** The lead is determined by the relationship, $l = np = 2 \times 4 = 8\,\text{mm}$.

5. **C** The torque is determined from the equation for T_L on page 114. Upon substituting the data this becomes:

$$T_L = \left(\frac{7.2 \times 10^3 \times 32 \times 10^{-3}}{2}\right)\frac{\pi \times 0.1\left(32 \times 10^{-3}\right) - 8 \times 10^{-3}}{\left(\pi \times 32 \times 10^{-3}\right) + 0.1\left(8 \times 10^{-3}\right)} + 0.5\left(7.2 \times 10^{-3}\right) \times 0.1\left(40 \times 10^{-3}\right)$$

$$= 16.7\ \text{N} \cdot \text{m}$$

6. **B** The torque required to raise the load is designated as T_R on page 114. The governing equation becomes

$$T_R = 7.2 \times 16 \times \frac{\pi \times 0.1 \times 32 + 8}{\left(\pi \times 32 \times 10^{-3}\right) + 0.1\left(8 \times 10^{-3}\right)} + 0.5\left(7.2 \times 10^{-3}\right) \times 0.1\left(40 \times 10^{-3}\right)$$

$$= 34.920\ \text{N} \cdot \text{m}$$

7. **A** The governing equation for efficiency is given on page 114. Namely

$$\eta = \frac{Fl}{2\pi T} = \frac{7.2 \times 10^3 \times 8}{2\pi \times 34.92} = 26.2\%$$

Example 5.5

A helical compression spring is manufactured from a cold drawn carbon steel wire with an ultimate tensile strength of 2170 MPa, a shear modulus of 80 GPa, and a diameter of 2 mm. The outside diameter of the spring is 11 mm. The ends are plain and there are 10 coils.

1. What is the yield strength?
   (A) 2032 MPa
   (B) 5631 GPa
   (C) 1693 MPa
   (D) 9813 MPa

2. What is the torsional yield strength?
   (A) 976.5 MPa
   (B) 798.3 GPa
   (C) 4356.5 MPa
   (D) 345.8 MPa

3. What is the shear stress if the applied load is 13.75 N?
   (A) 5078 GPa
   (B) 456.7 MPa
   (C) 35.7 MPa
   (D) 43.3 MPa

4. What is the spring rate?
   (A) 21.9 N/m
   (B) 43.5 kN/m
   (C) 38.9 N/m
   (D) 56.7 N/m

5. What is the deflection?
   (A) 34 mm
   (B) 0.34 mm
   (C) 56 mm
   (D) 62 mm

Solutions:

1. **C** The yield strength is obtained from page 114 based upon the ultimate strength. Hence

$$S_y = 0.78 S_{ut} = 0.78 \times 2170 \text{ MPa} = 1692.6 \text{ MPa}$$

2. **A** The torsional yield strength is also obtained from page 114. Thus

$$S_{sy} = 0.45 S_{ut} = 0.45 \times 2170 \text{ MPa} = 976.5 \text{ MPa}$$

3. **D** The shear stress is also determined from the equation given on page 115. Thus

$$\tau = \frac{K_s \times 8F \times D}{\pi d^3}$$

Now $D = 11 - 2 = 9$ mm, and $C = D/d = 9/2 = 5.035$. Therefore, the shear stress is

$$\tau = \left(\frac{2 \times 5.035 + 1}{2 \times 5.035}\right)\frac{8 \times 13.75 \times 9 \times 10^{-3}}{\pi\left(2 \times 10^{-3}\right)^3}$$

$$= 43\ 300\ 000 \text{ Pa} \quad \text{or} \quad 43.3 \text{ MPa}$$

4. **A** The spring rate is given on page 115. Thus

$$k = \frac{d^4 G}{8D^3 N} = \frac{\left(2 \times 10^{-3}\right)^4 \times 80 \times 10^6}{8 \times [9 \times 10^{-3}]^3 \times 10} = 21.94 \text{ N/m}$$

5. **D** The spring deflection relationship is given on page 115:

$$x = \frac{F}{k} = \frac{13.75}{219.4} = 62 \text{ mm}$$

Example 5.6

A torsion spring from a domestic appliance is manufactured from a cold-drawn carbon steel with an ultimate tensile strength of 2000 MPa and 5 total turns. The wire diameter is 1.8 mm and the maximum diameter is 15 mm.

1. What is the spring rate?
   (A) 5219 kN/rad
   (B) 3289 N/rad
   (C) 0.5219 N.m/rad
   (D) 0.3289 N/rad

2. What torque will cause yielding of the spring material?
   (A) 456 N.m
   (B) 0.805 N.m
   (C) 619 N.m
   (D) 0.619 kN.m

3. What is the corresponding angular rotation due to the torque of Problem 2?
   (A) 1.54 rad
   (B) 0.32 rad
   (C) 14.5 rad
   (D) 0.154 rad

Solutions:

1. **C** The spring rate for a helical torsion spring is given on page 115. Hence

$$k = \frac{d^4E}{64DN} = \frac{(1.8\times10^3)^4\times2.1\times10^{11}}{64\times13.2\times10^3\times5} = 0.5219\ \text{N}\cdot\text{m}$$

We used the mean diameter of the spring as $D = 15 - 1.8 = 13.2$ mm. The Young's modulus was obtained on page 32.

2. **B** The yield strength is obtained from page 115; thus

$$S_y = 0.78S_{ut} = 0.78\times2000\ \text{MPa} = 1560\ \text{MPa}$$

The quotient

$$C = D/d = 13.2/1.8 = 7.33$$

Thus the correction factor, K_i is 1.11. Replacing s by S_y in the limiting case of yielding, and rearranging the equation defines the torque to be,

$$Fr = \frac{S_y\pi d^3}{32K_i} = \frac{1560\times10^6\,\pi(1.8\times10^{-3})^3}{32\times1.11} = 0.80467\ \text{N}\cdot\text{m}$$

3. **A** The spring rate was found in No.1 above and the torque in No.2:

$$\theta = \frac{Fr}{k} = \frac{0.80467}{0.5219} = 1.5418\ \text{rad}$$

6. Control Systems

by Brian Feeny

The main goal in control design is to achieve a desirable system response. To this end, attention is on modeling, the steady-state response, its stability, and the manner in which the steady state is reached through transient dynamics. Other issues in control design are disturbance rejection and sensitivity to parameter changes. Fundamental control design involves feedback, which leads to a closed-loop system.

Most of the content of these questions on control systems can be found in the *NCEES Handbook* on pages 70–72. The required mathematical background concerns Laplace transforms, covered on page 17, and matrix algebra. Information on any questions which go beyond this scope can be found in any text on control systems.

6.1 Block Diagrams

Block diagrams are a standard tool for describing control systems. Each block in the diagram corresponds to the transfer function of a sub process, each of which has an input and an output. The entire block diagram has an input, labeled $U(s)$ in this chapter, and an output, labeled $Y(s)$. The inputs and outputs are in the Laplace-transform domain.

Example 6.1

For the block diagram below, the transfer functions are given as

$$G_1(s)\frac{1}{s}, \quad G_2(s) = \frac{1}{s+3}, \quad H(s) = 4$$

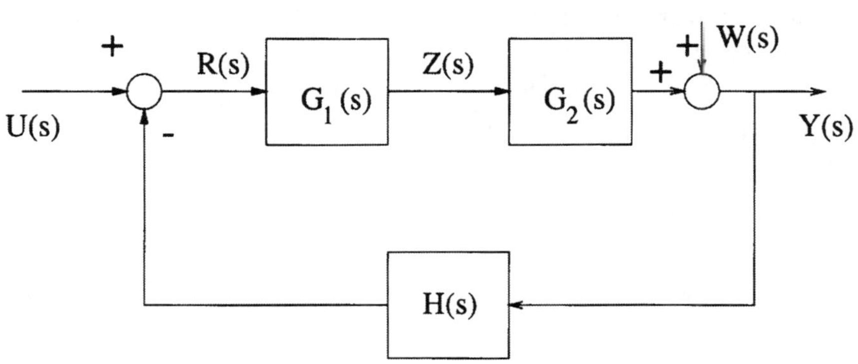

1. The value for the signal $Z(s)$ can be written as
   (A) $G_1(s)U(s)$
   (B) $G_1(s)[U(s)-H(s)Y(s)]$
   (C) $U(s)-H(s)Y(s)$
   (D) $U(s)-H(s)G_2(s)U(s)$

2. Find the transfer function between the input $U(s)$ and the output $Y(s)$, $G(s)=Y(s)U(s)$.
   (A) $\frac{1}{s+3}$
   (B) $\frac{1}{s(s+3)}$
   (C) $\frac{4}{s(s+3)}$
   (D) $\frac{1}{s^2+3s+4}$

3. The characteristic equation is given by
   (A) $s+3=0$
   (B) $s(s+3)=0$
   (C) $s=0$
   (D) $s^2+3s+4=0$

4. The open-loop transfer function is
   (A) $\frac{1}{s+3}$
   (B) $\frac{1}{s(s+3)}$
   (C) $\frac{4}{s(s+3)}$
   (D) $\frac{1}{s^2+3s+4}$

5. Find the transfer function between the disturbance $W(s)$ and the output $Y(s)$, $G_d(s)=Y(s)W(s)$.
   (A) 1
   (B) $\frac{1}{s+3}$
   (C) $\frac{4}{s(s+3)}$
   (D) $\frac{s^2+3s}{s^2+3s+4}$

Solutions:

1. **B** The signals entering the summation junction are $U(s)$ and $-H(s)Y(s)$. Therefore, the signal coming out of the summation junction is $U(s)-H(s)Y(s)$. This quantity is the input $R(s)$ to box $G_1(s)$, whose output
$$Z(s)=G_1(s)R(s)=G_1(s)(U(s)-H(s)Y(s)).$$

2. **D** The transfer function can be computed from
$$G(s)=\frac{G_1(s)G_2(s)}{1+H(s)G_1(s)G_2(s)}=\frac{\frac{1}{s}\frac{1}{s+3}}{1+4\frac{1}{s}\frac{1}{s+3}}=\frac{1}{s(s+3)+4}$$

3. **D** The characteristic equation can be obtained by setting the denominator of the transfer function to zero. Hence
$$s(s+3)+4=s^2+3s+4=0.$$

4. **C** The open-loop transfer function is defined as
$$G_1(s)G_2(s)H(s)=\frac{4}{s(s+3)}$$

5. **D** Similar to problem 1, the transfer function under the input of W is
$$G_d(s)=\frac{1}{1+H(s)G_1(s)G_2(s)}=\frac{1}{1+4\frac{1}{s}\frac{1}{s+3}}=\frac{s(s+3)}{s(s+3)+4}$$

6.2 State-Variable Modeling

Another common and useful way to describe dynamical systems is by state-variable models. State-variable models are particularly useful in analyzing multi-input, multi-output control systems. A state variable is one of the quantities necessary for describing the system at any time during its evolution. In this chapter, the state variables are listed in a vector **x**. The vector **y** represents the output, and the vector $\mathbf{u}(t)$ denotes the input. (Bold letters represent matrices and vectors.)

Example 6.2

Given the system $\ddot{x}+3\dot{x}+4x=6u(t)$, with the output given by $y=5x$, a state-variable model can be constructed as $\dot{\mathbf{x}}=\mathbf{Ax}+\mathbf{Bu}$, $\mathbf{y}=\mathbf{Cx}+\mathbf{Du}$. The states are defined as $x_1=x$ and $x_2=\dot{x}$.

1. The matrix **A** is

(A) $\begin{bmatrix}1 & 3\\ 4 & 6\end{bmatrix}$

(B) $\begin{bmatrix}3 & 4\end{bmatrix}$

(C) $\begin{bmatrix}0 & 1\\ -4 & -3\end{bmatrix}$

(D) $\begin{bmatrix}4\\ 3\end{bmatrix}$

2. The matrix **B** is

(A) $\begin{bmatrix} 0 \\ 6 \end{bmatrix}$

(B) $\begin{bmatrix} 6 & 5 \end{bmatrix}$

(C) $\begin{bmatrix} 0 & 0 \\ 0 & 6 \end{bmatrix}$

(D) $\begin{bmatrix} 6 & 0 \\ 0 & 5 \end{bmatrix}$

3. The matrix **C** is

(A) $\begin{bmatrix} 0 \\ 5 \end{bmatrix}$

(B) $\begin{bmatrix} 5 & 0 \end{bmatrix}$

(C) $\begin{bmatrix} 5 & 0 \\ 0 & 0 \end{bmatrix}$

(D) $\begin{bmatrix} 6 & 0 \\ 0 & 5 \end{bmatrix}$

Solutions:

Since $x_1 = x$ and $x_2 = \dot{x} = x_1$, the first state-variable equation is $\dot{x}_1 = x_2$. The second equation is obtained from $\dot{x}_2 = \ddot{x} = -3\dot{x} - 4x + 6u = -4x_1 - 3x_2 + 6u$. The output equation can be written as $y = 5x_1$. We can write all of this in matrix form, such that

$$\begin{pmatrix} \dot{x}_1 \\ \dot{x}_2 \end{pmatrix} = \begin{bmatrix} 0 & 1 \\ -4 & -3 \end{bmatrix} \begin{pmatrix} x_1 \\ x_2 \end{pmatrix} + \begin{bmatrix} 0 \\ 6 \end{bmatrix} u(t)$$

$$y = \begin{bmatrix} 5 & 0 \end{bmatrix} \begin{pmatrix} x_1 \\ x_2 \end{pmatrix}$$

From here, we can pick out the matrices for the problems.

1. **C** The matrix $\mathbf{A} = \begin{bmatrix} 0 & 1 \\ -4 & -3 \end{bmatrix}$

2. **A** The matrix $\mathbf{B} = \begin{bmatrix} 0 \\ 6 \end{bmatrix}$

3. **C** The matrix $\mathbf{C} = \begin{bmatrix} 5 & 0 \end{bmatrix}$

Example 6.3

Suppose we have $\dot{\mathbf{x}} = \mathbf{Ax} + \mathbf{Bu}$, $\mathbf{y} = \mathbf{Cx} + \mathbf{Du}$, with

$$\mathbf{A} = \begin{bmatrix} 2 & 1 \\ 1 & 0 \end{bmatrix}$$

$$\mathbf{B} = \begin{bmatrix} 1 \\ 0 \end{bmatrix}$$

$$\mathbf{C} = \begin{bmatrix} 0 & 1 \end{bmatrix}$$

$$\mathbf{D} = \begin{bmatrix} 0 \end{bmatrix}$$

The initial conditions are zero.

1. Determine the characteristic equation.
   (A) $s^2 - 1 = 0$
   (B) $2s - 1 = 0$
   (C) $s^2 - 2s - 1 = 0$
   (D) $s = 0$

2. The transfer function is:
   (A) $\dfrac{1}{s(s-2)-1}\begin{bmatrix} s & 1 \\ 1 & s-2 \end{bmatrix}$
   (B) $\begin{bmatrix} s-2 & -1 \\ -1 & s \end{bmatrix}$
   (C) $\dfrac{1}{s(s-2)-1}\begin{bmatrix} s \\ 1 \end{bmatrix}$
   (D) $\dfrac{2s-2}{s(s-2)-1}$

Solutions:

1. **C** The characteristic equation can be expressed as $|s\mathbf{I} - \mathbf{A}| = 0$. Since

$$s\mathbf{I} - \mathbf{A} = \begin{bmatrix} s-2 & 1 \\ 1 & s \end{bmatrix}$$

we have

$$|s\mathbf{I} - \mathbf{A}| = s(s-2) - 1 = 0$$

for the characteristic equation.

2. **D** When the initial conditions are zero, the transfer function can be obtained from

$$Y(s) = \{\mathbf{C}\Phi(s)\mathbf{B}\}U(s),$$

where

$$\Phi(s) = [s\mathbf{I} - \mathbf{A}]^{-1} = \frac{1}{s(s-2)-1}\begin{bmatrix} s & 1 \\ 1 & s-2 \end{bmatrix}$$

By matrix multiplication, we obtain

$$\frac{Y(s)}{U(s)} = \mathbf{C}\Phi(s)\mathbf{B} = \frac{1}{s^2 - 2s - 1}$$

6.3 Response

The response of a control system is particularly important, since it is the entity which we would like to control. This section focuses on response characteristics of first- and second-order plants. The order of a system is the degree of its characteristic equation. The order also coincides with the number of poles in the system, and the number of states required for the state-variable model.

Example 6.4

Given the first-order plant of the form

$$G(s) = \frac{5}{s+2} = \frac{Y(s)}{U(s)}$$

1. The location of the pole is
   (A) −2
   (B) 5
   (C) 1/2
   (D) There is no pole.

2. The time constant is
   (A) undefined
   (B) 5
   (C) 1/2
   (D) 2

3. If $u(t)$ is an impulse, i.e., $U(s) = 1$, find $y(t)$.
   (A) $\delta(t)$
   (B) e^{-2t}
   (C) $5\,e^{-2t}$
   (D) $5 - 5\,e^{-2t}$.

4. If $u(t)$ is a unit step function, i.e., $U(s) = 1/s$, find the steady-state value of $y(t)$, i.e., find $\lim_{t\to\infty} y(t)$.
   (A) 5/2
   (B) 5
   (C) 1/2
   (D) 1

Solutions:

1. **A** A pole is a root of the characteristic equation, or $s+2=0$. Hence, $s=-2$ locates the single pole.

2. **C** The time constant τ is a quantity that defines the decay rate of an exponential function of the form $e^{-t/\tau}$. If there are several exponentials in a response, the largest value of τ defines the speed of the response, and hence the time constant of the response. Responses in this system contain terms such as e^{-2t}. Thus $\tau = 1/2$.

3. **C** If $U(s)=1$, then

$$Y(s) = G(s)U(s) = 5/s(s+2)$$

According to the Laplace transform table, the inverse transform of $1/(s+2)$ is e^{-2t}. Thus, $y(t) = 5e^{-2t}$.

4. **A** If $U(s) = 1/s$, then

$$Y(s) = 5/s(s+2)$$

By the final-value theorem,

$$\lim_{t\to\infty} y(t) = [sY(s)]_{s=0} = 5/2$$

Example 6.5

Suppose we have a second-order plant of the nondimensional form

$$G(s) = \frac{8}{s^2+s+4}$$

and the time is in seconds (s has units of per second).

1. Find the undamped natural frequency.
   (A) 4 rad/s
   (B) 2 rad/s
   (C) 8 rad/s
   (D) 2.8284 rad/s

2. The frequency of damped oscillation is:
   (A) 1.9365 rad/s
   (B) 1.8708 rad/s
   (C) 2.6016 rad/s
   (D) 4.25 rad/s

3. The peak time during a step response is:
   (A) 1.6223 s
   (B) 3.2446 s
   (C) 3.1416 s
   (D) ∞

4. Find the percent overshoot.
   (A) 44.4 %
   (B) 1.44 %
   (C) 144 %
   (D) There is no overshoot

5. The time constant is:
   (A) 1/2 s
   (B) 1 s
   (C) 2 s
   (D) 4 s

Solutions:

1. **B** The standard form of second-order systems is

$$G(s) = \frac{k\omega_n^2}{s^2 + 2\zeta\omega_n s + \omega_n^2}$$

where ω_n^2 is the undamped natural frequency. Here, $\omega_n^2 = 4\ \text{rad}^2/\text{s}^2$ and therefore $\omega_n = 2$ rad/s

2. **A** The damped natural frequency is given by

$$\omega_d = \omega_n\sqrt{1-\zeta^2}$$

We can compute ζ from $\zeta\omega_n = 1$, and $\omega_n = 2$ rad/s. Thus, $\zeta = 1/4$, and $\omega_d = 1.9365$ rad/s.

3. **A** The peak time during a step response for this system turns out to be half of the damped period of response, and is thus $T_p = \pi/\omega_d = 1.6223$ s.

4. **A** The percent overshoot can be computed from the overshoot quantity, C_p. Accounting for the constant in the numerator, $k = 2$, we have

$$P.O. = \frac{C_p - C_\infty}{C_\infty} \times 100\%$$

where

$$C_\infty = \lim_{t\to\infty} C(t) = \lim_{s\to 0} sY(s) = G(0)$$

by the final value theorem. Thus, the percent overshoot can be derived as

$$P.O. = e^{-\pi\zeta\sqrt{1-\zeta^2}} \times 100\% = 44.4\%$$

5. **A** The transient response consists of terms of the form

$$e^{-\zeta\omega_n t}\cos(\omega_d t + \phi)$$

Thus, the time constant is $\tau = 1/\zeta\omega_n = 1/2$ s.

Example 6.6

A second-order plant has complex poles at $s = -3 \pm 4i$.

1. The undamped natural frequency ω_n is:
   (A) 5 rad/s
   (B) 4 rad/s
   (C) 3 rad/s
   (D) 1/3 rad/s

2. The damped natural frequency ω_d is:
   (A) 5 rad/s
   (B) 4 rad/s
   (C) 3 rad/s
   (D) 1/3 rad/s

3. The time constant τ is
   (A) 5 s^{-1}
   (B) 4 s^{-1}
   (C) 3 s^{-1}
   (D) 1/3 s^{-1}

Solutions:

1. **A** For second-order systems with complex-conjugate poles, the undamped natural frequency is the distance from the poles to the origin of the complex *s* plane. This distance is $\sqrt{3^2 + 4^2} = 5$

2. **B** The damped natural frequency is the distance from the poles to the real axis of the complex *s* plane. In this case, the distance is 4.

3. **D** The time constant is the reciprocal of minus the real part of the pole which is furthest to the left in the complex plane (most negative). Here, that quantity is 1/3.

6.4 Closed-Loop Systems

Fundamental control methods implement feedback control. When feedback is used to control a system, the result is a closed-loop system. Closed-loop systems are typically more robust, more stable, and less sensitive to modeling errors than the corresponding open-loop system. Thus, using feedback is a standard way of applying control strategies. Common controllers are the PID controllers, and the lead-lag controllers. The concerns in a control system are the steady state, its stability, and the manner in which the steady state is reached through transient dynamics.

Example 6.7

For the block diagram below, the plant is given by $G(s) = \frac{1}{s+a}$, and the compensator is given by k.

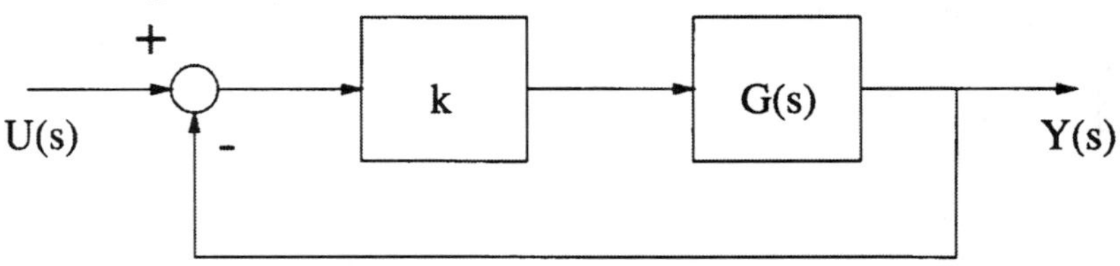

1. Find the range of values of k for stability.
   (A) $k < 0$
   (B) $k > 0$
   (C) $k > a$
   (D) $k > -a$
2. The steady-state tracking error to a unit step input is:
   (A) 0
   (B) k
   (C) $a/(k+a)$
   (D) ∞
3. The steady-state tracking error to a ramp input is:
   (A) 0
   (B) k
   (C) $a/(k+a)$
   (D) ∞
4. The system type is:
   (A) 0
   (B) 1
   (C) 2
   (D) 3

5. The controller type is:
   (A) P
   (B) PI
   (C) PD
   (D) PID

6. Which of the following controllers would achieve zero tracking error to a step input?
   (A) P
   (B) PI
   (C) PD
   (D) It is not possible

Solutions:

1. **D** The closed-loop transfer function is

$$G(s) = \frac{kG_1(s)}{1 + kG_1(s)} = \frac{k}{s + a + k}$$

The pole is therefore located at $s = -a - k$. For stability, the pole must be in the left half plane. This requires $k > -a$ for stability.

2. **C** Given a unit step input, $U(s) = 1/s$, the final value theorem dictates a steady-state response to be $G(0) = k/(k + a)$. The error is thus $1 - k/(k + a) = a/(k + a)$.

3. **D** Since the steady-state tracking error to a unit step is nonzero but finite, the system type can be determined, and the steady-state tracking error to a ramp will be unbounded.

4. **A** Since the steady-state tracking error to a unit step is nonzero but finite, it is by definition a type-0 system. The system type is the power p of the term s^p in the denominator of the open-loop transfer function.

5. **A** The compensator consists of a constant, or proportional, term only. Therefore it is a P controller.

6. **B** The system type increases as we introduce integrators to the open-loop system. If we add an integrator to the compensator, making it a PI controller, the system type increases by one. We would then have a type-1 system, which tracks a step with zero steady-state error. To see this, replace k with $k + K/s = (ks + K)/s$. Then, the transfer function becomes

$$G(s) = \frac{ks + K}{s(s + a) + ks + K} \quad \text{and } G(0) = 1$$

6.5 Stability

One of the usual goals of a control system is to achieve a stable steady-state response.

Example 6.8

A closed-loop system has poles located at $s=-3$, $s=-2\pm i$, and $s=+1$. The system is

(A) Stable
(B) Unstable
(C) Marginally stable
(D) Neutrally stable

Solution:

A system is bounded-input bounded-output stable if and only if all of the poles are in the left half plane. Since one pole has a positive real part, it is in the right half plane, and the system is unstable. The answer is B.

Example 6.9

The characteristic equation of a closed loop system is given by

$$s^3+6s^2-2s+k=0$$

For what values of k is this system stable?

(A) $k<0$
(B) $k>0$
(C) all values k
(D) no values of k

Solution:

Since not all coefficients of the characteristic polynomial are nonzero with the same sign, regardless of k, there will be at least one root in the right half plane. Thus the system is unstable for all k. The answer is D.

Example 6.10

Suppose the characteristic equation of a closed-loop system with gain k is

$$s^3+3s^2+2s+k=0$$

The range of k for which the system is stable is:

(A) $k>0$
(B) $0<k<6$
(C) $0<k<3$
(D) all k

Solution:

If we construct a Routh array for $a_0s^3 + a_1s^2 + a_2s + a_3$, for which $a_0 = 1$, $a_1 = 3$, $a_2 = 2$, and $a_3 = k$, then the first column of the Routh array would consist of the quantities

$$a_0,\ a_1,\ b_1 = \frac{a_0a_2 - a_0a_3}{a_1} = \frac{6-k}{3}$$

and $c_1 = (b_1a_3 - 0a_1)/b_1 = k$. For all of these quantities to be of the same sign, and nonzero, we need $k > 0$ and $k < 6$. The answer is **B**.

6.6 Frequency-Domain Analysis

A powerful approach for control design is to use the open-loop frequency response. The compensator can be designed to manipulate the frequency response in an appropriate way. A simple test for stability is the *Nyquist criterion,* in which the number of clockwise encirclements of –1 in the complex plane by the polar frequency-response plot are checked.

Example 6.11

The open-loop frequency response $G(s)$ of a system is given below.

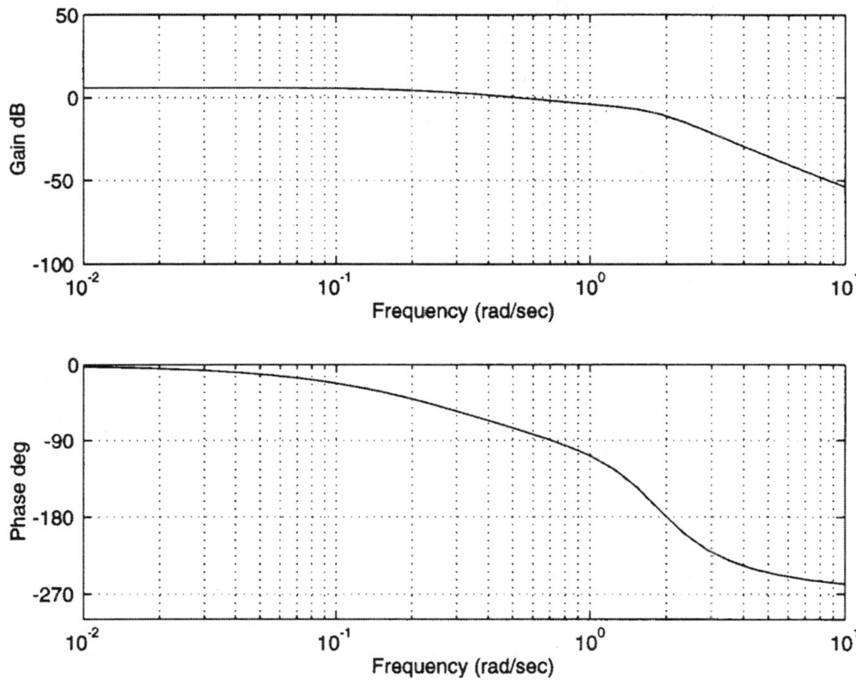

1. The phase margin is approximately:
   (A) –75 degrees
   (B) 105 degrees
   (C) 0 degrees
   (D) undefined

2. The gain margin is approximately:
   (A) 10 dB
   (B) 1 dB
   (C) undefined
   (D) ∞

3. The system can be destabilized with a constant-gain compensator of gain:
   (A) 10 dB
   (B) 1 dB
   (C) already not stable
   (D) always stable

4. What can be said about the number of poles?
   (A) nothing can be said
   (B) 1 pole
   (C) 2 poles
   (D) at least 3 poles

Solutions:

1. **B** If ω_0 is the frequency for which the magnitude of the frequency response is unity or 0 dB ($20 \log |G(\omega_0)| = 0$), then the phase margin is 180 degrees plus the angle of the frequency response at ω_0. In other words, the phase margin $P.M. = 180 + \angle G(\omega_0)$ degrees. Here, $\omega_0 \approx 0.5$ rad/s. The corresponding phase in the frequency response is $\angle G(\omega_0) \approx$ –75 degrees, yielding a phase margin $\approx$ 105 degrees.

2. **A** If ω_c is the frequency for which the phase of the frequency response is –80 degrees, then the gain margin *G.M.* is minus the magnitude of the frequency response at ω_c. Hence, $G.M. = -20\log|G(\omega_c)|$. Here $\omega_c \approx 1$ rad/s. Hence, gain margin is $G.M. \approx 10$ dB by examination of the graph.

3. **A** The gain margin defines the gain which, when added to the open-loop path, destabilizes the system. Thus, any gain larger than 10 dB will destabilize the system.

4. **D** For each pole, the high-frequency phase response increments by –90 degrees. Similarly, for each zero, the high-frequency phase response increments by +90 degrees. Since the phase response shown gets below 180 degrees, we know there must be at least three poles. However, it is unknown from this plot whether the phase might get further below –270 degrees for larger frequencies, or whether there are zeros influencing the plot.

Mechanical Engineering Practice Test

This practice exam has been developed to help test-takers prepare for the afternoon Discipline-Specific test in Electrical Engineering. In addition, if test-takers are undecided, taking this test may also help them determine whether they should take the afternoon DS test or the afternoon General test.

The subjects tested are those subjects that the NCEES states will make up the exam. We have placed the same number of problems from each area of Electrical Engineering that can be expected to be on the actual exam. The difficulty level can, of course, be quite different on the actual exam; we have, however, attempted to create an exam with essentially the same level of difficulty as one would expect on the actual exam.

If you find the problems on this exam to be more difficult than the problems on the afternoon General test, you may wish to take the General test in the afternoon. About one half of all Electrical Engineering examinees take the afternoon General test. This practice exam should help you make that decision before the day of the test.

cut here!

FUNDAMENTALS OF ENGINEERING EXAM

Afternoon Session—Mechanical Exam

(Simulated answer form with topical breakout and scoring grid.)

BE SURE EACH MARK IS DARK AND COMPLETELY FILLS THE INTENDED SPACE AS ILLUSTRATED HERE: ●.

THERMO	FANS, PUMPS, COMPRESSORS	DYNAMIC SYSTEMS	DESIGN	MATERIALS
1 Ⓐ Ⓑ Ⓒ Ⓓ	16 Ⓐ Ⓑ Ⓒ Ⓓ	28 Ⓐ Ⓑ Ⓒ Ⓓ	40 Ⓐ Ⓑ Ⓒ Ⓓ	52 Ⓐ Ⓑ Ⓒ Ⓓ
2 Ⓐ Ⓑ Ⓒ Ⓓ	17 Ⓐ Ⓑ Ⓒ Ⓓ	29 Ⓐ Ⓑ Ⓒ Ⓓ	41 Ⓐ Ⓑ Ⓒ Ⓓ	53 Ⓐ Ⓑ Ⓒ Ⓓ
3 Ⓐ Ⓑ Ⓒ Ⓓ	18 Ⓐ Ⓑ Ⓒ Ⓓ	30 Ⓐ Ⓑ Ⓒ Ⓓ	42 Ⓐ Ⓑ Ⓒ Ⓓ	54 Ⓐ Ⓑ Ⓒ Ⓓ
4 Ⓐ Ⓑ Ⓒ Ⓓ	19 Ⓐ Ⓑ Ⓒ Ⓓ	31 Ⓐ Ⓑ Ⓒ Ⓓ	43 Ⓐ Ⓑ Ⓒ Ⓓ	
5 Ⓐ Ⓑ Ⓒ Ⓓ	20 Ⓐ Ⓑ Ⓒ Ⓓ	32 Ⓐ Ⓑ Ⓒ Ⓓ	44 Ⓐ Ⓑ Ⓒ Ⓓ	
6 Ⓐ Ⓑ Ⓒ Ⓓ	21 Ⓐ Ⓑ Ⓒ Ⓓ	33 Ⓐ Ⓑ Ⓒ Ⓓ	45 Ⓐ Ⓑ Ⓒ Ⓓ	
Score: ______	Score: ______	Score: ______	Score: ______	Score: ______

FLUIDS	STRESS ANALYSIS	INSTRUMENTATION	HEAT TRANSFER	CONTROLS
7 Ⓐ Ⓑ Ⓒ Ⓓ	22 Ⓐ Ⓑ Ⓒ Ⓓ	34 Ⓐ Ⓑ Ⓒ Ⓓ	46 Ⓐ Ⓑ Ⓒ Ⓓ	55 Ⓐ Ⓑ Ⓒ Ⓓ
8 Ⓐ Ⓑ Ⓒ Ⓓ	23 Ⓐ Ⓑ Ⓒ Ⓓ	35 Ⓐ Ⓑ Ⓒ Ⓓ	47 Ⓐ Ⓑ Ⓒ Ⓓ	56 Ⓐ Ⓑ Ⓒ Ⓓ
9 Ⓐ Ⓑ Ⓒ Ⓓ	24 Ⓐ Ⓑ Ⓒ Ⓓ	36 Ⓐ Ⓑ Ⓒ Ⓓ	48 Ⓐ Ⓑ Ⓒ Ⓓ	57 Ⓐ Ⓑ Ⓒ Ⓓ
10 Ⓐ Ⓑ Ⓒ Ⓓ	25 Ⓐ Ⓑ Ⓒ Ⓓ	37 Ⓐ Ⓑ Ⓒ Ⓓ	49 Ⓐ Ⓑ Ⓒ Ⓓ	
11 Ⓐ Ⓑ Ⓒ Ⓓ	26 Ⓐ Ⓑ Ⓒ Ⓓ	38 Ⓐ Ⓑ Ⓒ Ⓓ	50 Ⓐ Ⓑ Ⓒ Ⓓ	
12 Ⓐ Ⓑ Ⓒ Ⓓ	27 Ⓐ Ⓑ Ⓒ Ⓓ	39 Ⓐ Ⓑ Ⓒ Ⓓ	51 Ⓐ Ⓑ Ⓒ Ⓓ	
Score: ______	Score: ______	Score: ______	Score: ______	Score: ______

HVAC				COMPUTERS
13 Ⓐ Ⓑ Ⓒ Ⓓ				58 Ⓐ Ⓑ Ⓒ Ⓓ
14 Ⓐ Ⓑ Ⓒ Ⓓ				59 Ⓐ Ⓑ Ⓒ Ⓓ
15 Ⓐ Ⓑ Ⓒ Ⓓ				60 Ⓐ Ⓑ Ⓒ Ⓓ
Score: ______				Score: ______

Mechanical Discipline Exam

Questions 1 - 6

A reheat Rankine cycle is sketched. Properties are found in appropriate tables in the Handbook provided and others are given.

$$h_1 = 3468 \text{ kJ/kg}$$

$$s_1 = 7.4317 \text{ kJ/kg}\cdot\text{K}$$

$$\left(c_p\right)_{\text{water}} = 4.18 \text{ kJ/kg}\cdot\text{K}$$

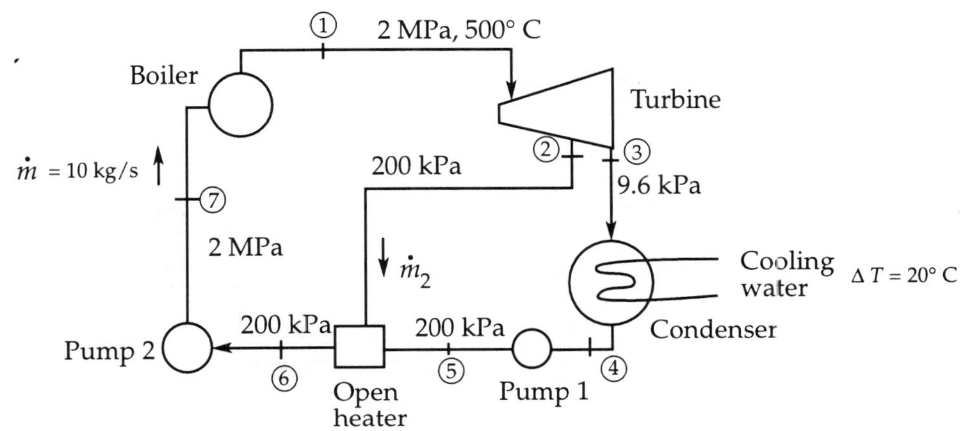

1. If pump 2 is 85% efficient, its power requirement is:
   (A) 13.2 kW
   (B) 21.2 kW
   (C) 25.7 kW
   (D) 36.1 kW

2. Estimate the rate of coal burned in the boiler, assuming a boiler efficiency of 86%(use $HV_{\text{coal}} = 28\ 000$ kJ/kg):
   (A) 1690 kg/hr
   (B) 2190 kg/hr
   (C) 3220 kg/hr
   (D) 4430 kg/hr

3. Estimate $\dot{m}_2$.
   (A) 1.2 kg/s
   (B) 1.8 kg/s
   (C) 2.2 kg/s
   (D) 2.4 kg/s

4. The maximum power output of the turbine is most nearly:
   (A) 5 MW
   (B) 8 MW
   (C) 10 MW
   (D) 13 MW

5. What is the cooling water mass flux?
(A) 280 kg/s
(B) 190 kg/s
(C) 110 kg/s
(D) 78 kg/s

6. Estimate the maximum possible cycle efficiency.
(A) 33.1%
(B) 35.8%
(C) 37.6%
(D) 39.2%

Questions 7 - 12

The 10-cm-diameter, wrought iron pipe transports 180 m^3/hr of water. The nozzle exit diameter is 4 cm.

$\mu = 10^{-3}\ N\cdot s/m^2$

$\rho = 1000\ kg/m^3$

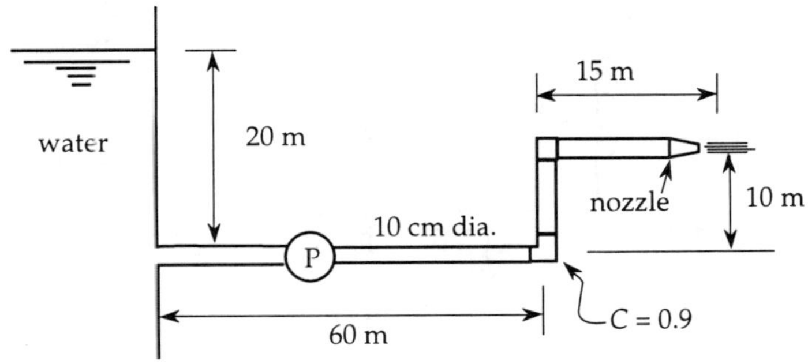

7. The Reynolds number in the pipe is most nearly:
(A) 640
(B) 3800
(C) 38 000
(D) 640 000

8. The head loss for the entire pipe (excluding the pump) is:
(A) 25 m
(B) 30 m
(C) 35 m
(D) 40 m

9. Find the pressure just before the nozzle.
(A) 770 kPa
(B) 680 kPa
(C) 560 kPa
(D) 420 kPa

10. The minimum power required by the pump is:
(A) 43 kW
(B) 52 kW
(C) 64 kW
(D) 75 kW

11 Estimate how far the energy grade line is above the pipe exit.
(A) 2 m
(B) 20 m
(C) 41 m
(D) 81 m

12. If the pipe had a long horizontal section, what parameter would limit how far the pump could be located from the water reservoir?
(A) Froude number
(B) Surface tension
(C) Vapor pressure
(D) Friction factor

Questions 13 - 15

One hundred cubic meters per minute of atmospheric air at 35°C and 70% relative humidity is to be conditioned to 24°C and 40% humidity by using a chiller and a reheater. $R_{air} = 0.287$ kJ/kg·K.

13. Estimate the volume of water removed each hour.
(A) 120 L
(B) 100 L
(C) 80 L
(D) 60 L

14. The reheating required is most nearly:
(A) 50 MJ/hr
(B) 60 MJ/hr
(C) 80 MJ/hr
(D) 100 MJ/hr

15. What cooling rate is required by the chiller?
(A) 142 kJ/s
(B) 126 kJ/s
(C) 105 kJ/s
(D) 91 kJ/s

Questions 16 - 18

An Otto cycle operates between 15°C and 1500°C with a compression ratio of 8. Atmospheric air at 100 kPa enters the cycle.

16. Calculate the efficiency of this ideal cycle.
(A) 48%
(B) 52%
(C) 56%
(D) 60%

17. The pressure just before combustion is most nearly:
(A) 800 kPa
(B) 1270 kPa
(C) 1840 kPa
(D) 2260 kPa

18. The maximum pressure is most nearly:
(A) 6 MPa
(B) 5 MPa
(C) 4 MPa
(D) 3 MPa

Questions 19 - 21

Atmospheric air at 100 kPa and 20°C enters a compressor through a 10-cm-diameter inlet. The pressure ratio is 6.

19. The exiting temperature, assuming an ideal compressor, is:
(A) 216°C
(B) 291°C
(C) 356°C
(D) 489°C

20. If the power requirement is 10 kW, what is the mass flow rate through the ideal compressor?
(A) 1 kg/min
(B) 2 kg/min
(C) 3 kg/min
(D) 4 kg/min

21. If the compressor efficiency is 80%, what is the exit temperature?
(A) 227°C
(B) 265°C
(C) 362°C
(D) 538°C

Questions 22 - 27

Point A is at the very top of the aluminum shaft of radius 4 cm.

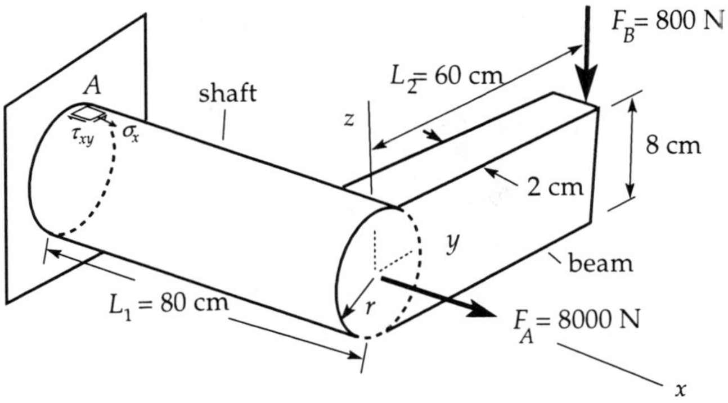

22. The moment diagram drawn for the shaft resembles a:
(A) rectangle
(B) triangle
(C) parabola
(D) circle

23. The normal stress σ_x on an element oriented as shown at point A is:
(A) $F_A / \pi r^2 + F_B L_1 r / I_{shaft}$
(B) $F_A / \pi r^2 + F_B L_2 r / I_{beam}$
(C) $F_A / \pi r^2$
(D) $F_B (L_1 + L_2) r / I_{shaft}$

24. The shearing stress on an element oriented as shown at point A is:
(A) 1.2 MPa
(B) 2.4 MPa
(C) 3.2 MPa
(D) 4.8 MPa

25. The maximum shearing stress at point A is:
(A) 4.8 MPa
(B) 7.1 MPa
(C) 8.6 MPa
(D) 10.2 MPa

26. The maximum normal stress at point A is:
(A) 8.6 MPa
(B) 10.2 MPa
(C) 15.8 MPa
(D) 19.0 MPa

27. The far end of the rigid beam deflects a distance of:
(A) 3 mm
(B) 4 mm
(C) 5 mm
(D) 6 mm

Questions 28 - 33

The coefficient of dynamic friction between the mass and the inclined plane is represented by μ. Given: $m = 4$ kg and $k = 100$ N/m.

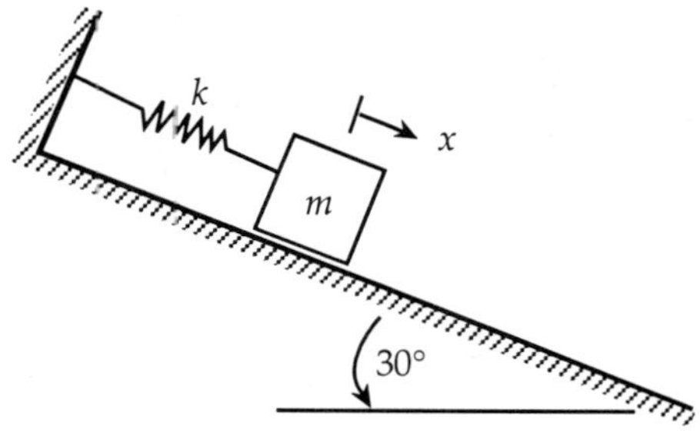

28. Find the potential energy stored in the spring at the static equilibrium position shown when the frictional force is zero.
(A) 0.96 N·m
(B) 1.92 N·m
(C) 2.46 N·m
(D) 3.84 N·m

29. If the coefficient of friction were zero (well greased), what would be the frequency of oscillation?
(A) 5 rad/s
(B) 5 hertz
(C) 25 rad/s
(D) 25 hertz

30. For a frictionless surface, calculate the maximum velocity if the mass is moved 20 cm down the plane from the equilibrium position and released from rest.
(A) 0.2 m/s
(B) 0.4 m/s
(C) 1.0 m/s
(D) 5.0 m/s

31. If a forcing function $F(t) = 10\sin\omega t$ acts down the frictionless plane in the x-direction, what frequency ω would lead to large oscillations?
(A) 10 rad/s
(B) 10 hertz
(C) 5 rad/s
(D) 5 hertz

32. If $\mu = 0.1$ with no forcing function, what differential equation describes motion down the plane?
(A) $\ddot{x} + 25x - 0.85 = 0$
(B) $\ddot{x} - 25x - 0.85 = 0$
(C) $\ddot{x} + 25x + 0.85 = 0$
(D) $\ddot{x} - 25x + 0.85 = 0$

33. The differential equation describing the motion with friction but no forcing function is:
(A) Linear, homogeneous
(B) Linear, nonhomogeneous
(C) nonlinear, homogeneous
(D) nonlinear, nonhomogeneous

Questions 34 – 36

A full strain gage bridge is shown below. The bridge consists of four strain gages R_1, R_2, R_3 and R_4. The strain gages are to be mounted on the top and bottom of a beam to generate E_{out} sensitive to vertical bending strains while remaining relatively insensitive to axial strain of the beam.

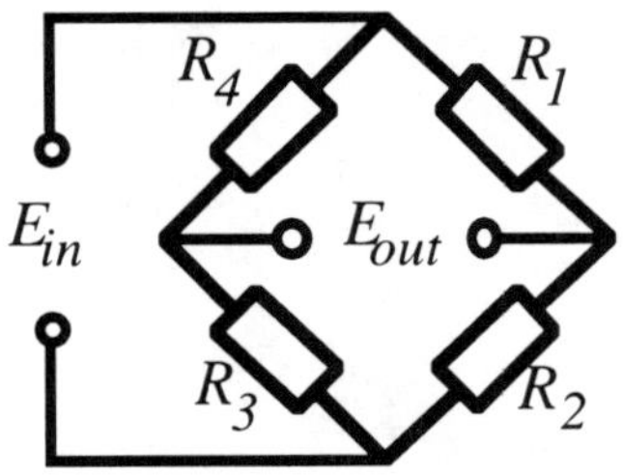

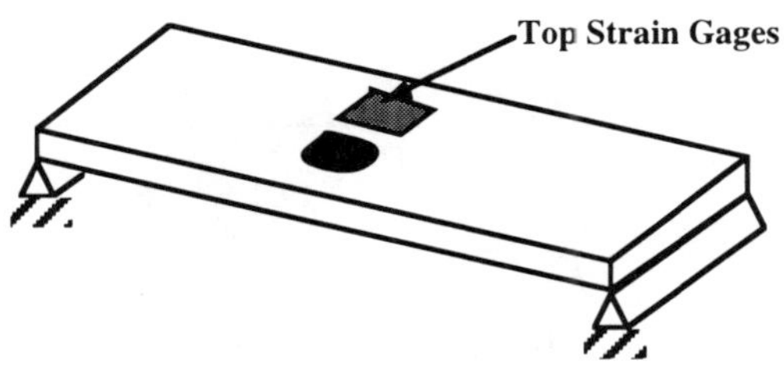

34. Which combinations of strain gage locations would provide the best bending strain sensitivity?
(A) All four strain gages on the top of the beam
(B) Gages R_1 and R_2 on the top of the beam with gages R_3 and R_4 on the bottom of the beam
(C) Gages R_1 and R_3 on the top of the beam with gages R_2 and R_4 on the bottom of the beam
(D) Gages R_1 and R_4 on the top of the beam with gages R_2 and R_3 on the bottom of the beam

35. If the input voltage E_{in} = 10 volts, all gages have a 330 Ohm nominal resistance and gage R_1 increases resistance by 1%, what is the bridge output voltage E_{out} ?
(A) 0.000 volt
(B) 0.010 volt
(C) 0.020 volt
(D) 0.025 volt

36. If the input voltage E_{in} = 10 volts, $R_1 = R_3 = R_A$ = 340 Ohm and $R_2 = R_4 = R_B$ = 320 Ohm, what is the bridge output voltage E_{out} ?
(A) 0.15 volt
(B) 0.30 volt
(C) 0.50 volt
(D) 0.0 volt

Questions 37 – 39

When subjected to a sudden 1 psi change in pressure, a pressure sensor has the time response shown below.

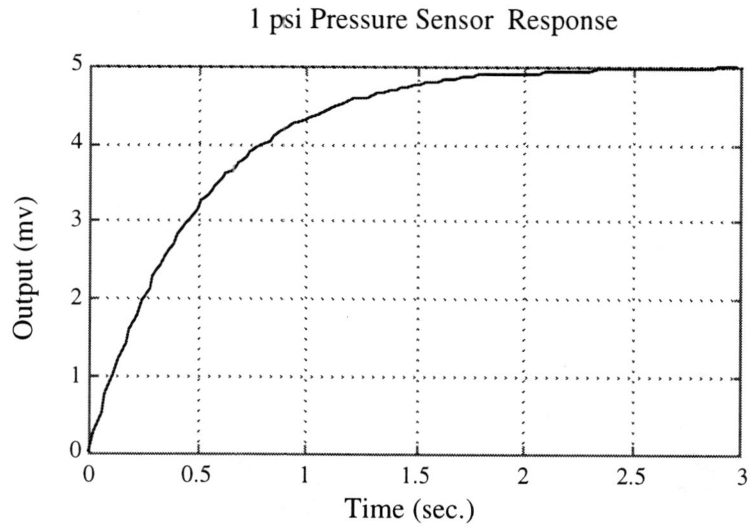

37. What is the approximate time constant of the sensor?
    (A) 0.25 seconds
    (B) 0.5 seconds
    (C) 1.0 seconds
    (D) 3.0 seconds

38. What is the approximate gain of the sensor?
    (A) 0.005 volts/psi
    (B) 4.3 volts/psi
    (C) 5 volts/psi
    (D) 5000 volts/psi

39. What is the largest frequency range over the gain remains within 3 dB of the above value?
    (A) below 0.1 Hz
    (B) below 0.3 Hz
    (C) below 1.0 Hz
    (D) below 10 Hz

Questions 40 – 45

A helical compression spring from a packaging machine has a wire diameter of 1.0 mm and is manufactured in ASTM A228 music wire. The outer diameter is 12 mm and the spring has 10 coils with the ends plain and ground.

40. The minimum tensile strength S_{ut} of the spring material is nearest:
   (A) 1000 MPa
   (B) 2000 MPa
   (C) 3000 MPa
   (D) 4000 MPa

41. The torsional yield strength for this class of cold-drawn carbon steel wire is:
   (A) 780 MPa
   (B) 870 MPa
   (C) 930 MPa
   (D) 1090 MPa

42. What load can the spring withstand when the shear stress equals the yield strength?
   (A) 72 N
   (B) 64 N
   (C) 48 N
   (D) 32 N

43. Calculate the spring rate.
   (A) 653 N/m
   (B) 772 N/m
   (C) 866 N/m
   (D) 931 N/m

44. The axial deflection caused bt the load of Problem 42 is nearest:
   (A) 36 mm
   (B) 24 mm
   (C) 12 mm
   (D) 6 mm

45. Determine what the length of the spring should be to avoid permanent deformation when an overload causes the spring to be compressed solid.
   (A) 66 mm
   (B) 56 mm
   (C) 46 mm
   (D) 36 mm

Questions 46 – 51

A building wall of height 3.5 m and length 8 m is composed of 2 cm of plywood (thermal conductivity of 0.12 W/m·K), 1.5 cm polystyrene (thermal conductivity of 0.027 W/m·K), and 1.2 cm rock sheet (thermal conductivity of 0.12 W/m·K). The convective heat transfer coefficient from the inside wall surface to the inside air is 1.5 W/m^2·K. The inside air temperature is 20°C and the outside air temperature is 2°C. The wind outside is at 2 m/s parallel to the length of the wall. The kinematic viscosity of the outside air may be taken to be 1.6×10^{-5} m^2/s , its thermal conductivity is 0.026 W/m·K, with a Prandtl number of 0.7.

46. What is the convective heat transfer coefficient between the outside wall surface and the outside air?
    (A) 10.5 W/m^2·K
    (B) 6.7 W/m^2·K
    (C) 1.9 W/m^2·K
    (D) 630 000 W/m^2·K

47. What is the thermal resistance due to conduction through the wall?
    (A) 34 K/W
    (B) 0.001 K/W
    (C) 0.82 K/W
    (D) 0.029 K/W

48. If the outside convective heat transfer coefficient were 10 W/m^2·K, what would be the total heat loss from the inside air to the outside air?
    (A) 32 W
    (B) 153 W
    (C) 317 W
    (D) 567 W

49. If the heat flow through the wall is 450 W, what is the inside wall surface temperature?
    (A) 9°C
    (B) 19°C
    (C) 29°C
    (D) 4°C

50. If the outside wall has an emissivity of 0.75 and a temperature of 4°C, what is the radiation loss from the wall to the outside? You may assume that the outside is at 2°C.
    (A) 400 W
    (B) 4 W
    (C) 20 W
    (D) 200 W

51. If the wall receives a solar flux of 200 W/m^2 and has a solar absorptivity of 0.55, what is the total heat gain from the sun by the wall?
(A) 1020 W
(B) 3080 W
(C) 5600 W
(D) 4020 W

Questions 52 – 54

Select the true statement from the following sets of statements for a medium-carbon steel?

52. (A) The hardness can be increased by annealing
(B) The yield stress can be increased by tempering
(C) Higher strength can be achieved by making the grain size smaller
(D) The primary microstructure, after rapid quenching, is Austenite

53. (A) The fatigue strength is not influenced by the microstructure
(B) Below the endurance limit, fatigue is an important failure mode
(C) Highly polished and corrosion free surfaces improve fatigue life
(D) The fracture surface of a sample, which failed due to fatigue, is very rough and jagged

54. (A) A major deformation mode of steel at elevated temperature is twinning
(B) Line defects, called dislocations, interact with carbon atoms during deformation giving rise to strain aging
(C) Maximum hardness in steel is obtained by a heat treatment called tempering
(D) Carburizing is a process in which a sample is carefully annealed to produce a uniform distribution of carbon

Questions 55 – 57

Respond to questions 55–57 by referring to the following diagram.

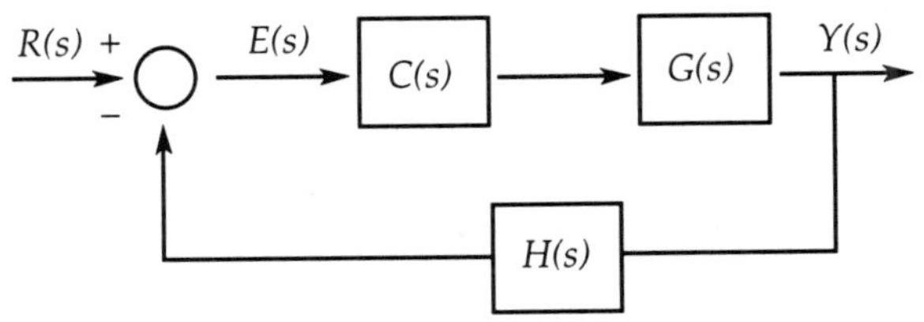

55. Find $E(s)$ in terms of $R(s)$.
(A) $E(s) = R(s) - Y(s)$
(B) $E(s) = C(s)G(s)R(s)$
(C) $E(s) = \dfrac{G(s)R(s)}{1 + C(s)H(s)}$
(D) $E(s) = \dfrac{R(s)}{1 + C(s)G(s)H(s)}$

56. Suppose $C(s)=\frac{k}{s}$, $G(s)=\frac{1}{s^2+1}$, and $H(s)=2$. For what values of k is the system stable?

(A) $k>0$
(B) $k<1$
(C) $k>1$
(D) $0<k<1$

57. Suppose $C(s)=k>0$ and the Bode diagram of the open-loop system is as follows:

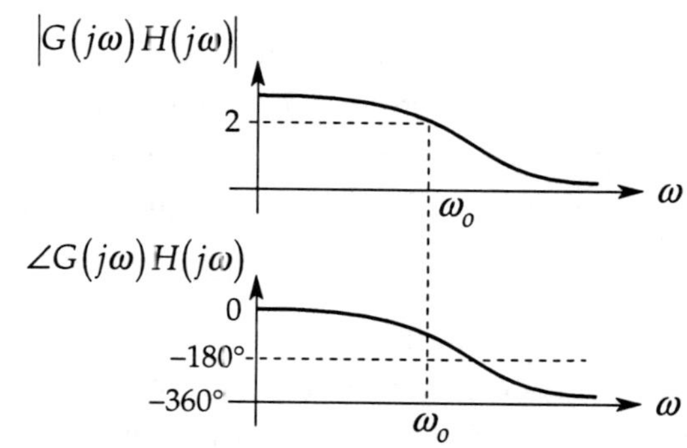

For what range of positive k is the closed-loop system stable?

(A) $k<1/2$
(B) $k>1/2$
(C) $k<2$
(D) $k>2$

58. A spreadsheet has cell $B2$ set to 4. Then cell $B3$ is set to the formula $\$B\$2+B2$. This formula is copied into cells $B4$, $B5$, $B6$, and $B7$ in order. The number in $B6$ is:

(A) 20
(B) 24
(C) 32
(D) 16

59. The pseudo-code segment is given the following list of integer values as data: 17, 15, 12, 18, 19, 25. After execution, what values will be displayed?

```
a := 100;  b := 0;  c := 0;
for i := 1 to 5
   read( x );
   if( x  a)
     a := x;
   if( x  b)
     b := x;
   c := c + x;
endfor
print( a, b, c/5 );
```

(A) 12, 25, 21
(B) 12, 17, 16
(C) 12, 19, 16
(D) 12, 25, 21

60. A file contains the integers: 3, 2, -4, 5, 1, -2, ... which are processed by the pseudo-code program:

```
  a := 0;
  i := 0;
  while( i  3)
    read(b);
   if( b  0 )
     a = a + b*b;
  i := i + 1;
endwhile;
print( a/i);
```

What will be the approximate value printed?

(A) 11
(B) 11.25
(C) 7.8
(D) 9.5

Answers to the Mechanical Exam

Thermodynamics

1. B
2. D
3. D
4. C
5. A
6. B

Fluids

7. D
8. C
9. A
10. B
11. D
12. C

HVAC

13. A
14. D
15. B

Fans, Pumps, Compressors

16. C
17. C
18. B
19. A
20. C
21. B

Stress Analysis

22. B
23. A
24. D
25. C
26. C
27. A

Dynamic Systems

28. B
29. A
30. C
31. D
32. C
33. B

Instrumentation

34. C
35. D
36. B
37. B
38. A
39. B

Design

40. B
41. C
42. D
43. C
44. A
45. C

Heat Transfer

46. B
47. D
48. C
49. A
50. D
51. B

Materials

52. C
53. C
54. B

Controls

55. D
56. D
57. A

Computers

58. A
59. C
60. D

tear out answer sheet

Solutions to the Mechanical Exam

1. **B** The energy equation applied to the pump is

$$\dot{W}_P = \frac{Q\Delta P}{\eta} = \dot{m}\frac{\Delta p}{\eta\rho} = 10\times\frac{2000-200}{0.85\times 1000} = 21.2 \text{ kW}$$

2. **D** The enthalpy change across a pump is quite small and is negligible compared to the enthalpy change across the boiler:

$h_7 \cong h_6 = 504$ kJ/kg ($h_6 \cong h_f$ at 0.1985 MPa – see p.48)

Energy: $\dot{m}_{coal}\times HV\times\eta = \dot{m}(h_1-h_7)$

$28\,000\dot{m}_{coal}\times 0.86 = 10(3468-504)$. $\therefore \dot{m} = 1.23$ kg/s or 4430 kg/hr

3. **D** First we must find h_2. Assume constant entropy from state 1 to state 2:

$s_2 = s_1 = 7.4317$, $P_2 = 0.2$ MPa. $\therefore h_2 = 2837$ kJ/kg

Heater: $\dot{m}_2h_2 + \dot{m}_5h_5 = \dot{m}_6h_6$. (Note: $h_5 \cong h_4 = h_f$ at 9.6 kPa.)

$\dot{m}_2\times 2837 + (10-\dot{m}_2)\times 188 = 10\times 504$. $\therefore \dot{m}_2 = 1.19$ kg/s

4. **C** The ideal turbine is isentropic (constant entropy):

$s_3 = s_1 = 7.4317$, $P_3 = 9.6$ kPa. $7.4317 = 0.6387 + x_3(7.5261)$

$\therefore x_3 = 0.903$ and $h_3 = 188 + 0.903(2395) = 2350$ kJ/kg

$$\dot{W}_T = \dot{m}(h_1-h_2) + (1-\dot{m})(h_2-h_3)$$
$$= 10(3468-2837) + (10-1.19)(2837-2350) = 10\,600 \text{ kW}$$

5. **A** The energy lost by the steam is gained by the water:

Energy: $\dot{m}_3(h_3-h_4) = \dot{m}_{water}c_p\Delta T_{water}$

$(10-1.19)(2350-188) = \dot{m}_{water}\times 4.18\times 20$. $\therefore \dot{m}_{water} = 278$ kg/s

6. **B** The cycle efficiency is output over input (we neglect the small power requirement of the pump):

$$\eta_{cycle} = \frac{\dot{W}_T}{\dot{Q}_B} = \frac{10\,600}{10(3468-504)} = 0.358 \text{ or } 35.8\%$$

7. **D** First, calculate the average velocity:

$$V = \frac{Q}{A} = \frac{180/3600}{\pi\times 0.05^2} = 6.37 \text{ m/s}. \quad \text{Re} = \frac{VD}{\nu} = \frac{6.37\times 0.1}{10^{-6}} = 6.37\times 10^5$$

8. **C** We include both minor losses and pipe friction losses:

$$h_L = \Sigma C\frac{V^2}{2g} + f\frac{L}{D}\frac{V^2}{2g} = \left[0.5 + 2\times 0.9 + 0.0173\frac{85}{0.1}\right]\frac{6.37^2}{2\times 9.8} = 35.2 \text{ m}$$

where we used $e/D = 0.046/100 = 0.00046$. From p. 41, $f = 0.0173$.

9. **A** Bernoulli's equation across the nozzle (there are negligible losses through a nozzle) is used:

$$0 = \frac{V_2^2 - V_1^2}{2g} + \frac{p_2 - p_1}{\gamma}. \qquad V_2 = V_1 \frac{A_1}{A_2} = 6.37 \frac{\pi \times 0.05^2}{\pi \times 0.02^2} = 39.8 \text{ m/s}$$

$$0 = \frac{39.8^2 - 6.37^2}{2 \times 9.8} + \frac{0 - p_1}{9800}. \qquad \therefore p_1 = 772\,230 \text{ Pa} \quad \text{or} \quad 772 \text{ kPa}$$

10 **B** Apply the energy equation from the surface to the pump exit:

$$h_P = \frac{V_2^2 - V_1^2}{2g} + \frac{p_2 - p_1}{\gamma} + z_2 - z_1 + h_L$$

$$= \frac{39.8^2 - 0}{2 \times 9.8} + 0 + 10 - 20 + 35.2 = 106 \text{ m}$$

$$\dot{W}_P = Q\gamma h_P = \frac{180}{3600} \times 9800 \times 106 = 51\,950 \text{ W} \quad \text{or} \quad 52 \text{ kW}$$

11. **D** The energy grade line is the pressure head plus the velocity head above the pipe:

$$\frac{V_2^2}{2g} + \frac{p_2}{\gamma} = \frac{39.8^2}{2 \times 9.8} = 80.8 \text{ m}$$

12. **C** The minimum pressure at the pump inlet cannot be less than the vapor pressure. If it is, the water would vaporize (cavitate).

13. **A** First find the mass flux of the atmospheric air:

$\dot{m}_a = \frac{100}{60} \times \frac{100}{0.287 \times 308} = 1.885$ kg/s where we used $P_{atm} = 100$kPa.

Referring to p.51 we find $\omega_1 = 0.0253$, $\omega_2 = 0.0075$ kg water/kg air.

$\dot{m}_w = \dot{m}_a(\omega_1 - \omega_2) = 1.885(0.0253 - 0.0075) = 0.0335$ kg/s.

Volume $= 0.0335 \times 3600/1000 = 0.121 \text{ m}^3$ or 121 L

14. **D** Reheating occurs at constant humidity ratio from the 100% humidity line to the final design state (use the psychrometric chart):

$$\dot{Q}_{heat} = \dot{m}_a \Delta h$$

$$= 1.885(43.5 - 29) = 27.3 \text{ kJ/s} \quad \text{or} \quad 98\,400 \text{ kJ/hr}$$

15. **B** The chiller reduces the humidity ratio of the air to the humidity ratio required at the final design state (use the psychrometric chart):

$$\dot{Q}_{chiller} = 1.885(96 - 29) = 126 \text{ kJ/s}$$

16. **C** The efficiency depends only on the compression ratio:

$$\eta = 1 - r^{1-k}$$

$$= 1 - 8^{1-1.4} = 0.565 \quad \text{or} \quad 56.5\%$$

17. **C** An isentropic process occurs between intake and when the piston is at top-dead-center:

$$P_2 = P_1 \left(\frac{v_1}{v_2} \right)^k = 100 \times 8^{1.4} = 1838 \text{ kPa}$$

18. **B** Maximum pressure occurs after combustion while the piston is assumed to remain at top-dead-center. First we find the temperature just before combustion:

$$T_2 = T_1\left(\frac{v_1}{v_2}\right)^{k-1} = 288 \times 8^{0.4} = 662 \text{ K}$$

$$P_3 = P_2\frac{T_3}{T_2} = 1838\frac{1500+273}{662} = 4920 \text{ kPa}$$

19. **A** Flow through an ideal compressor is assumed to be isentropic:

$$T_e = T_i\left(\frac{P_e}{P_i}\right)^{k-1/k} = 293 \times 6^{0.4/1.4} = 489 \text{ K} \quad \text{or} \quad 216°\text{C}$$

20. **C** We neglect the kinetic energy change across a compressor:

$$\dot{W}_C = \dot{m}\Delta h = \dot{m}c_p\Delta T$$

$$10 = \dot{m} \times 1.0(489-293). \qquad \therefore \dot{m} = 0.051 \text{ kg/s}$$

21. **B** Efficiency is defined as the ratio of enthalpy changes. Since enthalpy change in a gas is directly proportional to the temperature change, we can use

$$T_e = T_i + (T_{es} - T_i)/\eta = 293 + (489-293)/0.8 = 538 \text{ K} \quad \text{or} \quad 265°\text{C}$$

22. **B** The vertical shear diagram is a rectangle, a force F_B up at the left end and a force F_B down at the right end. The moment diagram is triangular, F_BL_1 at the left end and zero at the right end.

23. **A** Two normal stresses add at point A: F/A and Mc/I. The moment is $M = F_BL_1$. Hence,

$$\sigma_x = \frac{F_A}{A} + \frac{F_BL_1r}{I_{shaft}} = \frac{F_A}{\pi r^2} + \frac{F_BL_1r}{I_{shaft}}$$

24. **D** The shearing stress on the element in the position shown is

$$\tau = \frac{Tr}{J} = \frac{F_BL_2r}{J}$$

$$= \frac{800 \times 0.6 \times 0.04}{\pi \times 0.04^4/2} = 4.77 \times 10^6 \text{ Pa} \quad \text{or} \quad 4.77 \text{ MPa}$$

25. **C** The normal stress σ_x is

$$\sigma_x = \frac{F_A}{A} + \frac{F_BL_2r}{I_{shaft}}$$

$$= \frac{8000}{\pi \times 0.04^2} + \frac{800 \times 0.8 \times 0.04}{\pi \times 0.04^4/4} = 14.32 \times 10^6 \text{ Pa}$$

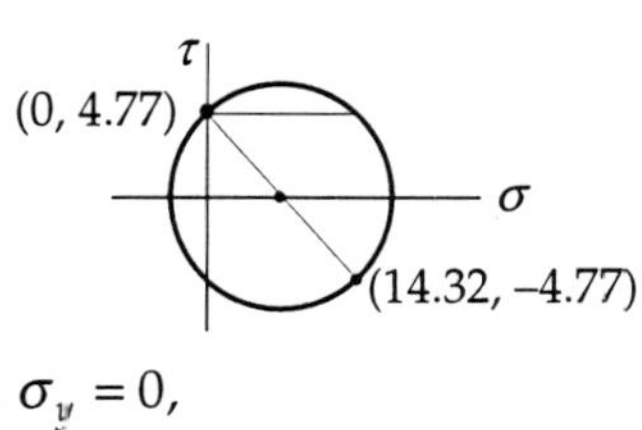

The maximum shearing stress is then, using Mohr's circle with $\sigma_y = 0$,

$$\tau_{max} = \sqrt{(\sigma_x/2)^2 + \tau^2} = \sqrt{(14.32/2)^2 + 4.77^2} = 8.61 \text{ MPa}$$

26. C The maximum normal stress is (se Mohr's circle above)

$$\sigma_{max} = \frac{\sigma_{max}}{2} + \tau_{max} = \frac{14.32}{2} + 8.61 = 15.77 \text{ MPa}$$

27. A Since the beam is assumed to be rigid it will only rotate:

$$\delta = \delta_{shaft} + \phi_{shaft} L_2 = \frac{F_B L_1^3}{3EI} + \frac{(F_B L_2) L_1}{JG} \times L_2$$

$$= \frac{800 \times 0.8^3}{3 \times 6.9 \times 10^{10} \times \pi \times 0.04^4 / 4} + \frac{800 \times 0.6 \times 0.8}{(\pi \times 0.04^4 / 2) \times 2.8 \times 10^{10}}$$

$$= 0.00098 + 0.00205 = 0.00303 \text{ m}$$

28. B Sum forces with zero frictional force:

$$mg \sin 30^\circ = k\delta. \qquad \therefore \delta = \frac{4 \times 9.8 \times 0.5}{100} = 0.196 \text{ m}$$

$$\text{P.E.} = \frac{1}{2} k\delta^2 = \frac{1}{2} \times 100 \times 0.196^2 = 1.92 \text{ N} \cdot \text{m}$$

29. A The differential equation would be (x is positive down the plane)

$$\sum F_x = ma_x$$

$$-k(x + \delta) + mg \sin 30^\circ = m\ddot{x}$$

From Problem 28, $-k\delta + mg \sin 30^\circ = 0$, so that

$$m\ddot{x} + kx = 0$$

The angular frequency is $\omega_n = \sqrt{k/m} = \sqrt{100/4} = 5 \text{ rad/s}$

30. C The differential equation is $4\ddot{x} + 100x = 0$ with $x(0) = 0.2$ and $\dot{x}(0) = 0$. The solution is

$$x(t) = C_1 \cos 5t + C_2 \sin 5t$$

$$x(0) = 0.2 = C_1$$

$$\dot{x}(0) = 0 = 5C_2$$

This gives $x(t) = 0.2 \cos 5t$ and $\dot{x}(t) = -\sin 5t$. Finally,

$$\dot{x}_{max} = 1.0 \text{ m/s}$$

31. D If the frequency of the forcing function is equal to the natural frequency of the system, resonance occurs and large oscillations result. From Problem 29, $\omega = 5$ rad/s. (In real life there is always some damping, however small. Resonance–i.e., large amplitude displacements–may still occur if $\omega \cong \omega_n$)

32. C The frictional force opposes motion and is equal to $\mu N = \mu mg \cos 30^\circ$. Referring to Problem 29,

$$-kx - \mu mg \cos 30^\circ = m\ddot{x}$$

Thus,

$$4\ddot{x} + 100x + 3.39 = 0, \quad \text{or} \quad \ddot{x} + 25x + 0.85 = 0$$

33. **B** All the differential equations listed in Problem 32 are linear and nonhomogeneous, nonhomogeneous because they have a nonzero term that does not contain the dependent variable.

34. **C** For the largest bending sensor signal, gage pair R_1 and R_3 must respond in the same way and in the opposite sense to the gage pair R_2 and R_4. The correct response is gages R_1 and R_3 on the top of the beam with gages R_2 and R_4 on the bottom of the beam. In this way when the beam is bent downwards, gages R_1 and R_3 on the top of the beam will be compressed decreasing their resistance together while gages R_2 and R_4 on the bottom of the beam will be elongated increasing their resistance together. Axial strains will identically affect all gages yielding no output voltage.

35. **D** The voltage E_{out} is computed from the voltage difference

$$E_{out} = E_{in}\left[\frac{R_3}{R_3+R_4} - \frac{R_2}{R_1+R_2}\right]$$

$$= E_{in}\left[\frac{R}{R+R} - \frac{R}{1.01R+R}\right]$$

$$= 10\left[\frac{1}{1+1} - \frac{1}{1.01+1}\right] = 10\left[\frac{1}{2} - \frac{1}{2.01}\right] = 0.025 \text{ volts}$$

36. **B** Similarly, the voltage E_{out} is computed from the voltage difference

$$E_{out} = E_{in}\left[\frac{R_3}{R_3+R_4} - \frac{R_2}{R_1+R_2}\right]$$

$$= E_{in}\left[\frac{R_A}{R_A+R_B} - \frac{R_B}{R_A+R_B}\right]$$

$$= 10\left[\frac{R_A-R_B}{R_A+R_B}\right] = 10\left[\frac{340-320}{340+320}\right] = 10\left[\frac{20}{660}\right] = 0.3 \text{ volts}$$

37. **B** The response after time equal to one time constant is about $1/e = 63\%$ of the final value. The final value is 5 mv and 63% of 5 mv is 3.2 mv. The response crosses 3.2 mv after 0.5 seconds.

38. **A** The gain of the sensor is the ratio of the change in input to the change in output:

$$G = 5 \text{ mv}/1\text{psi} = 5 \text{ mv/psi} = 0.005 \text{ volt/psi}$$

39. **B** The bandwidth of the first order actuator at –3 dB is

$$\omega = 1/\tau = 1/0.5 = 2 \text{ rad/sec} = 2/2\pi \text{ Hz} = 0.3 \text{ Hz}$$

40. **B** From p.115 of the Handbook, the minimum tensile strength of common spring steel is

$$S_{ut} = \frac{A}{d^m}$$

From the tabulated data for music wire, we have

$$S_{ut} = \frac{A}{d^m} = \frac{2060}{1^{0.163}} = 2060 \text{ MPa}$$

41. C From p.115, the torsional yield strength is

$$S_{sy} = 0.45S_{ut}$$
$$= 0.45 \times 2060 = 927 \text{ MPa}$$

42. D The governing equation is found on p.115, column A:

$$\tau = K_s \frac{8FD}{\pi d^3}$$

We must determine K_s. The mean spring diameter is given by

$$D = D_{outer} - d = 12 - 1 = 11 \text{ mm}$$

The spring index is

$$C = \frac{D}{d} = \frac{11}{1} = 11$$

The shear stress correction factor is defined on p.115 as

$$K_s = \frac{2C+1}{2C} = \frac{2\times 11+1}{2\times 11} = 1.045$$

The load is then found using $\tau = S_{st}$ to be

$$F = \frac{S_{sy}\pi d^3}{8DK_s} = \frac{927\times 10^6 \pi \times (10^{-3})^3}{8\times 0.011\times 1.045} = 31.7 \text{ N}$$

43. C The spring rate is calculated using the equation on p.115:

$$k = \frac{d^4 G}{8D^3 N}$$

where N is the number of active coils. Since the ends of the spring are plain and ground, and the number of coils is 10, the number of active coils is N = 10 – 1 = 9 using the information in the table on p.115. Using the value $G = 8.3\times 10^{10}$ Pa from p.32 we find

$$k = \frac{0.001^4 \times 8.3\times 10^{10}}{8\times 0.011^3 \times 9} = 866 \text{ N/m}$$

44. A The axial deflection is related to the applied load by the spring constant. Therefore at yielding, from p.115,

$$x = \frac{F}{k} = \frac{31.7}{866} = 0.0366 \text{ m} \quad \text{or} \quad 36.6 \text{ mm}$$

45. C The length L_o of the spring to avoid yielding is the sum of the solid length L_s and the length x associated with the load required to initiate yielding. From the table on p.115

$$L_s = N_t d = 10\times 1 = 10 \text{ mm}$$

Therefore,

$$L_o = L_s + x = 10 + 36.6 = 46.6 \text{ mm}$$

46. **B** The outside heat transfer coefficient may be calculated from the Nusselt number correlation for flow parallel to a constant temperature flat plate. We first calculate the Reynolds number as

$$\text{Re} = \frac{uL}{\nu} = \frac{(2)(8)}{1.6 \times 10^{-5}} = 1.0 \times 10^{6}$$

Then from the Handbook we find that

$$\text{Nu} = 0.0366\text{Re}^{0.8}\text{Pr}^{1/3} = 2050$$

The outside convective heat transfer coefficient is then given by

$$h_o = \frac{\text{Nu} \cdot k}{L} = \frac{(2050)(0.026)}{8} = 6.7 \text{ W/m}^2 \cdot \text{K}$$

47. **D** The conduction resistance is given by

$$R_{cond} = \left[\left(\frac{d}{kA}\right)_{plywood} + \left(\frac{d}{kA}\right)_{polystyrene} + \left(\frac{d}{kA}\right)_{rock\ board}\right] = 0.029 \text{ K/W}$$

48. **C** The total heat transfer is given by

$$\dot{q} = UA\left[T_{inside} - T_{outside}\right]$$

where

$$UA = \left[\left(\frac{1}{h_o A}\right) + \left(\frac{d}{kA}\right)_{plywood} + \left(\frac{d}{kA}\right)_{polystyrene} + \left(\frac{d}{kA}\right)_{rock\ board} + \left(\frac{1}{h_i A}\right)\right]^{-1}$$
$$= 17.6 \text{ W/K}$$

which then gives a heat loss of 317 W.

49. **A** The inside wall surface temperature is given by

$$T_{i,wall} = T_i - \frac{\dot{q}}{h_i A} = 9°\text{C}$$

50. **D** The outside wall radiation loss may be calculated from

$$\dot{q}_{rad} = \varepsilon\sigma A\left[T_{w,o}^4 - T_o^4\right]$$
$$= 200 \text{ W}$$

51. **B** The solar heat gain is given by $\dot{q}_{sun} = \alpha A \dot{q}''_{solar} = 3080$ W.

52. **C**

53. **C**

54. **B**

55. **D**

56. **D**

57. **A**

58. **A**

59. **C**

60. **D**

The D/S Review for Electrical Engineering

edited by M. C. Potter

Chapters

Overview of the EE/DS Topics

What You Need to Know about the NCEES Handbook

Most of the problems on the Electrical Engineering test of the afternoon session can be worked using pages 95 to 100 of the NCEES Handbook. Many of the problems, however, on the afternoon Electrical Engineering test will come from the material in pages 1 to 81 of the NCEES Handbook, material used for the General tests of the morning and afternoon sessions. Since all engineers must take the General morning exam, it is assumed that you are familiar with that material. You must realize, however, that you will have to make reference to equations, tables, and charts located in those first 81 pages of the NCEES Handbook when you are working problems in the afternoon Electrical Engineering session. The general engineering sections that you should be particularly familiar with are the sections on Mathematics, Electric Circuits, and Control Systems; a review of problems on Control Systems can be found in our Mechanical Engineering Part of this book. A review of general engineering material is found in our *Fundamentals of Engineering*, a review book written specifically for the FE/EIT exam.

Do I Have to Take the DS Test?

If, after you study this Electrical Engineering review section, and take the Electrical Engineering Discipline Practice Exam, you feel that you could more easily pass the General exam of the afternoon session, you should register to take the afternoon General test. If you have already registered for the Electrical Engineering afternoon test, you may contact your state board and request a change. Make sure your state board recognizes a *pass* on the General afternoon test for your particular discipline before you make that decision; the vast majority of the state boards will recognize a pass on any of the afternoon tests by any registrant.

This Review is Focused on the NCEES Handbook

In order to provide a very efficient, yet effective review, this DS Review has been designed to review only the material associated with the formulas and figures presented in the NCEES Handbook. The Handbook is the only material allowed in the exam. Thus it makes an efficient base to study around. It is assumed that you have a copy of the NCEES Handbook dated 1996 or later. If you do not, you can obtain one from Great Lakes Press by calling 1-800-837-0201.

GLP Offers a Concise A.M. Session Review & Other Study Aides

We also have an effective review for the General tests, both morning and afternoon, entitled *Fundamentals of Engineering*. It is used at hundreds of engineering schools across the nation. We also distribute the HP-48G and GX calculators, which are pre-programmed with many of the equations you'll need when working problems on the FE/EIT exam. Also, we've developed *Jump Start your HP-48G/X*, a 150-page book designed to efficiently help engineers get the most out of the HP48G/X. You can order these items by calling 1-800-837-0201.

1. Electromagnetic Fields

by Roland Zapp

The equations used to solve problems involving electromagnetic fields are listed on page 95 of the NCEES Handbook. Related information may be needed from pages 66 to 69 of the Handbook.

Example 1.1

The electric field for a dipole located at the origin of a spherical coordinate system is given by

$$\mathbf{E} = \frac{2k\cos\theta\,\mathbf{a}_r}{4\pi\varepsilon_0 r^3} + \frac{k\sin\theta\,\mathbf{a}_\theta}{4\pi\varepsilon_0 r^3}$$

1. A thin spherical surface of radius $r = 25$ cm is located at the origin of the given spherical coordinate system. It was determined that Maxwell's equations yield only the dipole field given. The potential on the surface of the sphere is:
   (A) zero
   (B) constant
   (C) $A/(r^2\cos\theta)$
   (D) $V_0\cos\theta$ where V_0 is a constant

2. A Gaussian surface of radius r_0 encloses the dipole field given. The surface integral $\oint_A \mathbf{E}\cdot d\mathbf{A}$ gives a value of:
   (A) $r_0 k$
   (B) $r_0 k/2$
   (C) k (where $k = Qd$ in the limit $d \to 0$)
   (D) None of the above

3. Applications of Maxwell's equations

   $$\nabla\times\mathbf{E} = -\frac{\partial\mathbf{B}}{\partial t} \text{ and } \nabla\cdot\mathbf{E} = \frac{\rho}{\varepsilon_0}$$

   to the given dipole field would yield
   (A) $\mathbf{B} = 0,\ \rho = 0$ everywhere
   (B) The curl and divergence are indeterminate
   (C) The equations are well behaved except at the origin
   (D) None of the above

Solutions:

1. **D** $V_0 \cos\theta$.
   General potential of a dipole is $\frac{P_0 \cos\theta}{4\pi\varepsilon_0 r^2}$ which at r = constant gives $K \cos\theta$.

2. **D** By Gauss's Law, the surface integral of **E** *over* any surface gives the charge enclosed. Since a dipole represents a zero net charge, the answer is zero.

3. **C** The curl and divergence are zero everywhere except at $r = 0$, where the derivatives are undefined.

Example 1.2

Given the electric field:

$$\mathbf{E} = 2x^2\mathbf{a}_x + 2y^2\mathbf{a}_y + 2z^2\mathbf{a}_z$$

1. Select one of the following:
   (A) $\nabla \cdot \mathbf{E} \neq 0$
   (B) $\oint_L \mathbf{E} \cdot d\mathbf{L} = 0$
   (C) $\oint_s \mathbf{E} \cdot d\mathbf{s} \neq 0$
   (D) All of the above

2. The work done in moving a negative charge $-|Q|$ from (0,0,0) to (1,1,1) in the electric field described is:
   (A) zero
   (B) two $|Q|$
   (C) four $|Q|$
   (D) None of the above

3. The potential associated with the given electric field is:
   (A) $V = -4(x + y + z)$
   (B) $V = -2(x^3 + y^3 + z^3)/3$
   (C) $V = -2(x^3y^3z^3)/3$
   (D) None of the above

Solutions:

1. **D** All of the above.

2. **B** Two $|Q|$.

3. **B** $V = -2(x^3 + y^3 + z^3)/3$.
   Note $\mathbf{E} = -\nabla V$. Can also get the answer to Prob. 2 by noting $W = QV$.

Example 1.3

Assume an electric flux density of $\mathbf{D} = 2x^2\mathbf{a}_x$. The total charge enclosed in a cube of side = 2 located in the first quadrant with $x = y = z = 0$ as one of its corners is:

(A) zero
(B) four
(C) sixteen
(D) thirty-two

Solution:

$\nabla \cdot \mathbf{D} = \rho. \quad \int \rho dv = Q = 32.$

The solution is **D**.

Example 1.4

A circular wire ring of radius a centered at the origin of a cylindrical coordinate system has a total positive charge Q uniformly distributed over the wire. The electric field at a location $z = z_0$ on the z-axis is given by:

(A) $\dfrac{Q}{4\pi\varepsilon_0(a^2 + z_0^2)}\mathbf{a}_z$

(B) $\dfrac{Q}{4\pi\varepsilon_0(a^2 + z_0^2)}\mathbf{a}_r$

(C) $\dfrac{Q}{4\pi\varepsilon r^2}\mathbf{a}_r$

(D) None of the above.

Solution:

By simply applying integration to the point charge field $\mathrm{E} = \dfrac{Q}{4\pi\varepsilon r^2}\mathbf{a}_r$ the field becomes

$$\mathbf{E} = \frac{Qz_0}{4\pi\varepsilon_0(a^2 + z_0^2)^{3/2}}\mathbf{a}_z$$

The solution is **D**.

Example 1.5

A discontinuity on a transmission line is detected by launching a known step voltage and observing the echo. This is done as shown below.

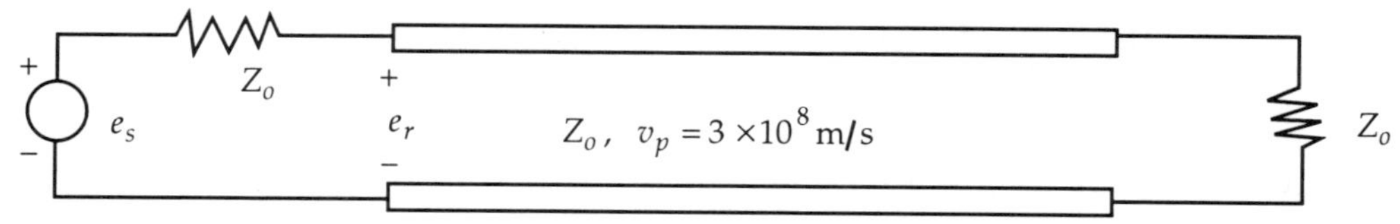

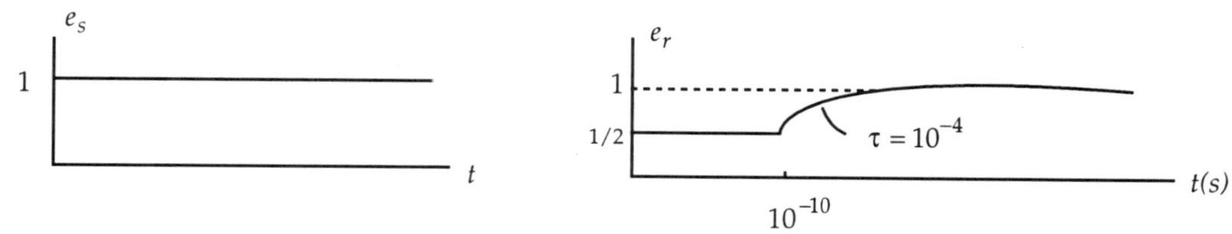

The following can be determined from this exercise:

(A) A discontinuity exists 3 cm from the line input.

(B) An open circuit exists on the line.

(C) An inductor of value 5×10^{-3} is present on the line.

(D) None of the above.

Solution:

A capacitive discontinuity exists 1.5 cm down the line (with $C = 2\mu\text{F}$).

The answer is **D**.

Example 1.6

Current flows axially with uniform density $\mathbf{J} = J_0\, \mathbf{a}_z$ amp/m^2 along a cylindrical conductor of radius a and length **L** with uniform conductivity σ_0, by the application of a potential difference V_0 between the ends of the conductor. Determine the magnetic field **H** inside the cylinder.

(A) $\mathbf{H} = \dfrac{J_0 r}{2}\mathbf{a}_\phi$

(B) $\mathbf{H} = \dfrac{J_0 V_0}{2\mathbf{L}r}\mathbf{a}_\phi$

(C) $\mathbf{H} = \dfrac{J_0}{2\pi r}\mathbf{a}_\phi$

(D) $\mathbf{H} = \dfrac{J_0 a^2}{2r}\mathbf{a}_\phi$

Solution:

Using Maxwell's equation, $\oint \mathbf{H} \cdot d\mathbf{L} = \int \mathbf{J} \cdot d\mathbf{s}$ and symmetry.

The answer is **A**.

Example 1.7

The power dissipated by the cylinder of Example 1.6 is:

(A) $P_d = J_0^2 / \sigma_0$

(B) $P_d = \dfrac{V_0^2 \sigma_0 a^2 \pi}{L}$

(C) $P_d = \dfrac{L J_0^2 a^2 \pi}{2\sigma_0}$

(D) None of the above

Solution:

$$P_d = \int \mathbf{J} \cdot \mathbf{E} dv = \frac{L}{\sigma_0} \int_{\varphi=0}^{2\pi} \int_{r=0}^{a} J_0^2 r dr d\varphi = \frac{L}{\sigma_0} J_0^2 a^2 \pi$$

where $J_0 = \sigma_0 \dfrac{V_0}{L}$.

The answer is **B**.

Example 1.8

The electric field **E** in free space is given by:

$$\mathbf{E} = \mathbf{a}_y E_0 \sin(\omega t - kz)$$

Determine the magnetic field **B** associated with this electric field.

(A) $\mathbf{B} = \mathbf{a}_x [-\dfrac{E_0 k}{\omega} \cos(\omega t - kz)]$

(B) $\mathbf{B} = \mathbf{a}_x [-\dfrac{E_0 k}{\omega} \sin(\omega t - kz)]$

(C) $\mathbf{B} = \mathbf{a}_y [-\dfrac{E_0 k}{\omega} \cos(\omega t - kz)]$

(D) $\mathbf{B} = \mathbf{a}_y [-\dfrac{E_0 k}{\omega} \sin(\omega t - kz)]$

Solution:

This is solved using $\nabla \times \mathbf{E} = -\dfrac{\partial \mathbf{B}}{\partial t}$. The answer is **B**.

2. Electric Power

by Dennis Wiitanen

Page 95 of the NCEES Handbook provides some information on electric power, listing the fundamental relationships for ac Machines and balanced three phase systems. In addition, a knowledge of ac circuits as outlined on pages 68 and 69 is required to work the basic problems in power.

2.1 Balanced Three Phase Systems

Three phase systems are employed because of ease of generation and transmission efficiency. Solving balanced three phase problems requires the application of the equations on page 95 in conjunction with the ac circuits relationships on pages 68 and 69.

Example 2.1 Wye Connected Load

A three phase load of wye connected impedances of 4.0 + j3.0 ohms each is supplied by a balanced three phase source with a line voltage of 207.8 volts.

1. What is the magnitude of the voltage across each impedance?
   (A) 120 V
   (B) 207.8 V
   (C) 360 V
   (D) 440 V

2. What is the power factor of the load?
   (A) 0.20
   (B) 0.40
   (C) 0.80
   (D) 36.8

3. What is the magnitude of the line current?
   (A) 10 A
   (B) 24 A
   (C) 38 A
   (D) 41.6 A

4. What is the real power delivered to the load?
   (A) 2.30 kW
   (B) 2.88 kW
   (C) 3.99 kW
   (D) 6.91 kW

5. What is the reactive power deliverd to the load?
   (A) 1.73 kVAR
   (B) 2.88 kVAR
   (C) 5.18 kVAR
   (D) 8.64 kVAR

Solutions:

1. **A** $|V_P| = |V_L|/\sqrt{3} = 207.8/\sqrt{3} = 120.0$

2. **C** p.f. = $\cos\theta_P = \cos(\tan^{-1} 3/4) = .80$

3. **B** $|I_P| = |V_P|/|Z_P|$; for a wye system, $|V_P| = |V_L|/\sqrt{3}$ and $I_P = I_L$.
   Therefore, $|I_L| = (|V_L|/\sqrt{3})/|Z_P|$; also, $|Z_P| = (R^2 + X^2)^{0.5}$
   Finally,
   $|I_L| = 207.8/\sqrt{3(4.0^2 + 3.0^2)} = 24.0$ A

4. **D** $P = \sqrt{3}|V_L||I_L|\cos\theta_P = \sqrt{3}(207.8)(24)\cos(\tan^{-1} 3/4) = 6910$ W = 6.91 kW

5. **C** $Q = \sqrt{3}|V_L||I_L|\sin\theta_P = \sqrt{3}(207.8)(24)\sin(\tan^{-1} 3/4) = 5{,}180$ W = 5.18 kVAR

Example 2.2 General Load Specification

A balanced three phase load supplied by a 460 volt balanced three phase source draws 50 kVA at an 0.9 lagging power factor.

1. What is the real power delivered to the load?
   (A) 12.3 kW
   (B) 15.0 kW
   (C) 32.8 kW
   (D) 45.0 kW

2. What is the reactive power deliverd to the load?
   (A) 21.8 kVAR
   (B) 32.6 kVAR
   (C) 45.0 kVAR
   (D) 50.0 kVAR

3. What is the magnitude of the line current the load draws from the source?
(A) 12.6 A
(B) 19.1 A
(C) 41.6 A
(D) 62.8 A

4. If the load were represented by a set of delta connected impedances, what would be the required impedance per phase?
(A) $1.41 \angle 25.8^0\ \Omega$
(B) $2.44 \angle 25.8^0\ \Omega$
(C) $4.23 \angle 25.8^0\ \Omega$
(D) $12.7 \angle 25.8^0\ \Omega$

5. If the load were represented by a set of wye connected impedances, what would be the required impedance per phase?
(A) $1.41 \angle 25.8^0\ \Omega$
(B) $2.44 \angle 25.8^0\ \Omega$
(C) $4.23 \angle 25.8^0\ \Omega$
(D) $7.33 \angle 25.8^0\ \Omega$

Solutions:

1. **D** $VA = P + jQ;\ P = \text{Re}\{VA\} = VA \cos \theta_P$; p.f. = $\cos \theta_P$; therefore

$P = VA(\text{p.f.}) = 50{,}000(.9) = 45{,}000$ W = 45.0 kW

2. **A** $VA = P + jQ;\ Q = \text{Im}\{VA\} = VA \sin \theta_P$; p.f. = $\cos \theta_P$; so

$Q = VA(\sin \cos^{-1}\text{p.f.}) = 50{,}000(\sin (\cos^{-1}.9)) = 50{,}000(\sin \pm 25.8^0)$ (positive angle for lagging p.f.)

Therefore,

$Q = 50{,}000(\sin 25.8^0) = 21{,}800$ VAR = 21.8 kVAR

3. **D** $VA = \sqrt{3}\,|V_L|\,|I_L|$; which gives

$|I_L| = VA/(\sqrt{3}\,|V_L|) = 50{,}000/(\sqrt{3}\text{x}460) = 62.8$ A

4. **D** $|Z_P| = |V_P|/|I_P|$; in a delta load $V_P = V_L$ and $|I_P| = |I_L|/\sqrt{3}$

Therefore

$|Z_\Delta| = |V_L|/(|I_L|/\sqrt{3}) = 460/(62.8/\sqrt{3}) = 12.7$ ohm

Also, p.f. = $\cos \theta_P$, so $\theta_P = \cos^{-1}$ (p.f.) = $\cos^{-1}$ (.9) = $\pm 25.8^0$ (positive angle for lagging p.f.) so

$Z_\Delta = |Z_\Delta| \angle 25.8^0 = 12.7 \angle 25.8^0$

5. **C** $|Z_P| = |V_P|/|I_P|$; in a wye load $|V_P| = |V_L|/\sqrt{3}$ and $I_P = I_L$.

Therefore $|Z_Y| = |V_L|/\sqrt{3}/|I_L| = (460/\sqrt{3})/62.8 = 4.23$ ohm.

Also, p.f. = cos θ_P , so θ_P = cos^{-1} (p.f.) = cos^{-1} (.9) = $\pm 25.8^0$ (positive angle for lagging p.f.) so

$Z_Y = |Z_Y| \angle 25.8^0 = 4.23 \angle 25.8^0$

2.2 AC Machines

The two dominant ac machines in use are the synchronous machine and the induction machine. The synchronous machine is used primarily in the generation of electric power, while most applications converting electrical to mechanical power employ induction motors. The primary equations relating shaft speed to applied frequency are given on page 95.

Example 2.3 Induction Motor

A 6-pole induction motor operating under load drawing 9.35 kW from a 460 volt three phase source delivers 10.0 horsepower at a slip of 5.0 percent.

1. What is the synchronous speed of the motor?
   (A) 1,200 rpm
   (B) 1,800 rpm
   (C) 3,258 rpm
   (D) 7,540 rpm

2. What is the shaft speed of the motor?
   (A) 900 rpm
   (B) 1,140 rpm
   (C) 1,200 rpm
   (D) 1,710 rpm

3. What torque is being delivered to the load?
   (A) .56 n·m
   (B) 1.04 N·m
   (C) 9.76 N·m
   (D) 62.5 N·m

4. What is the efficiency of the motor at this load?
   (A) zero
   (B) 50%
   (C) 80%
   (D) 91%

Solutions:

1. **A** $n_s = 120\,f/P = 120(60)/6 = 1{,}200$ rpm
2. **B** $n_r = (1-s)n_s = (1-s)120\,f/P = (1-.05)120(60)/6 = 1140$ rpm
3. **D** $T = P(\text{watts})/\omega(\text{rad/sec})$; 1 h.p. = 746 watts

   Therefore,

   $$T = \frac{10 \times 746}{1140 \times 2\pi/60} = 62.5 \text{ N·m.}$$
4. **C** Efficiency = $P_{out}/P_{in} = (746)(10.0)/9350 = .80 = 80\%$

3. Communications and Signal Processing

by Roland Zapp

The material in the NCEES Reference Handbook used to solve problems in Communications and Signal Processing is found on page 95, with some continuation on page 96. There may also be some reference required to information contained on pages 66 through 69 of the Handbook.

Example 3.1

A linear-time -invariant (LTI) filter with impulse response

$$x(t) = e^{-t} \quad t \geq 0$$
$$= 0 \quad t < 0$$

has an input

$$x(t) = e^{-2t} \quad t \geq 0$$
$$= 0 \quad t < 0$$

1. The output response is given by:

   (A) $y(t) = e^{-2t} \quad t \geq 0$

   (B) $y(t) = e^{-t} \quad t \geq 0$

   (C) $y(t) = 2e^{-t}(1 - e^{-t}) \quad t \geq 0$

   (D) $y(t) = e^{-t}(1 - e^{-t}) \quad t \geq 0$

2. The LTI filter has a system function given by:

   (A) $H(\omega) = \dfrac{1}{1 + j\omega} \quad \omega \geq 0$

   (B) $H(\omega) = \dfrac{1}{1 + \omega^2} - j\dfrac{\omega}{1 + \omega^2} \quad \omega \leq 0$

   (C) $|H(\omega)| = \dfrac{\sqrt{1 + \omega^2}}{1 + \omega^2} \quad -\infty \leq \omega \leq 0$

   (D) None of the above

Solutions:

1. **D** Using $y(t) = \int_0^t x(\tau)h(t-\tau)d\tau$ yields $y(t) = e^{-t}(1-e^{-t})$.

2. **C** Based on the Fourier transform:

$$H(\omega) = \frac{1}{1+j\omega} = \frac{1-j\omega}{1+\omega^2} \qquad -\infty \le \omega \le \infty$$

$$\therefore |H(\omega)| = \frac{\sqrt{1+\omega^2}}{1+\omega^2} \qquad -\infty \le \omega \le \infty$$

Example 3.2

Find the impulse response $h(t)$ of a filter which yields the output to the input:

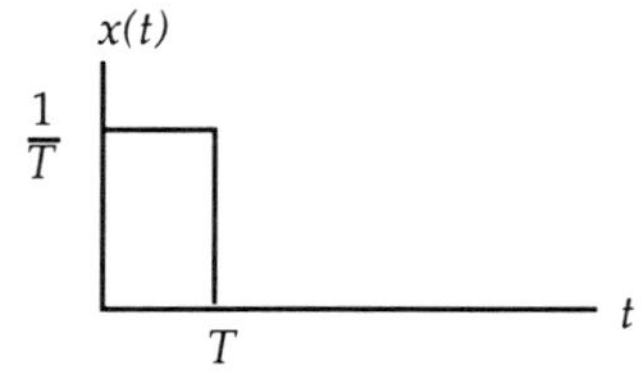

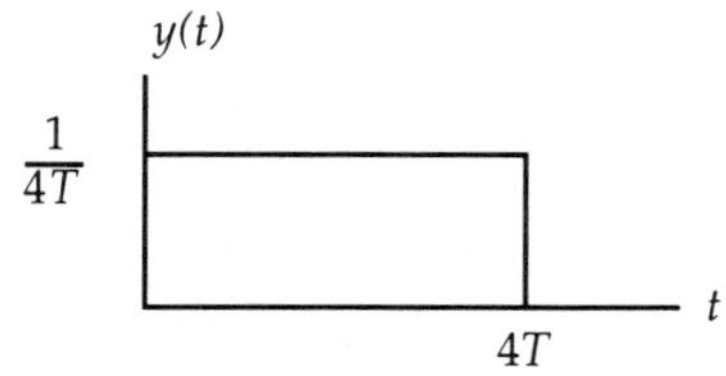

(A) $h(t) = 1 \quad 0 \le t \le 4T$

(B) $h(t) = 4 \quad 0 \le t \le 4T$

(C) $h(t) = \frac{1}{4T^2}t \quad 0 \le t \le T$

$= \frac{1}{4T^2} \quad T \le t \le 4T$

$= \frac{1}{4T^2}(5T - t) \quad 4T \le t \le 5T$

(D) None of the above

Solution:

By convolution, a train of 4 impulses, separated by T and of amplitude 1/4 will give the desired output.

The answer is **D**.

Example 3.3

The periodic impulse train shown is passed through the idealized low-pass filter shown. What is the average output power:

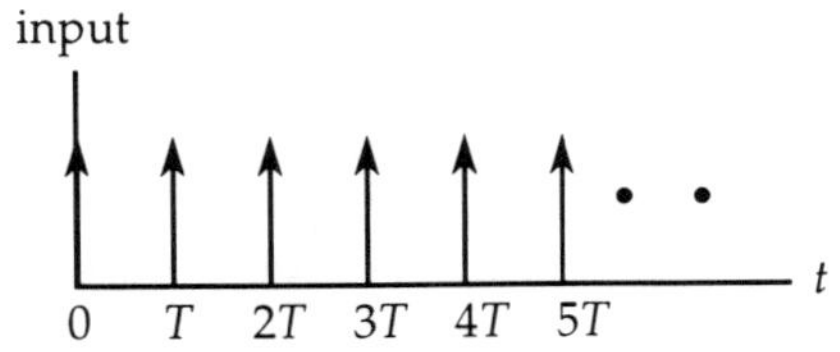

unit area impulses

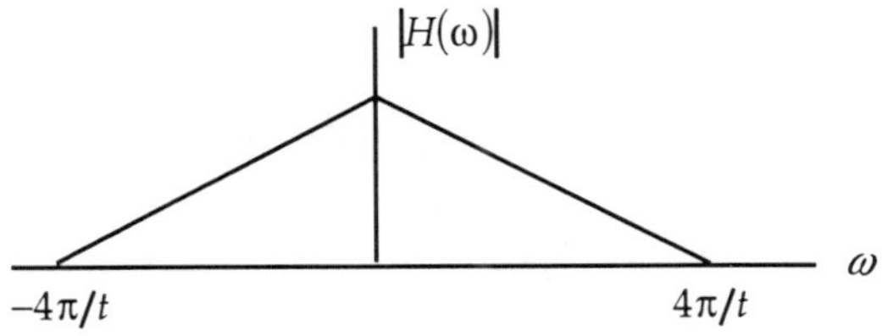

(A) $\langle P\rangle = \dfrac{5}{4T^2}$

(B) $\langle P\rangle = \dfrac{1}{2T^2}$

(C) $\langle P\rangle = \dfrac{3}{2T^2}$

(D) None of the above

Solution:

Input $= \sum_{k=-\infty}^{\infty} c_k e^{jk\omega_0 T}$ where $\omega_0 = \dfrac{2\pi}{T}$ and $c_k = \dfrac{1}{T}$ for all k.

Filter allows only c_0, c_{-1}, c_{+1} or $c_0^2 + 2\left[\dfrac{c_1}{2}\right]^2 = \dfrac{3}{2T^2}$.

The answer is **C**.

Example 3.4

The Fourier series $x(t) = \sum c_n e^{j2\pi f_0 nt}$ for the signal shown is:

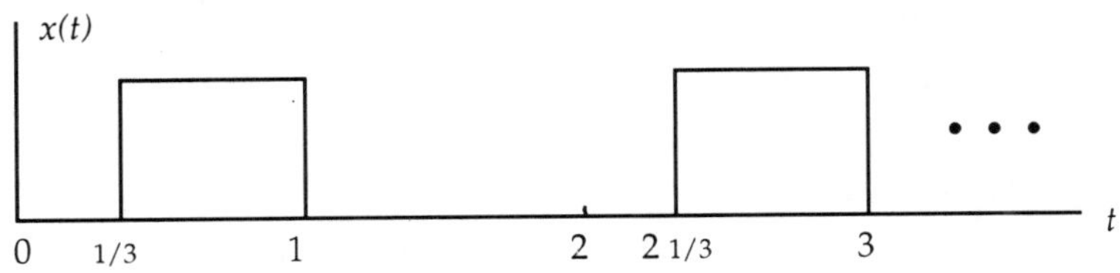

(A) $c_n = \dfrac{2e^{-j2\pi n/3}}{\pi n}\sin\left(\dfrac{\pi n}{3}\right)$

(B) $c_n = \dfrac{2}{3}\operatorname{sinc}\left(\dfrac{\pi n}{3}\right)e^{-2\pi n f_0/3}$

(C) $c_n = \operatorname{sinc}\left(\dfrac{\pi n}{3}\right)e^{-4\pi n f_0/3}$

(D) None of the above

Solution:

Use a Fourier series expansion. See page 16 of the NCEES Handbook.

The answer is **A**.

Example 3.5

The energy contained in an aperiodic signal is given by:

$$E = \int_{-\infty}^{\infty} f^2(t)dt$$

If $f(t) = \dfrac{\sin(t-\tau)}{t-\tau}$ the energy is

(A) infinite

(B) one

(C) π

(D) 2π

Solution:

Using the Parseval power theorem (or energy theorem):

$$\int_{-\infty}^{\infty} f^2(t)dt = \int_{-\infty}^{\infty} |F(f)|^2 df$$

$$= \int_{-1/2\pi}^{1/2\pi} \pi^2 df = \pi$$

The answer is **C**.

Example 3.6

A tone modulating AM signal is given by:

$$A_c[1+\cos\omega_m t]\cos\omega_c t$$

where $\omega_c = 600\pi \times 10^3$ and $\omega_m = 6\pi \times 10^3$.

1. This will have positive frequency spectral components located at:
   (A) 300 KHz and 303 KHz
   (B) 300 KHz and 297 KHz
   (C) 297 KHz and 303 KHz
   (D) None of the above

2. The AM signal is transmitted by USSB-SC (upper single side band suppressed carrier). The expression for this signal is:
   (A) $\frac{A_c}{2}\cos(606 \times 10^3 \pi t)$
   (B) $A_c \cos(6 \times 10^3 \pi t)\cos(600 \times 10^3 \pi t)$
   (C) $A_c \cos(600 \times 10^3 \pi t)$
   (D) None of the above

Solutions:

1. **D** Components will be located at all 3 frequencies: 297, 300, 303 KHz.
2. **A** Only the upper spectral frequency is transmitted.

Example 3.7

Consider the tone-modulated FM signal

$$x(t) = \cos(2\times 10^8 \pi t + \sin 400\pi t)$$

applied to an ideal Band pass filter of bandwidth 500 Hz and centered at 10^8 Hz. The output from this filter has sinusoidal components at frequencies:

(A) 10^8 Hz plus $(10^8 + 200)$ Hz plus $(10^8 + 400)$ Hz

(B) 10^8 Hz plus $(10^8 - 200)$ Hz plus $(10^8 - 400)$ Hz

(C) 10^8 Hz plus $(10^8 - 200)$ Hz plus $(10^8 + 200)$ Hz

(D) None of the above

Solution:

The FM spectrum contains the carrier plus components at every harmonic of 200 Hz. The filter allows only the carrier and components at ±200 Hz to pass. The answer is **C**.

Example 3.8

Given the periodic signal

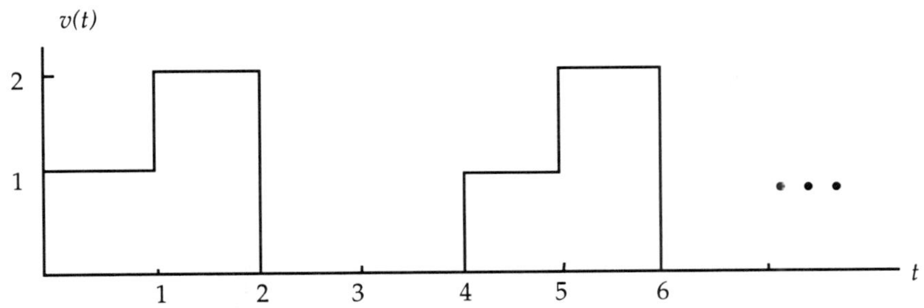

The exponential Fourier series for this signal is:

(A) $v(t) = \frac{1}{4}\sum_n \left(\frac{\sin n\pi f_0}{n\pi f_0}\right) e^{-j\pi n f_0} e^{j2\pi n f_0 t}$

(B) $v(t) = \frac{1}{2}\sum_n \left(\frac{\sin n\pi f_0}{n\pi f_0}\right) e^{-j3\pi n f_0} e^{j2\pi n f_0 t}$

(C) $v(t) = \frac{1}{2}\sum_n \left(\frac{\sin n\pi f_0}{n\pi f_0}\right) e^{-j\pi n f_0} \left(\frac{1}{2} + e^{-2j\pi n f_0}\right) e^{j2\pi n f_0 t}$

(D) None of the above

Solution:

Using $\sum_n c(n) e^{j2\pi n b f_0 t}$ calculate $c(n)$ or use transform techniques. The answer is **C**.

4. Solid State Devices

by Bong Ho

The material and equations needed to work problems involving Solid State devices is found on pages 96–99 of the NCEES Reference Handbook. It will also be necessary to use equations found in the Electric Circuits section on pages 66–70. The several examples that follow illustrate the devices presented in the NCEES Reference Handbook.

Example 4.1 Semiconductor Material

A bar of intrinsic (pure) germanium is 2 cm long and has a cross-section area of 0.01 cm^2. A 10-volt battery is connected across it at room temperature.

Properties of germanium:

$$n_i = 2.4 \times 10^{13}\,\text{cm}^{-3}$$

$$\mu_n = 3900\,\text{cm}^2/\text{V}\cdot\text{sec}$$

$$\mu_p = 1900\,\text{cm}^2/\text{V}\cdot\text{sec}$$

1. What is the conductivity of the semiconductor material?

   (A) $3.5\times10^{-2}(\Omega\cdot\text{cm})^{-1}$

   (B) $2.2\times10^{-2}(\Omega\cdot\text{cm})^{-1}$

   (C) $4.7\times10^{-2}(\Omega\cdot\text{cm})^{-1}$

   (D) $1.1\times10^{-3}(\Omega\cdot\text{cm})^{-1}$

2. What is the current in the circuit?

   (A) 1.1×10^{-3} amp

   (B) 2×10^{-3} amp

   (C) 1.3×10^{-2} amp

   (D) 7.2×10^{-4} amp

3. How long does it take for an electron to drift the length of the bar?
   (A) 0.1 sec
   (B) 0.1 msec
   (C) 0.1 μsec
   (D) 10 μsec

4. How much power is dissipated in the bar?
   (A) 1.1 W
   (B) 2.1 W
   (C) 0.11 W
   (D) 0.011 W

Solutions:

1. **B** For an intrinsic semiconductor: $n_i = n = p$

$$\begin{aligned}\sigma &= q\left(n\mu_n + p\mu_p\right)\\ &= qn_i\left(\mu_n + \mu_p\right)\\ &= 1.6\times10^{-19}\times2.4\times10^{13}(3900+1900)\\ &= 2.2\times10^{-2}(\Omega\cdot\text{cm})^{-1}\end{aligned}$$

2. **A** Electric field intensity:

$$E = \frac{V}{l} = \frac{10}{2} = 5\frac{\text{V}}{\text{cm}}$$

Current density:

$$J = \sigma E = 2.2\times10^{-2}\times5 = 1.1\times10^{-1}\ \text{A/cm}^2$$

Current:

$$I = JA = 1.1\times10^{-1}\times0.01 = 1.1\times10^{-3}\ \text{A}$$

3. **B** Electron velocity:

$$v = \mu_n E = 3900\times5 = 1.95\times10^{4}\ \text{cm / sec}$$

The time is then

$$t = \frac{l}{v} = \frac{2}{1.95\times10^{4}} = 0.1\times10^{-3}\ \text{sec} = 0.1\ \text{msec}$$

4. **D** Power:

$$P = VI = 10\times1.1\times10^{-3} = 0.011\ \text{W}$$

Example 4.2 P-N Diode

A silicon *P-N* diode has a cross-section area of 10^{-3} cm^2. Both *P* and *N* sides are 0.1 cm long. At room temperature, the properties of Si are:

$$n_i = 1.5\times10^{10}\ \text{cm}^{-3},\quad \mu_n = 1.5\times10^{3}\ \text{cm}^2/\text{V}\cdot\text{sec},\quad \mu_p = 6\times10^{2}\ \text{cm}^2/\text{V}\cdot\text{sec}$$

$$\rho_p = 10^{-3}\ \Omega\cdot\text{cm},\quad \rho_N = 10^{-1}\ \Omega\cdot\text{cm}$$

1. What is the majority carrier concentration on the *P* side?
   (A) 1×10^{17} cm^{-3}
   (B) 1×10^{18} cm^{-3}
   (C) 1×10^{19} cm^{-3}
   (D) 1×10^{20} cm^{-3}

2. What is the minority carrier concentration on the *P* side?
   (A) 22 cm^{-3}
   (B) 2.2×10^{3} cm^{-3}
   (C) 2.2×10^{4} cm^{-3}
   (D) 2.2×10^{5} cm^{-3}

3. What is the built-in potential of the diode?
   (A) 0.09 V
   (B) 0.45 V
   (C) 0.7 V
   (D) 0.9 V

4. Which side is at a higher (more positive) potential?
   (A) N side
   (B) P side
   (C) Both sides are at the same potential
   (D) Insufficient data

5. If the width of the junction is 0.1 μm $(1\times10^{-7}\ \text{meter})$, what will be the order of magnitude of the electric field intensity *E* inside the junction?
   (A) 10^3 V/m
   (B) 10^4 V/m
   (C) 10^5 V/m
   (D) 10^6 V/m

6. If the saturation current of the diode is $I_s = 4.13\times10^{-15}$ amp, what will be the diode current if 0.5 V is put across the diode?
   (A) 1×10^{-6} amp
   (B) 1×10^{-7} amp
   (C) 1×10^{-8} amp
   (D) 1×10^{-9} amp

Solutions:

1. **C** $\rho_p = \dfrac{1}{\sigma_p} = \dfrac{1}{p_p q \mu_p}$

$$\therefore p_p = \frac{1}{\rho_p q \mu_p} = \frac{1}{10^{-3} \times 1.6 \times 10^{-19} \times 6 \times 10^{2}} = 1 \times 10^{19}\ \text{cm}^{-3} = N_a$$

2. **A** $n_p = \dfrac{n_i^2}{p_p} = \dfrac{\left(1.5 \times 10^{10}\right)^2}{0.1 \times 10^{20}} = 22\ \text{cm}^{-3}$

3. **D** $V_0 = \dfrac{kT}{q} \ln \dfrac{N_a N_d}{\eta_i^2} = \dfrac{1.38 \times 10^{-23}}{1.6 \times 10^{-19}} \ln \dfrac{4.17 \times 10^{16} \times 1 \times 10^{19}}{2.25 \times 10^{20}} = 0.9\ \text{V}$

   where we used

$$N_d = n_n = \frac{1}{\rho_N q \mu_n} = \frac{1}{10^{-1} \times 1.6 \times 10^{-19} \times 1.5 \times 10^{3}} = 4.17 \times 10^{16}\ \text{cm}^{-3}$$

4. **A** Donor ions are positive on the N side.

5. **D** $E = \dfrac{V_0}{d} = \dfrac{0.9}{0.1 \times 10^{-6}} = 9 \times 10^{6}\ \text{V/m}$

6. **A** For S_i, $n = 1$:

$$i_D = I_s \left[e^{\frac{V_D}{nV_T}} - 1 \right]$$

$$= 4.13 \times 10^{-15} \left[e^{\frac{0.5}{0.026}} - 1 \right] = 1 \times 10^{-6}\ \text{amp}$$

Example 4.3 NPN Bipolar Junction Transistor BJT

A common-emitter BJT amplifier circuit is given below.

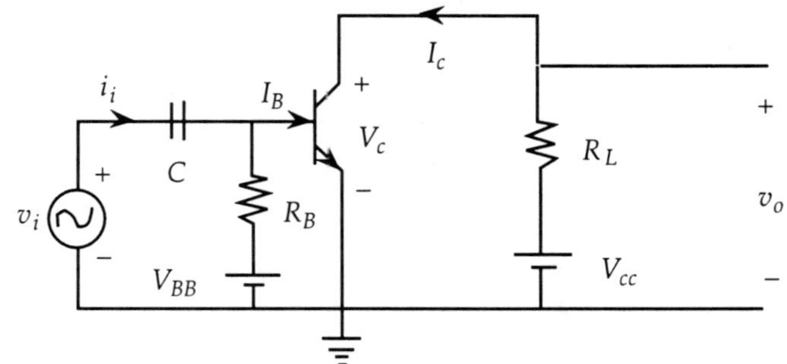

$V_{CC} = 12$ V
$V_{BB} = 3$ V
$R_B = 3.75$ kΩ
$R_L = 150$ Ω
$\beta = 90$
$r_\pi = 1$ kΩ
C = very large

1. Use the DC equivalent circuit to find the base current I_B.
   (A) 0.3 mA
   (B) 0.8 mA
   (C) 0.4 mA
   (D) 0.61 mA

2. Use the DC equivalent circuit to find the collector voltage V_C.
   (A) 3.76 V
   (B) 6 V
   (C) 4.21 V
   (D) 1.52 V

3. With an input signal $v_i = 0.1\sin\omega t$ volt, use the low-frequency small signal equivalent circuit to find the voltage gain $A_V = \dfrac{V_o}{V_i}$.
   (A) –6.6
   (B) 15.2
   (C) –13.5
   (D) –37.8

4. Find the current gain $A_I = \dfrac{i_c}{i_i}$ for Problem 3.
   (A) 71
   (B) 52
   (C) –83
   (D) –31

Solutions:

1. **D** $I_B = \dfrac{V_{BB} - 0.7}{R_B} = \dfrac{3 - 0.7}{3.75 \times 10^3} = 0.61 \text{ mA}$

2. **A** $I_C = \beta I_B = 90 \times 0.61 \times 10^{-3} = 54.9 \text{ mA}$

   $V_C = V_{CC} - I_C R_L = 12 - 54.9 \times 10^{-3} \times 150 = 12 - 8.235 = 3.76 \text{ V}$

3. **C** The AC equivalent circuit is

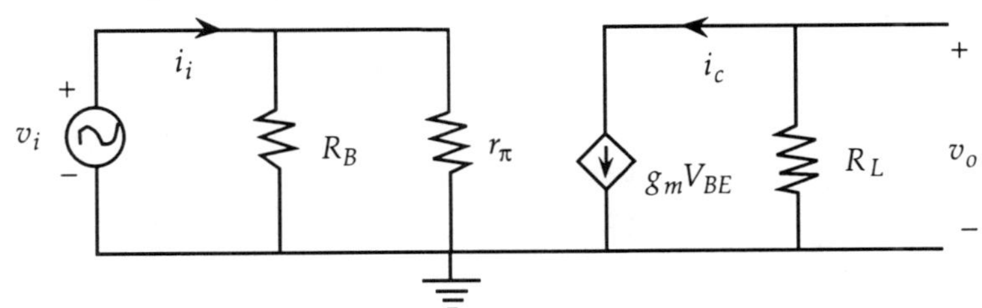

$$g_m = \frac{\beta}{r_\pi} = \frac{90}{10^3} = 9 \times 10^{-2} \text{ A/V}$$

$$V_{BE} = v_i$$

$$v_o = -g_m V_{BE} R_L = -9 \times 10^{-2} \times 0.1 \times 150 = -1.35 \sin\omega t \text{ volt}$$

Voltage Gain: $A_V = \dfrac{v_o}{v_i} = \dfrac{-1.35 \sin\omega t}{0.1 \sin\omega t} = -13.5$

4. **A** Refer to the AC equivalent above:

$$i_i = \frac{v_i}{R_B r_\pi/(R_B + r_\pi)} = \frac{0.1 \sin \omega t}{0.79 \times 10^3} = 0.126 \sin \omega t \text{ mA}$$

$$i_c = g_m V_{BE} = g_m v_i = 9 \times 10^{-2} \times 0.1 \sin \omega t = 9 \sin \omega t \text{ mA}$$

Current Gain: $$A_I = \frac{i_c}{i_i} = \frac{9 \times 10^{-3} \sin \omega t}{0.126 \times 10^{-3} \sin \omega t} = 71.4$$

Example 4.4 N-Channel Field-Effect Transistor FET

Given the FET amplifier circuit below with

$I_{Dss} = 10$ mA, $V_P = -4$ V, $g_m = 2.5 \times 10^{-3}$ A/V and $r_d = 100$ kΩ

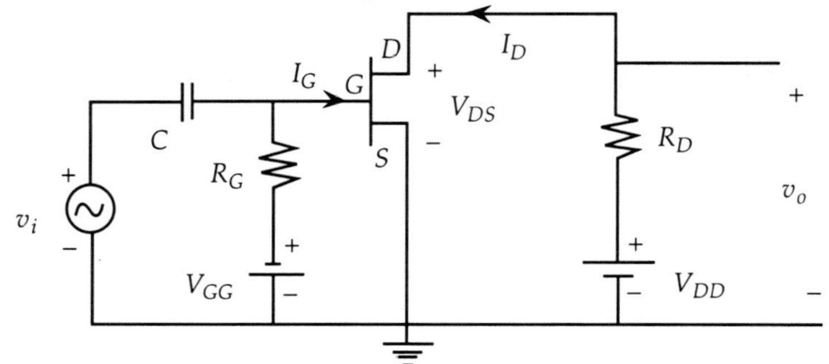

$V_{DD} = 10$ V
$V_{GG} = -2$ V
$R_G = 100$ kΩ
$R_D = 1$ kΩ
C = very large

The FET is operating in the saturation region.

1. What is the gate current I_G?
   (A) –0.02 mA
   (B) 0.02 mA
   (C) 0.013 mA
   (D) 0.0 mA

2. What is the drain voltage V_{DS}?
   (A) 7.5 V
   (B) 5 V
   (C) 3.5 V
   (D) 2.5 V

3. What is the voltage gain of the amplifier? (Using the AC equivalent circuit given.)
   (A) 5
   (B) –2.5
   (C) 2.5
   (D) –5

Solutions:

1. **D** Gate is open circuited by either reverse biasing the gate P-N junction in the case of JFET or by silicon dioxide layer in the MOSFET.

2. **A** Using the drain current expression in the saturation region,

$$i_D = I_{DSS}\left(1-\frac{v_{GS}}{V_p}\right)^2$$

$$= 10\times10^{-3}\left(1-\frac{-2}{-4}\right)^2 = 2.5\times10^{-3} \text{ amp}$$

Then,

$$v_{DS} = V_{DD} - I_D R_D$$

$$= 10 - 2.5\times10^{-3}\times10^3 = 7.5 \text{ V}$$

3. **B** The AC equivalent circuit is

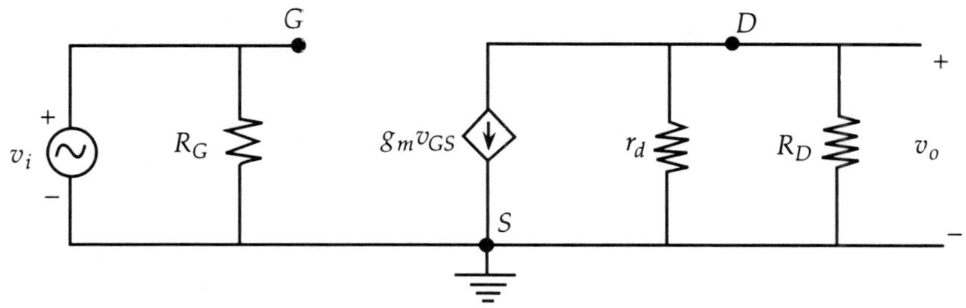

$$g_m = \frac{2\sqrt{I_{DSS}I_D}}{|V_p|} = \frac{2\left[10\times10^{-3}\times2.5\times10^{-3}\right]^{\frac{1}{2}}}{4} = 2.5\times10^{-3} \text{ A/V}$$

$$v_o \approx -g_m v_{GS} R_D = -g_m R_D v_i \qquad (v_{GS} = v_i)$$

Voltage Gain: $A_v = \frac{v_o}{v_i} = -g_m R_D = -2.5\times10^{-3}\times10^3 = -2.5$

5. Computer Engineering

by R. Lal Tummala

Many of the questions relating to Computer Engineering can be answered using the equations and material found on page 99 of the NCEES Handbook. So make sure you are familiar with that material. This material is reviewed in the following sections. Additional discussion is provided whenever it is appropriate to complement the material in the Handbook.

5.1 Number Systems and Codes

This section deals with converting numbers from one system to another.

Example 5.1

What is the decimal equivalent of octal number 555 ?

(A) 556

(B) 458

(C) 365

(D) 326

Solution:

The base for the octal number is 8. Decimal equivalent of octal number 555 = $5 \times 8^2 + 5 \times 8^1 + 5 \times 8^0 = 320 + 40 + 5 = 365$. The answer is **C**.

Example 5.2

What is the decimal equivalent of the hexadecimal number, *FE*?

(A) 290

(B) 255

(C) 128

(D) 254

Solution:

The base for the hexadecimal number is 16. The decimal equivalent of $FE = F \times 16^1 + E \times 16^0 = 15 \times 16 + 14 = 240 + 14 = 254$. The answer is **D**.

Example 5.3

What is the decimal equivalent of the following binary coded decimal number:

0001 1001 0110 0111 ?

(A) 1965
(B) 4058
(C) 1967
(D) 850

Solution:

Divide the number into groups of four starting from the right.

0001	1001	0110	0111
1	9	6	7

The answer is C.

Example 5.4

The hexadecimal equivalent of the 16 bit binary number 1110 1011 1000 0011 is:

(A) *EFE3*
(B) *EEFF*
(C) *EB85*
(D) *EB83*

Solution:

Group the binary digits into groups of four from right and replace each group by the equivalent hexadecimal digit. The answer is **D.**

Decimal Equivalent

Decimal equivalent of a number is obtained by multiplying with the positive powers of the base for the digits to the left of the radix point and negative powers of the base to the radix point.

Example 5.5

What is the decimal equivalent of binary number : 10011.101 ?

(A) 13.4
(B) 13.5
(C) 19.625
(D) 19.5

Solution:

The base for binary numbers is 2

$(1 \times 2^4 + 0 \times 2^3 + 0 \times 2^2 + 1 \times 2^1 + 1 \times 2^0) . (1 \times 2^{-1} + 0 \times 2^{-2} + 1 \times 2^{-3}) = 19.625$

The answer is C.

Conversion of Decimal Numbers

Conversion of decimal numbers to binary, octal or hexadecimal numbers is accomplished by "repeated division" by the base. The process is stopped when the quotient becomes zero. This is illustrated by the following example.

Example 5.6

Convert a decimal number 29 to a binary number.

(A) 1 1 1 1 0

(B) 1 0 1 1 1

(C) 1 1 0 1 1

(D) 1 1 1 0 1

Solution:

29/2	=	14 + 1	Remainder	=	1 ↑
14/2	=	7 + 0	Remainder	=	0
7/2	=	3 + 1	Remainder	=	1
3/2	=	1 + 1	Remainder	=	1
1/2	=	0 + 1	Remainder	=	1

The process stops because the quotient is zero. The remainder obtained in each step is the desired digit for the binary number in the reverse order shown by the arrow.

The corresponding binary number for decimal 29 is: 11101. The answer is **D**.

Alternate Method of Conversion

An alternate method for large numbers is to convert a decimal number into hexadecimal (octal) and replace each hexadecimal (octal) number by the corresponding binary number.

Example 5.7

Convert a decimal number 29 to binary number by first converting to an octal number.

Solution:

29/8	=	3 + 5	Remainder	=	5
3/8	=	0 + 3	Remainder	=	3

The octal number : 35 and the binary number: 011101. Since the left most zero has no meaning, it is usual practice to leave out the leftmost zeros. The answer = 11101.

Example 5.8

The binary number for a decimal number 982 is :

(A) 11101111

(B) 11101011

(C) 1112111

(D) 1111010110

Solution:

982/16 = 61 + 6	Remainder = 6
61/16 = 3 + 13	Remainder = 13 = D in hexadecimal number system
3/16 = 0 + 3	Remainder = 3

The hexadecimal number = 3D6

Replace each hexadecimal digit by four bit binary equivalent = 0011 1101 0110. Leave out the two leftmost zeros and the answer is 1111010110. The answer is **D**.

5.2 Binary Addition and Subtraction

The binary addition and subtraction is performed exactly like the decimal arithmetic. It is carried from right to left. The difference is that in binary arithmetic, we only have 1 and 0 and so the carry occurs whenever the sum in each column exceeds 2. Binary subtraction can be done similarly using the concept of *borrowing*, familiar in decimal arithmetic. In practice, however, subtraction is accomplished using 2's complement arithmetic. Negative numbers are represented as 2's complement of the corresponding positive number. In the computer the 2's complement of a binary number is obtained by complementing (replacing 1 by 0 and vice versa) and adding 1 to the result. So, to obtain $A - B$, we write it as : $A+(-B)$ where $-B$ represents the 2's complement of B.

Example 5.9

Given: $A = 00111$, $B = 00101$. Find $A + B$.

Solution:

A	0	0	1	1	1	
B	0	0	1	0	1	
			1	1		← Carry digit
$A+B=$	0	1	1	0	0	

Example 5.10

Two binary numbers, A = 00111, B = 00101 are given.

1. Find $A + B$
   (A) 10000
   (B) 10101
   (C) 01100
   (D) 00212

2. Find the 2's complement of B ?
   (A) 11000
   (B) 11011
   (C) 00112
   (D) 00111

3. Find $A - B$.
   (A) 00010
   (B) 11001
   (C) 00110
   (D) 00202

Solution:

1. **C**. (See Example 5.9)

2. **B**. Given B = 00101
   Complement each bit = 11010
   Add 1 to the result + 1

   2's complement of B = 11011 = $-B$

3. **A**. The subtraction is done by adding : $A + (-B)$ and ignoring the carry bit.

A		0	0	1	1	1
$-B$		1	1	0	1	1
$A + (-B)$	1	0	0	0	1	0

↑ Carry bit

5.3 Logic Operations and Boolean Algebra

Boolean logic deals with zeros (0) and ones (1). Three basic logic operations, "AND (.)" , "OR (+)" and "Exclusive-OR (⊕)" along with their symbols are given in the Handbook. Boolean algebra deals with the algebra of binary variables, i.e., variables that can take only two values: "0" (False) and "1" (True).

Logic circuits are divided into two classes. *Combinational Circuits* and *Sequential Circuits*. A combinational circuit consists of combinations of logic gates like AND, OR, Exclusive-OR, etc., whose output depends only on the present inputs. On the other hand, sequential circuits consist of flip-flops and logic gates, and their output depend not only on the present inputs but also on the previous outputs.

Designing of Combinational logic circuits starts with the mapping of the problem into a *truth table*. A truth table defines the output for all possible combinations of the input variables. The output is either "1" , "0" or don't cares, denoted by "X". Don't cares represent the outputs due to input combinations that cannot occur. This information from the truth table is used to define Boolean equations. The structure of the Boolean equation is of the form:

$$X = F(A, B, C ..)$$

where Boolean variable X represents the output and Boolean variables *A, B, C ..* represent the inputs to the logic circuit. For example, the following equation:

$$X = AB + AC + CDE$$

implies that X is *true* if A and B are true, OR A and C are true OR, C and D and E are true.

Boolean algebra is used to simplify these functions. Finally, logic gates are used to implement the simplified function. Some important theorems useful in simplification are given below.

In the following theorems, "+" represents an OR operation and "." represents an AND operation.

(1 = TRUE, 0 = FALSE.)

1. $A + 1 = 1,$
2. $A + 0 = A$
3. $A \,.\, 1 = A,$
4. $A \,.\, 0 = 0$
5. $A + A = A,$
6. $A \,.\, A = A$
7. $A + \overline{A} = 1,$
8. $A \,.\, \overline{A} = 0$
9. $A(B + C) = AB + AC$
10. $A + AB = A$
11. $A(A + B) = A$
12. DeMorgan's theorems: $\overline{AB} = \overline{A} + \overline{B}$; $\overline{A + B} = \overline{A}\,\overline{B}$

The above concepts are illustrated with the following example.

Example 5.10

Given the following truth table,

a) Write the logic function

c) Simplify the function

d) Implement the function using logic gates.

Truth Table

A	B	C	F
0	0	0	1
0	0	1	1
0	1	0	0
0	1	1	1
1	0	0	1
1	0	1	0
1	1	0	0
1	1	1	0

Solution:

1 **2** **3** **4**

a) $F = \overline{A}\,\overline{B}\,\overline{C} + \overline{A}\,\overline{B}C + \overline{A}BC + A\overline{B}\,\overline{C}$

b) Combining terms **1** and **4** & **2** and **3**, we get

$$F = (A + \overline{A})\overline{B}\,\overline{C} + \overline{A}C(B + \overline{B})$$

Using Theorem 7 above, we have :

$$F = \overline{B}\,\overline{C} + \overline{A}C$$

c) There are many ways to implement this function. Two different implementations are shown below.

i) AND-OR-NOT implementation:

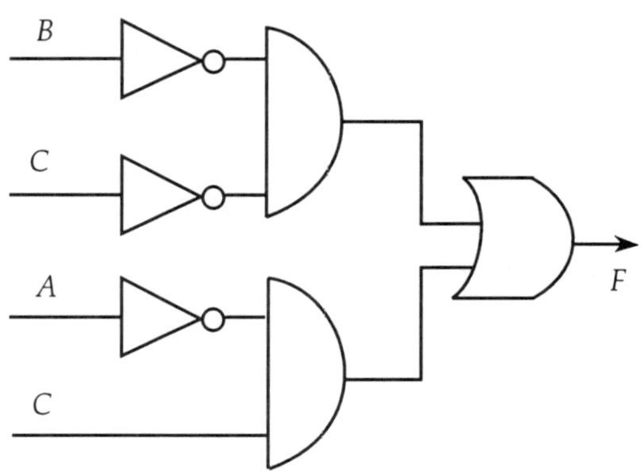

ii) NAND gate implementation:

$$F = \overline{B}\overline{C} + \overline{A}C$$

Applying DeMorgan's theorem:

$$\overline{\overline{F}} = \overline{\overline{\overline{B}\,\overline{C} + \overline{A}C}} = \overline{\overline{\overline{B}\,\overline{C}}.\overline{\overline{A}C}}$$

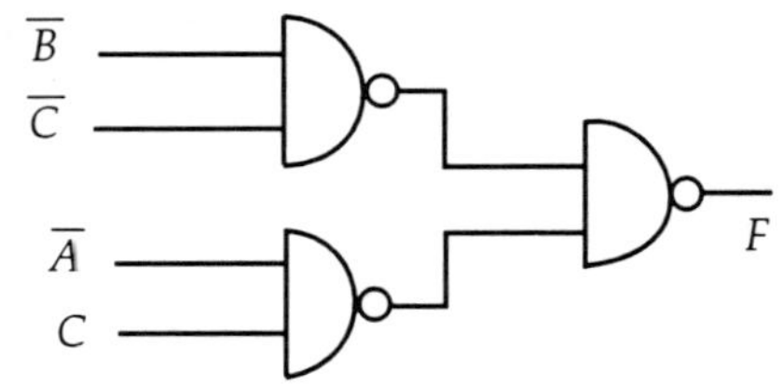

Example 5.12

Given the following digital circuit:

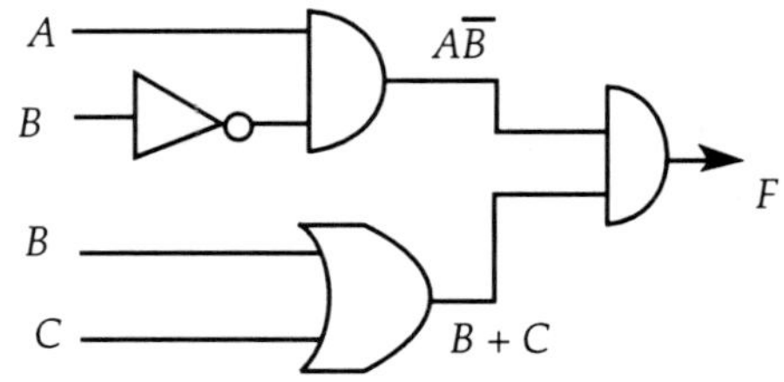

1. The equation for the function F after simplification is:
   (A) $A\ \overline{B}\ C$
   (B) $A\ B\ C$
   (C) $A\ \overline{B}\ +\ \overline{A}\ B$
   (D) $B\ \overline{C}$

2. The equation for $\overline{F}$ is :
   (A) $A\ \overline{B}\ C$
   (B) $\overline{A}\ + B +\ \overline{C}$
   (C) $A\ B +\ \overline{C}$
   (D) $A + B +\ \overline{C}$

Solutions:

1. **A.** $F = A\overline{B}\ (B + C) = A\,\overline{B}\,B + A\,\overline{B}\,C\ = 0 + A\,\overline{B}\,C\ = A\,\overline{B}\,C$
2. **B.** Using DeMorgan's theorem:

$$\overline{F} = \overline{A\overline{B}C} = \overline{A} + B + \overline{C}$$

Example 5.13

The truth table of a digital logic circuit is given below.

A	B	F
0	0	1
0	1	1
1	0	1
1	1	0

The expression for the simplified logic function F is :

(A) $F = A + \overline{B}$

(B) $F = \overline{A}\,\overline{B}$

(C) $F = A + B$

(D) $F = \overline{AB}$

Solution:

$$F = \overline{A}\,\overline{B} + \overline{A}B + A\overline{B}$$
$$= \overline{A}\,\overline{B} + \overline{A}\,\overline{B} + \overline{A}B + A\overline{B} = \overline{B}(A+\overline{A}) + \overline{A}(B+\overline{B}) = \overline{A} + \overline{B} = \overline{AB}$$

The answer is **D**.

5.4 Karnaugh Maps

A Karnaugh map (K-map) is another method used to simplify logic functions. Even though it can be used for any number of variables, it loses its usefulness beyond four variables. The K-maps for 2, 3 and 4 variables are shown below.

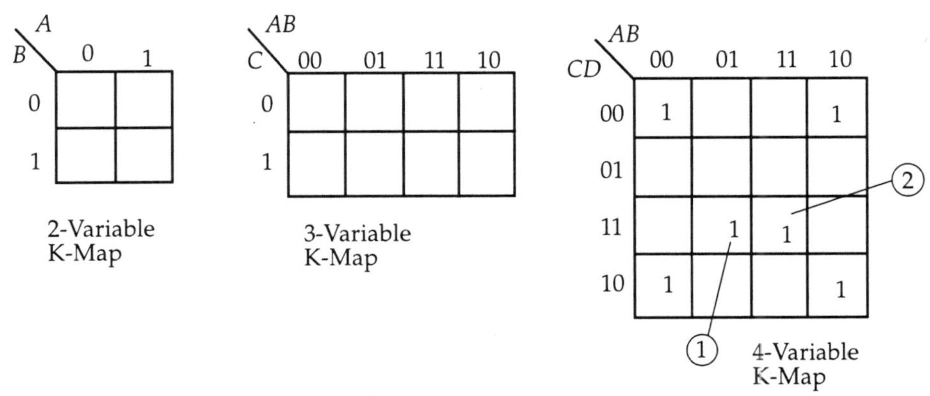

2-Variable K-Map

3-Variable K-Map

4-Variable K-Map

Each square of the map is identified by a certain combination of the variables. For example, the square marked "②" represents $ABCD$. Notice, by design, adjacent squares vertically or horizontally differ by only one variable change. These adjacent squares can be combined to eliminate variables. For example, in the 4-variable map, **1** and **2** can be combined to eliminate the variable A, resulting in a term BCD. Similarly four terms represented by "1"s in the corners of the 4-variable map can be combined to produce a term, $\overline{B}\,\overline{D}$. Entries in each square represent the output values corresponding to each row in the truth table. A map is created for each output variable. The plotted function in the K- map is simplified by encircling the largest adjacent squares of 1's. Six simple rules are given below. These rules are applied to each square containing the 1's in the K - map.

1. All encirclements should contain 2^N 1' s , where N is an integer.

2. Identify the adjacent squares containing "1"s that can be combined with a single other box in only one way. Encircle such two square combinations. Skip any square that can be combined more than one way.

3. Identify the squares that can be combined with three other squares in only one way. If all four squares so involved are not already covered in groupings of two, encircle these four squares. Skip any square that can be combined in a group of four in more than one way.

4. Repeat the procedure for groups of eight, etc.

5. After the above procedure, if there still remain some uncovered squares, they can be combined with each other or with other squares already used. If no square can be combined with the others, they appear in the final result without any simplification.

6. Don't cares may be used to simplify functions in the K-map, but are not included in the final logic equation.

Example 5.14

The following function is obtained from the truth table.

$$F = AB\overline{C} + A\overline{B}\,\overline{C} + \overline{A}\,\overline{B}\,\overline{C}$$
$$\quad\;\; 110 \qquad 100 \qquad 000$$

Plot the function on 3 variable K -map and simplify the function.

Solution:

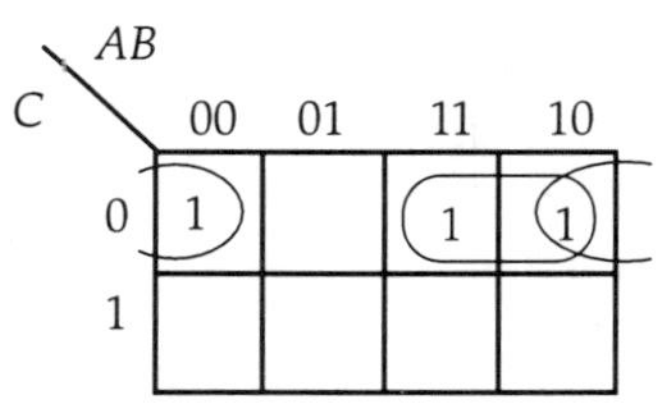

The simplified function is:

$$F = A\overline{C} + \overline{B}\,\overline{C}$$

Example 5.15

Simplify the following function given on 4- variable K-map.

The simplified function is given by:

(A) $A\overline{D}B + C\overline{D}$

(B) $\overline{A}D + \overline{C}D$

(C) $AD + \overline{C}D$

(D) $A\overline{B}D + C\overline{D}$

CD \ AB	00	01	11	10
00			X	
01	1	1	X	1
11	1	1		
10		X		

Solution:

The answer is **B.**

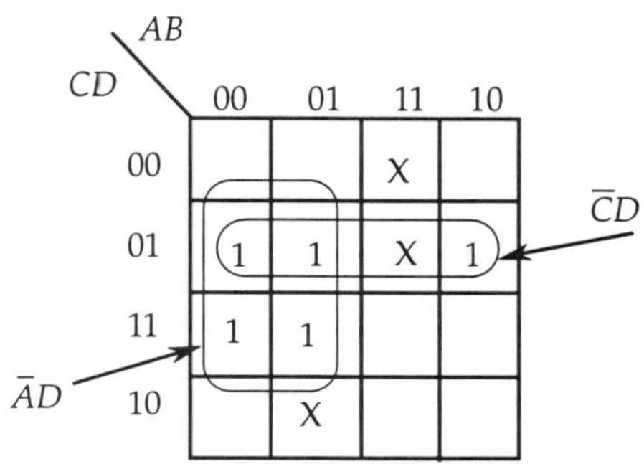

5.5 Flip-Flops

As mentioned earlier, *sequential circuits remember the past. The output of a sequential circuit depends not only on the present input but also on the previous outputs (caused by the past inputs).* Flip-flops along with the logic gates are used for this purpose. Three basic flip-flops are described in the NCEES handbook. They are *RS* flip-flop, *JK* flip-flop and *D* flip-flop. The flip-flop output is synchronized with a clock (CLK) signal. Q_n represents the output before the clock pulse is applied. Q_{n+1} represents the output of the flip-flop after the clock signal is applied. The output Q_{n+1} depends on the signal present on the input terminals (e.g., *J* and *K* for *JK* flip-flop) and the previous output, Q_n.. The truth tables are given in the NCEES Handbook on page 99.

The Composite Flip-Flop transition tables given in the NCEES Handbook on page 99 are used to determine the corresponding inputs required on the flip-flop inputs to accomplish the change in the output. Since this change also depends on the previous output, it has to be taken into consideration in determining the inputs. For example, if we want to change the output of *JK* flip flop from $Q_n = 1$ to $Q_{n+1} = 0$, then we should set $K = 1$ and $J = x$, where "*x*" represents don't care, i.e., *J* could be either "0" or "1" before the clock signal is applied.

Example 5.16

An Exclusive-OR logic gate is added to a *JK* flip-flop as shown below.

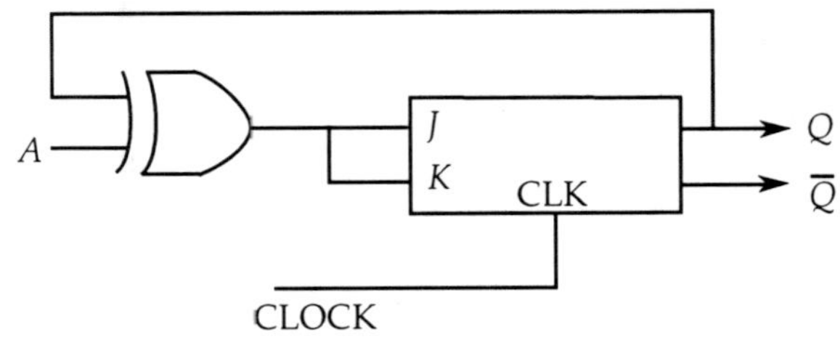

The new circuit represents a :

(A) *JK* Flip Flop
(B) *D* FF
(C) *SR* FF
(D) None of the above

Solution:

A	Q_n	Q_{n+1}
0	0	0
0	1	0
1	0	1
1	1	1

Type *D* flip-flop. The answer is **B**.

Example 5.17

A sequential circuit is shown below. In the circuit, Q^1 is the most significant bit and Q^0 is the least significant bit, i.e., state 10 means $Q^1 = 1$, $Q^0 = 0$.

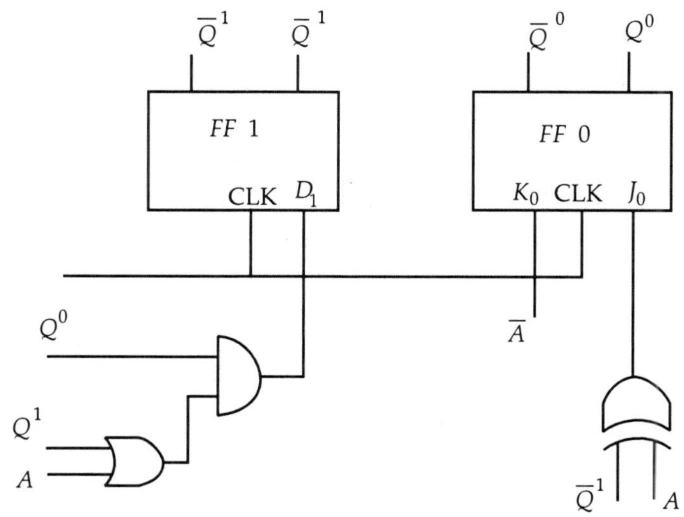

1. If the sequential circuit is in 00 state, what is the next state of the circuit if $A = 0$?
   (A) 01
   (B) 00
   (C) 10
   (D) 11

2. If the circuit's initial state is 01, what is the next state if $A = 1$?
   (A) 00
   (B) 01
   (C) 11
   (D) 01

3. If the sequential circuit is in 00 state, what input should we apply to the JK terminals and D terminals of the flip-flops to change it to 01?
   (A) $J = K = 0, D = 0$
   (B) $J = 1, K = X, D = 0$
   (C) $J = 1, K = 1, D = 1$
   (D) None of the above

4. If the sequential circuit is in 00 state, what should be the value of A to change it to 01?
   (A) $A = 0$
   (B) $A = 1$
   (C) $A = 01$
   (D) $A =$ don't care

5. The output of the circuit is defined by the following logic circuit. What is the output equation?

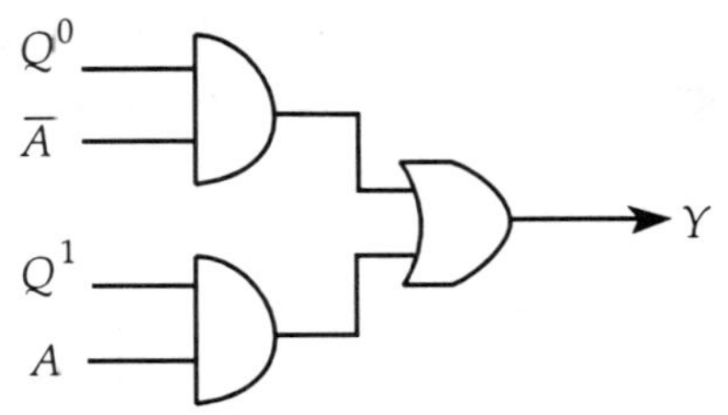

(A) $Y = Q^1 A$
(B) $Y = Q^0 \overline{A} + Q^1 A$
(C) $Y = Q^0 A + B$
(D) $Y = Q^1 + Q^0$

Solution:

The answers are easily obtained by writing the following logic equations from the circuit.

$$J_0 = A \oplus \overline{Q}^1, K_0 = \overline{A}$$
$$D_1 = (A + Q^1) \cdot Q^0$$

1. **A** $J_0 = 1, \; K_0 = 1, \; D_1 = 0 \; \rightarrow \; Q^0_{n+1} = 1, \; Q^1_{n+1} = 0$
2. **C** $J_0 = K_0 = 0, \; D_1 = 1$
3. **B**
4. **A**
5. **B**

Electrical Engineering Practice Test

This practice exam has been developed to help test-takers prepare for the afternoon Discipline-Specific test in Electrical Engineering. In addition, if test-takers are undecided, taking this test may also help them determine whether they should take the afternoon DS test or the afternoon General test.

The subjects tested are those subjects that the NCEES states will make up the exam. We have placed the same number of problems from each area of Electrical Engineering that can be expected to be on the actual exam. The difficulty level can, of course, be quite different on the actual exam; we have, however, attempted to create an exam with essentially the same level of difficulty as one would expect on the actual exam.

If you find the problems on this exam to be more difficult than the problems on the afternoon General test, you may wish to take the General test in the afternoon. About one half of all Electrical Engineering examinees take the afternoon General test. This practice exam should help you make that decision before the day of the test.

cut here!

FUNDAMENTALS OF ENGINEERING EXAM

Afternoon Session—Electrical Exam

(Simulated answer form with topical breakout and scoring grid.)

BE SURE EACH MARK IS DARK AND COMPLETELY FILLS THE INTENDED SPACE AS ILLUSTRATED HERE: ●.

POWER SYSTEMS

1 (A) (B) (C) (D)
2 (A) (B) (C) (D)
3 (A) (B) (C) (D)

Score: ________

SIGNAL PROCESSING

16 (A) (B) (C) (D)
17 (A) (B) (C) (D)
18 (A) (B) (C) (D)

Score: ________

ANALOG CIRCUITS

25 (A) (B) (C) (D)
26 (A) (B) (C) (D)
27 (A) (B) (C) (D)
28 (A) (B) (C) (D)
29 (A) (B) (C) (D)
30 (A) (B) (C) (D)

Score: ________

ELECTROMAGNETICS

34 (A) (B) (C) (D)
35 (A) (B) (C) (D)
36 (A) (B) (C) (D)
37 (A) (B) (C) (D)
38 (A) (B) (C) (D)
39 (A) (B) (C) (D)

Score: ________

NETWORKS

46 (A) (B) (C) (D)
47 (A) (B) (C) (D)
48 (A) (B) (C) (D)
49 (A) (B) (C) (D)
50 (A) (B) (C) (D)
51 (A) (B) (C) (D)

Score: ________

DIGITAL SYSTEMS

4 (A) (B) (C) (D)
5 (A) (B) (C) (D)
6 (A) (B) (C) (D)
7 (A) (B) (C) (D)
8 (A) (B) (C) (D)
9 (A) (B) (C) (D)

Score: ________

SOLID STATE

19 (A) (B) (C) (D)
20 (A) (B) (C) (D)
21 (A) (B) (C) (D)
22 (A) (B) (C) (D)
23 (A) (B) (C) (D)
24 (A) (B) (C) (D)

Score: ________

INSTRUMENTATION

31 (A) (B) (C) (D)
32 (A) (B) (C) (D)
33 (A) (B) (C) (D)

Score: ________

COMMUNICATIONS

40 (A) (B) (C) (D)
41 (A) (B) (C) (D)
42 (A) (B) (C) (D)
43 (A) (B) (C) (D)
44 (A) (B) (C) (D)
45 (A) (B) (C) (D)

Score: ________

COMPUTERS

52 (A) (B) (C) (D)
53 (A) (B) (C) (D)
54 (A) (B) (C) (D)
55 (A) (B) (C) (D)
56 (A) (B) (C) (D)
57 (A) (B) (C) (D)
58 (A) (B) (C) (D)
59 (A) (B) (C) (D)
60 (A) (B) (C) (D)

Score: ________

CONTROLS

10 (A) (B) (C) (D)
11 (A) (B) (C) (D)
12 (A) (B) (C) (D)
13 (A) (B) (C) (D)
14 (A) (B) (C) (D)
15 (A) (B) (C) (D)

Score: ________

Electrical Discipline Exam

Questions 1 – 3

Three single-phase, 7200 volt to 480 volt, 100 kVA transformers are connected in a step-down bank with the 7200 windings connected in wye and the 480 volt windings connected in delta.

1. What is the rated line-to-line voltage on the wye side?
   (A) 4,160 V
   (B) 7,200 V
   (C) 12,470 V
   (D) 14,400 V

2. What is the rated line current on the wye side?
   (A) 4.63 A
   (B) 13.9 A
   (C) 24.0 A
   (D) 41.6 A

3. What is the rated line current on the delta side?
   (A) 208 A
   (B) 361 A
   (C) 480 A
   (D) 625 A

Questions 4 – 6

A logic circuit is shown:

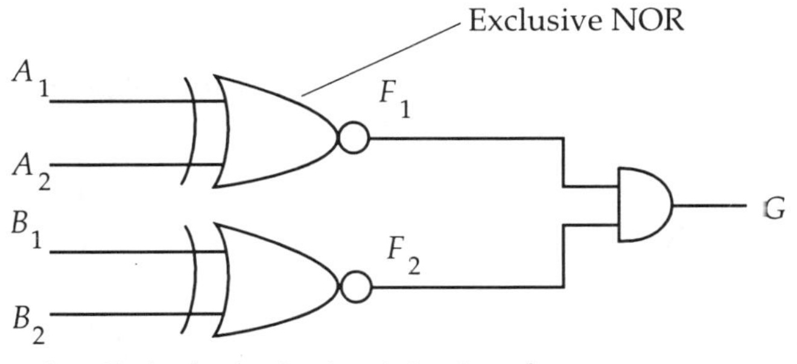

4. The function F_1 in the logic circuit is given by:
   (A) A_1A_2
   (B) $A_1A_2 + \overline{A}_1\overline{A}_2$
   (C) $A_1 + A_2$
   (D) $A_1 + A_1A_2$

5. If $A_1 = 1$ and $B_1 = 1$, then $G = 1$ if:
   (A) $A_2 = 1$ and $B_2 = 1$
   (B) $A_2 = 0$ and $B_2 = 1$
   (C) $A_2 = 0$ and $B_2 = 0$
   (D) $A_2 = 1$ and $B_2 = 0$

6. If $B_1 = 1$ and $B_2 = 1$, then F_2 will be equal to
(A) 0
(B) 1
(C) not defined
(D) F_1

Questions 7 – 9

Flip Flops A and B form a sequential circuit as shown:

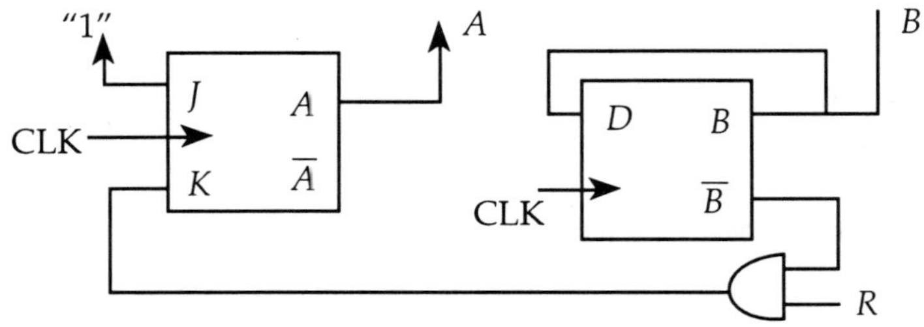

7. If $R = 1$, after the clock pulse, the binary count 10 changes to:
(A) 00
(B) 01
(C) 10
(D) 11

8. If $R = 0$, after the clock pulse the binary count 00 changes to:
(A) 00
(B) 01
(C) 10
(D) 11

9. If $R = 1$, after the clock pulse the binary count 01 changes to:
(A) 00
(B) 01
(C) 10
(D) 11

Questions 10 – 12

A servo system for a pen plotter is given by the following block diagram:

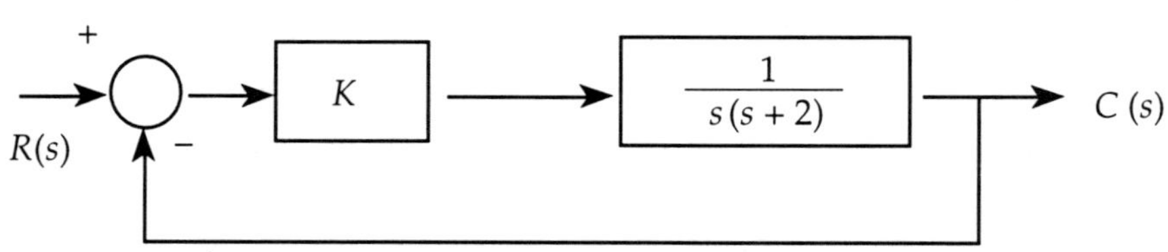

10. The value of K required to result in a closed-loop system characteristic equation of $P(s) = s^2 + 2s + 100$ is given by:
(A) 0
(B) 50
(C) 100
(D) 25

11. The value of K required to get the fastest response without any overshoot ($\zeta = 1$) is given by:
(A) 1
(B) 10
(C) 5
(D) 100

12. If K is set to 16, the damped natural frequency of oscillation to a step input is most nearly:
(A) 4 rad/s
(B) 16 rad/s
(C) 12 rad/s
(D) 10 rad/s

Questions 13 – 15

Given the following closed-loop system:

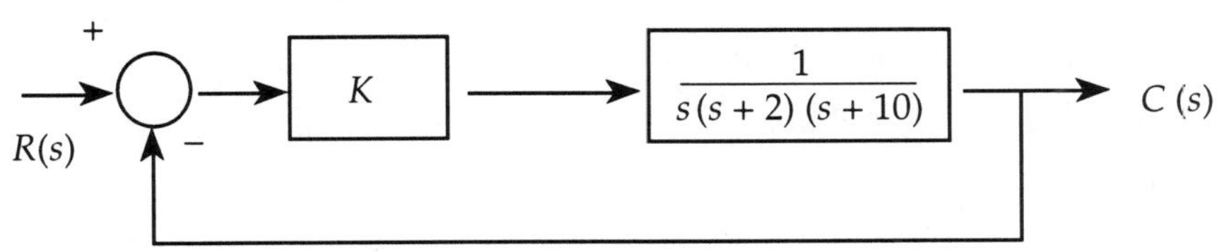

13. The value of K for which the system becomes marginally stable is nearly:
(A) 10
(B) 240
(C) 100
(D) 60

14. The frequency of oscillation when K is set to make the system marginally stable is :
(A) 0.7 Hz
(B) 10 Hz
(C) 1.2 Hz
(D) 2 Hz

15. If $K = 0$, the steady-state error to a step input of magnitude 10 is:
(A) 0
(B) 10
(C) 2
(D) 200

Questions 16 – 18

Two discrete-time signal processing systems are depicted below. Each system consists of a decimator and a linear time-invariant filter in series. The input to both systems is the discrete-time signal $x[n]$. The discrete-time impulse response of the linear time-invariant filter is denoted by $h[n]$. The decimator discards every odd sample from its input to produce its output; for example, the input $x[n]$ to the decimator in System I produces the output $x[2n]$.

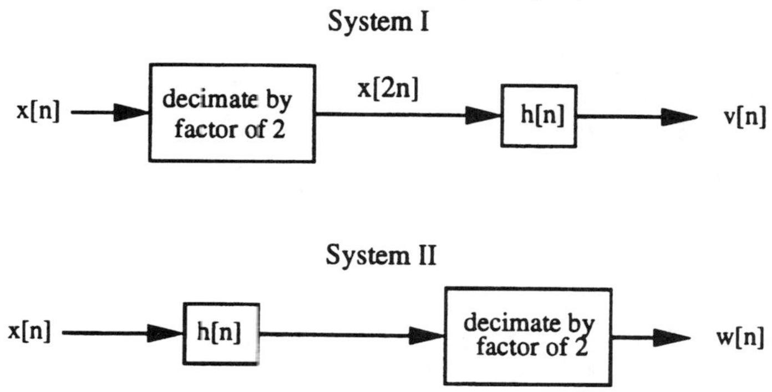

16. In general, with no special assumptions on $x[n]$ or $h[n]$,

    (A) $v[n]=w[n]$, and both systems are time-invariant.

    (B) $v[n]\neq w[n]$, and both systems are time-invariant.

    (C) $v[n]=w[n]$, and both systems are time-varying.

    (D) $v[n]\neq w[n]$, and both systems are time-varying.

17. If $h[n]=\delta[n]+0.5\delta[n-1]$, where $\delta[n]$ is a unit-sample (discrete-time impulse) function, and $x[n]=\cos \pi n$, then the output $v[n]$ of System I is:

    (A) $v[n]=0$, a constant signal

    (B) $v[n]=1.5$, a constant signal

    (C) $v[n]=-0.5\cos \pi n$

    (D) $v[n]=0.5^n u[n]$, where $u[n]$ is the unit-step function

18. If $h[n]=\delta[n]+\delta[n-1]$, where $\delta[n]$ is a unit-sample (discrete-time impulse) function, and $x[n]=\cos \pi n$, then the output $w[n]$ of System II is:

    (A) $w[n]=-0.5\cos \pi n$, a constant signal

    (B) $w[n]=1.5$, a constant signal

    (C) $w[n]=-0.5\cos \pi n$

    (D) $w[n]=0.5^n u[n]$, where $u[n]$ is the unit-step function

19. For a silicon *pn*-diode at room temperature, what should the applied diode voltage V_D be in order to cause the diode current i_D to reach 90% of the diode saturation current I_s?

    (A) 0.7 V

    (B) 0.5 V

    (C) 0.16 V

    (D) 0.016 V

20. Determine the base current I_B of the following BJT given $V_{BE} = 0.7$ V and $\beta = 100$.

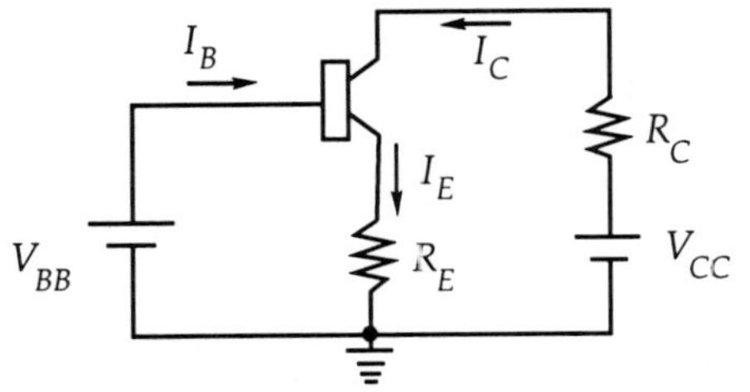

$V_{CC} = 10$ V
$V_{BB} = 5$ V

$R_C = 2$ kΩ
$R_E = 1$ kΩ

(A) 0.0426 mA
(B) 0.426 mA
(C) 0.746 mA
(D) 7.46 mA

21. The following JFET is operating in the saturation region. Find the gate-to-drain voltage V_{GD}.

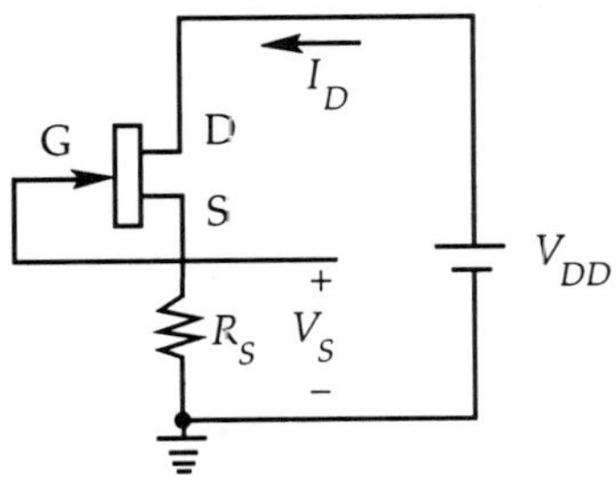

$V_{DD} = 5$ V
$R_S = 1$ kΩ

$I_{DSS} = 2$ mA
$V_p = -2$ V

(A) -2 V
(B) 2 V
(C) 0 V
(D) -3 V

22. Find the power dissipated in the JFET.

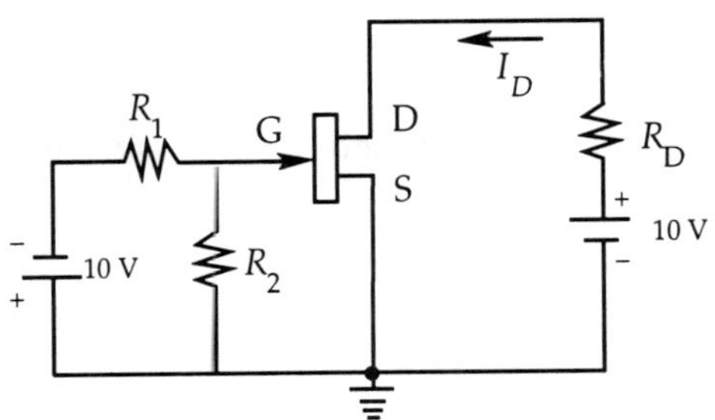

$R_1 = 90$ kΩ
$R_2 = 10$ kΩ
$R_D = 1$ kΩ
$I_{DSS} = 18$ mA
$V_p = -3$ V

(A) 0.016 W
(B) 0.16 W
(C) 1.6 W
(D) 16 W

Questions 23 – 24

Given the following BJT amplifier circuit.

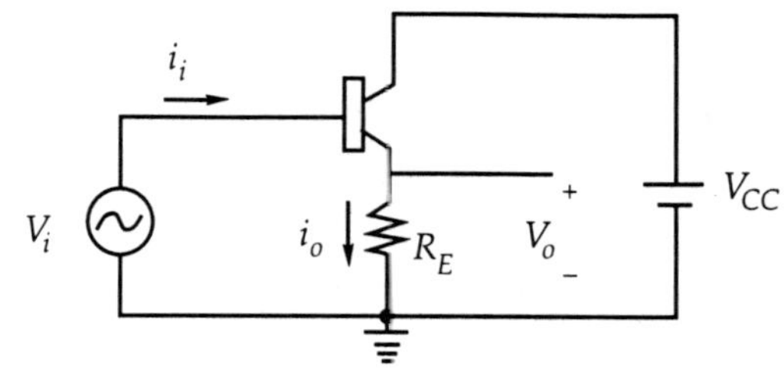

$V_{CC} = 12$ V
$R_E = 1$ kΩ
$\beta = 100$
$r_\pi = 2$ kΩ

23. Find the small-signal current gain $A_i = i_o / i_i$.
    (A) 101
    (B) 100
    (C) 99
    (D) 50

24. Find the small-signal voltage gain $A_V = V_o / V_i$.
    (A) 0.5
    (B) 0.98
    (C) 9.8
    (D) 18.8

Questions 25 – 28

For the cascade Op Amp configuration shown, respond to the following questions.

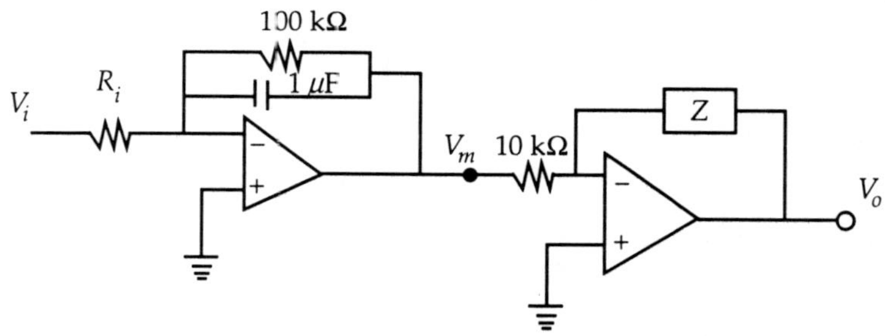

25. Determine the R_i needed to obtain an active filter with maximum amplitude gain of $V_m / V_i = 25$.
    (A) 25 kΩ
    (B) 10 Ω
    (C) 2.5 kΩ
    (D) none of the above

26. The filter bandwidth for V_m / V_i:
    (A) cannot be determined without R_i
    (B) is less than 2 Hz
    (C) is greater than 10 Hz but less than 100 Hz
    (D) is greater than 100 Hz

27. If a second-order low pass filter (2 pole filter) is desired, we can choose Z as:

(A) a series R–C impedance with time constant, $\tau = 10^{-1}$

(B) a parallel R–C impedance with time constant, $\tau = 10^{-1}$

(C) a parallel R–L impedance with time constant, $\tau = 10^{-1}$

(D) a series R–L impedance with time constant, $\tau = 10^{-1}$

28. If Z is a 1.0 henry inductor in parallel with a 10.0 Ω resistor and R_i is chosen as 1.0 kΩ, the configuration resembles:
(A) a low-pass filter with maximum amplitude gain of 100
(B) a high-pass filter with maximum amplitude gain of 1.0
(C) a band-pass filter with maximum amplitude gain of 0.05
(D) a high-pass filter with maximum amplitude gain of 100

Questions 29 – 30

Given the configuration shown.

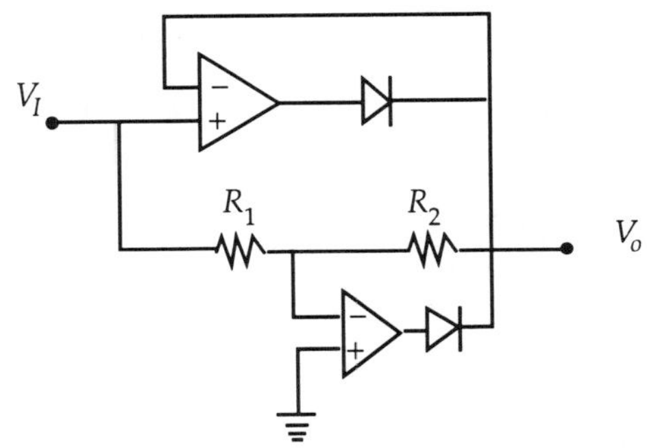

29. The configuration shown most nearly represents:
(A) a full-wave rectifier
(B) a half-wave rectifier
(C) a voltage follower
(D) a differential amplifier

30. If the resistor R_1 were to burn out (open circuit), the configuration would appear as:
(A) a full-wave rectifier
(B) a half-wave rectifier
(C) a voltage follower
(D) a non-inverting amplifier

Questions 31 - 33

An *npn* BJT transistor configuration is used as an instrumentation amplifier. The bias conditions are as shown. Assume $V_{BE} = 0.7$ V. This amplifier is to be used to linearly amplify signals in the range $V_i = 2.7 \pm 1.0\text{V}$ with $R_b = 2\ \text{k}\Omega$ and $R_C = 100\ \Omega$.

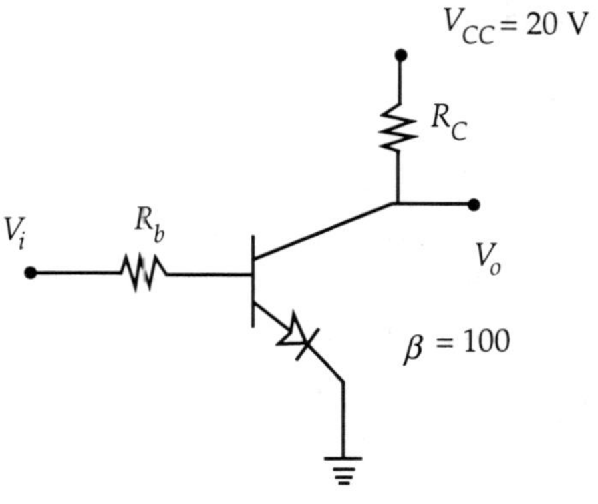

31. The value of V_o for $V_i = 1.7$ V is:
    (A) 0.5 V
    (B) 5.0 V
    (C) 10.0 V
    (D) 15.0 V

32. The value of V_o for $V_i = 3.7$ V is:
    (A) 0.5 V
    (B) 5.0 V
    (C) 10.0 V
    (D) 15.0 V

33. The value of i_e for $V_i = 2.7$ V is:
    (A) 99 mA
    (B) 100 mA
    (C) 101 mA
    (D) 102 mA

Questions 34 – 36

The field for a dipole is given by

$$\mathbf{E} = \frac{2p\cos\theta}{4\pi\varepsilon_o r^3}\mathbf{a}_r + \frac{p\sin\theta}{4\pi\varepsilon_o r^3}\mathbf{a}_\theta$$

34. A Gaussian surface of radius r_o is taken at the origin in order to evaluate the integral $\int_A \mathbf{E}\cdot d\mathbf{A}$ for the dipole described above. The result of this surface integral is:
    (A) too complex to evaluate
    (B) $|p|$ where $|p| = Qd$, the limit as $d \to 0$
    (C) zero
    (D) $Qd/2$

35. The potential on the surface of the sphere is:
(A) zero
(B) constant $\neq 0$
(C) indeterminant
(D) $V_0 \cos\theta$ where V_0 is a constant

36. Evaluation of $\nabla \times \mathbf{E}$ and $\nabla \cdot \mathbf{E}$ would yield:
(A) zero everywhere
(B) zero except at $r = 0$
(C) zero for the curl, and $|p| = Qd/\varepsilon_0$ for the divergence
(D) zero for the divergence, and $|p| = Qd/\varepsilon_0$ for the curl

Questions 37 – 39

Two uniform spherical surface charge layers of equal radii and equal but opposite surface charge densities of value $\pm\rho_0$, respectively, are touching in free space, as shown. Assume the potential is referenced to zero at infinity.

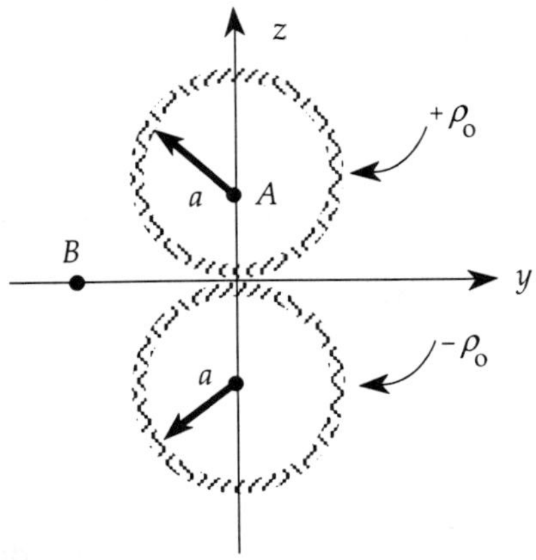

37. Find the exact potential at the point $A(0, 0, a)$.
(A) $V = 0$
(B) $V = -\dfrac{\rho_0 a}{2\varepsilon_0}$
(C) $V = \dfrac{\rho_0 a}{\varepsilon_0}$
(D) $V = \dfrac{\rho_0 a}{2\varepsilon_0}$

38. Find the exact potential at the point $B(0, -a, 0)$.
(A) $V = 0$
(B) $V = -\dfrac{\rho_0 a}{2\varepsilon_0}$
(C) $V = \dfrac{\rho_0 a}{\varepsilon_0}$
(D) $V = \dfrac{\rho_0 a}{2\varepsilon_0}$

39. Find the magnitude of the electric field at $B(0, -a, 0)$.

(A) $E = 0$

(B) $E = \dfrac{\sqrt{2}\rho_0}{2\varepsilon_0}$

(C) $E = \dfrac{\rho_0}{2\varepsilon_0}$

(D) $E = \dfrac{\rho_0}{\varepsilon_0}$

Questions 40 – 42

Given an AM system with modulation index $\mu = 1$ (or 100% modulation) and modulation voltage

$$v(t) = K[2\cos\omega_m t + 4\cos 2\omega_m t + 2\cos 3\omega_m t]$$

Assume $\omega_c = 2\pi f_c = 10\omega_m$.

40. The value of K for 100% modulation (without distortion) is:

(A) 1.0
(B) 0.5
(C) 0.25
(D) 0.125

41. The highest frequency carried by this AM signal is:

(A) 10 ω_m
(B) 12 ω_m
(C) 13 ω_m
(D) 14 ω_m

42. The approximate average power carried by the sidebands is:

(A) 0.5 W
(B) 0.1 W
(C) 0.6 W
(D) 1.0 W

Questions 43 – 45

Given the tone modulating signal

$$x(t) = \cos 6\pi \times 10^3 t$$

and the carrier

$$v(t) = A_c \cos 600\pi \times 10^3 t$$

with $\mu = 1$ (the modulation index).

43. The maximum frequency in this AM signal is:

(A) 300 kHz
(B) 600 kHz
(C) 2 kHz
(D) none of these

44. The power dissipated across a resistor R due to the modulated signal is:

(A) $3A_c^2/4R$

(B) $A_c^2/2R$

(C) A_c^2/R

(D) none of these

45. The signal resulting from USSB-SC (upper single sideband suppressed carrier) is:

(A) $0.5A_c \cos 606\pi \times 10^3 t$

(B) $A_c \cos 600\pi \times 10^3 t$

(C) $A_c \cos 6\pi \times 10^3 t$

(D) none of these

Questions 46 – 49

A switch is closed at $t = 0$ in the circuit shown.

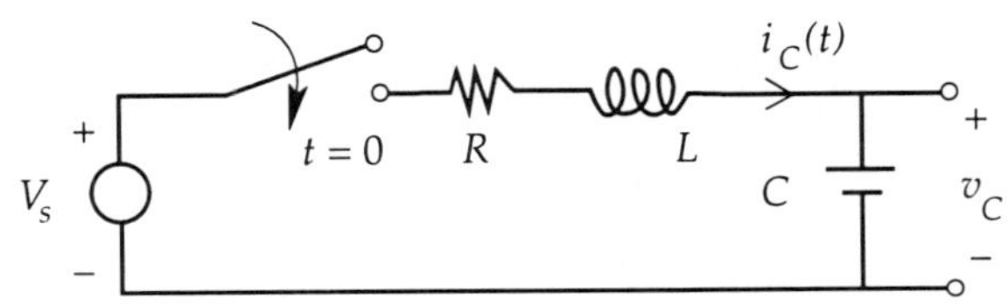

46. Assuming that the capacitor was initially uncharged, the current that flows at $t = 0$ after the switch is closed is:

(A) V_s/R

(B) V_s/L

(C) zero

(D) none of the above

47. The rate of change of current at $t = 0$ after the switch is closed, i.e., $i_C'(0^+)$ is:

(A) V_s/R

(B) V_s/L

(C) zero

(D) none of the above

48. The condition for an overdamped response is:

(A) $\left(\frac{R}{2L}\right)^2 = \frac{1}{LC}$

(B) $\left(\frac{R}{2L}\right)^2 > \frac{1}{LC}$

(C) $\left(\frac{R}{2L}\right)^2 < \frac{1}{\sqrt{LC}}$

(D) none of the above

49. The solution $i_c(t) = Ke^{-at}\cos(bt + \phi)$ where K, a, b, ϕ are constants:
(A) cannot possibly exist
(B) exists only for overdamped conditions
(C) would require that $\phi = \pi / 2$
(D) is the solution for the given circuit

Questions 50 – 51

A certain system has two measurements performed on a set of terminals. The first provides a voltage of 20 V across a 2 kΩ resistor placed across the terminals. The second provides 25 V across a 5 kΩ resistor placed across the terminals.

50. This system has the characteristic:
(A) no Thevenin equivalent is possible
(B) the system is nonlinear
(C) more information is required to make a decision
(D) a Norton equivalent is possible

51. A 4 kΩ resistor is placed across the terminals. Assuming the system is linear, this would give:
(A) a power dissipation of 144 mW in the resistor
(B) a Thevenin equivalent for the system of 30 V in series with 1 kΩ
(C) a Norton equivalent is possible
(D) all of the above

Questions 52 – 54

A simple byte oriented microcontroller has a 16-bit stack pointer (SP) register that points to the next empty memory location in the stack. A PUSH instruction loads an operand into the stack and then decrements the SP; a PULL instruction increments the SP and then unloads an operand from the stack.

52. Based on the above description, which of the following statements about the stack is true?
(A) The stack will fill up towards either lower memory or higher memory addresses depending on the number of items stored.
(B) The stack will always fill up towards higher memory addresses.
(C) The stack will always fill up towards lower memory addresses.
(D) This stack structure will not work since the stack pointer is sometimes incremented and sometimes decremented.

53. If the SP is set initially at memory location 0000_{16}, what will it contain after the serial execution of ten successive PUSH instructions?
(A) 0010_{16}
(B) $000A_{16}$
(C) $FF10_{16}$
(D) $FFF6_{16}$

54. Which of the following is NOT a common use for a memory stack of the type described above?
   (A) The stack can be used to store program counter return links for nested subprograms.
   (B) The stack can be used to store parameters passed among subprograms.
   (C) The stack can be used to store and retrieve randomly accessed database records.
   (D) The stack can be used to sequentially store synchronously received input data.

Questions 55 – 57

A special flip-flop, specified below, has been designed for a clocked sequential logic circuit.

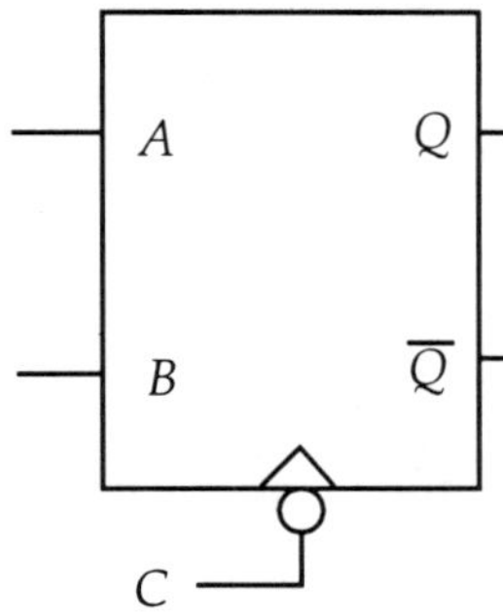

Circuit Icon

A	B	Q_{n-1}
0	0	$\overline{Q}_n$
0	1	1
1	0	0
1	1	Q_n

Truth Table

55. Which of the following is a true statement about when the flip-flop output state can change?
   (A) State changes are 1-level sensitive.
   (B) State changes are 0-level sensitive.
   (C) State changes are rising-edge sensitive.
   (D) State changes are falling-edge sensitive.

56. The specified flip-flop has the following characteristic equation:
   (A) $Q_{n+1} = \overline{A}\,\overline{Q}_n + BQ_n$
   (B) $Q_{n+1} = A\overline{Q}_n + \overline{B}Q_n$
   (C) $Q_{n+1} = AB + \overline{Q}_n$
   (D) Cannot be determined from the information given.

57. Determine the excitation table for the flip-flop.

(A)

Q_n	$\rightarrow Q_{n+1}$	A	B
0	0	X	1
0	1	X	0
1	0	0	X
1	1	1	X

(B)

Q_n	$\rightarrow Q_{n+1}$	A	B
0	0	1	X
0	1	0	X
1	0	X	0
1	1	X	1

(C)

Q_n	$\rightarrow Q_{n+1}$	A	B
0	0	1	1
0	1	0	0
1	0	1	0
1	1	0	1

(D) Cannot be determined from the information given.

58. A spreadsheet has cell $B2$ set to 4. Then cell $B3$ is set to the formula $\$B\$2 + B2$. This formula is copied into cells $B4$, $B5$, $B6$, and $B7$ in order. The number in $B6$ is:

(A) 20
(B) 24
(C) 32
(D) 16

59. The pseudo-code segment is given the following list of integer values as data: 17, 15, 12, 18, 19, 25. After execution, what values will be displayed?

```
a := 100; b := 0; c := 0;
for i := 1 to 5
  read( x );
  if( x a)
    a := x;
  if( x b)
    b := x;
  c := c + x;
endfor
print( a, b, c/5 );
```

(A) 12, 25, 21
(B) 12, 17, 16
(C) 12, 19, 16
(D) 12, 25, 21

60. A file contains the integers 3, 2, -4, 5, 1, -2, ... which are processed by the pseudo-code program:

```
a := 0;
i := 0;
while( i 3)
  read(b);
 if( b 0 )
   a = a + b*b;
 i := i + 1;
endwhile;
print( a/i);
```

What will be the approximate value printed?

(A) 11

(B) 11.25

(C) 7.8

(D) 9.5

Answers to the Electrical Exam

Power Systems

1. C
2. B
3. B

Digital Systems

4. B
5. A
6. B
7. A
8. C
9. D

Controls.

10. C
11. A
12. A
13. B
14. A
15. A

Signal Processing

16. D
17. B
18. A

Solid State

19. D
20. A
21. D
22. A
23. A
24. B

Analog Circuits

25. D
26. B
27. B
28. C
29. A
30. B

Instrumentation

31. D
32. B
33. C

Electromagnetics

34. C
35. D
36. B
37. D
38. A
39. B

Communications

40. D
41. D
42. B
43. D
44. A
45. A

Networks

46. C
47. B
48. B
49. C
50. C
51. D

Computers

52. C
53. D
54. C
55. D
56. A
57. B
58. A
59. C
60. D

tear out answer sheet

Solutions to Electrical Test

1. **C** For a wye connection,

$$V_{line} = \sqrt{3} \times V_{phase}$$

which is

$$V_{line} = 1.732 \times 7,200 = 12,470 \text{ V}$$

2. **B** We know in a three phase circuit,

$$\text{Total rated } VA = \sqrt{3} \times V_{line} \times I_{line}$$

from which

$$I_{line} = \frac{VA}{\sqrt{3} \times V_{line}}$$

and also that the Total rated $VA = 3 \times$ rated VA of each transformer:

$$\text{Total rated } VA = 3 \times 100,000 = 300,000$$

from which

$$I_{line} = 300,000 / (\sqrt{3} \times 12,470) = 13.9 \text{ A}$$

3. **B** This is identical to Problem 2, but we must realize that on the delta side,

$$V_{line} = V_{phase}, \qquad \text{or} \quad V_{line} = 480$$

Then

$$I_{line} = 300,000 / (\sqrt{3} \times 480) = 361 \text{ A}$$

4. **B** F_1 represents the output of the Exclusive-Nor gate (Exclusive-OR followed by a NOT gate). We can obtain the answer by complementing the output in the truth table for Exclusive-OR gate. The function from the truth table is

$$F_1 = A_1A_2 + \overline{A}_1\overline{A}_2$$

5. **A** If $A_1 = 1$, then $F_1 = A_2$ and if $B_1 = 1$, then $F_2 = B_2$. Therefore, $G = F_1F_2 = A_2B_2$.

6. **B**

7. **A** The inputs are $J = 1$, $K = 1$ and $D = 0$. Therefore, the binary count changes to 00.

8. **C**

9. **D**

10. **C** The closed-loop characteristic equation is given by

$$1+\frac{K}{s(s+2)}=0 \quad \text{or} \quad s(s+2)+K=0$$

Therefore, K should be set to 100.

11. **A** $2\zeta\omega_n = 2$ and $\omega_n^2 = K$. If $\zeta = 1$ then $K = 1$.

12. **A** With K = 16 the characteristic equation is $s^2 + 2s + 16 = 0$. Therefore, $\omega_n = 4$ and $\zeta = 0.25$. The damped frequency of oscillation, which is $\omega_n\sqrt{1-\zeta^2}$, is close to 4 rad/s.

13. **B** The characteristic equation is given by

$$s(s+2)(s+10)+K=0 \quad \text{or} \quad s^3+12s^2+20s+K=0$$

Apply the Routh-Horwitz criterion:

s^3	1	20
s^2	12	K
s	$(240-K)/12$	0
s^0	K	

For the system to be stable, there should be no sign changes in the first column of the above table. This is possible if the value of $0 < K < 240$. The system is marginally stable if K = 240.

14. **A** If K = 240, the frequency of oscillation is obtained by solving the auxiliary equation

$$12s^2 + 240 = 0$$

which gives the frequency of oscillation of $\sqrt{20}$ rad/s or 0.71 Hz.

15. **A** Since this is a type 1 system, the steady-state error to a step input is zero and independent of gain K..

16. **D** The decimator is a time-varying operation. Therefore, the output produced by decimating the linear time-invariant (LTI) filtering will be different from the output obtained by LTI filtering followed by decimation, and the two systems are not equivalent. To see this, check the condition for time-invariance. If we shift the input to the decimator, is the output simply shifted by the same amount? In general, the answer is no. However, if the even and odd samples of the signal are identical, then shifting the input does produce a shifted output.

17. **B** The signal $x[n] = \cos(\pi n) = (-1)^n$, a sequence of alternating 1's and -1's. Hence $x[2n] = 1$ for all n, and the output of the decimator is a constant signal. The LTI filter s impulse response is $h[n] = \Delta[n] - 0.5\Delta[n-1]$, so the output of the filter in response to the constant $x[2n] = 1$ signal is also constant with value 1+0.5 = 1.5, so $v[n] = 1.5$ for all n.

18. **A** The convolution between the signal $x[n] = \cos(\pi n) = (-1)^n$ and filter $h[n] = \Delta[n] + \Delta[n-1]$ is zero for all time. This is easy to verify directly by computing the convolution sum. Another way to see this is to note that the z-transform of $h[n]$ has a zero at -1, and therefore this filter annihilates the frequency component at digital frequency π. Decimating the constant zero signal produces zero output, so the output of the overall system is $w[n] = 0$.

19. **D** From the diode equation

$$i_D = I_s[e^{V_D/nV_T} - 1]$$

For Si, $n = 1$ and the thermal voltage $V_T = kT/q = 0.025$ V. When i_D reaches 90% of I_s,

$$0.9I_s = I_s[e^{V_D/0.025} - 1]. \quad \therefore e^{V_D/0.025} = 1.9 \text{ and } V_D = 0.01605 \text{ V}$$

20. **A** KLV for the base-emitter loop:

$$V_{BB} = V_{BE} + I_E R_E$$
$$5 = 0.7 + I_E \times 10^3. \quad \therefore I_E = 4.3 \text{ mA}$$

But

$$I_E = I_C + I_B$$
$$= \beta I_B + I_B = (\beta + 1) I_B. \quad \therefore I_B = \frac{I_E}{1+\beta} = \frac{4.3}{101} = 0.0426 \text{ mA}$$

21. **D** Since the gate and the source are tied together, $V_{GS} = 0$:

$$I_D = I_{DSS}\left(1 - \frac{V_{GS}}{V_p}\right)^2 = I_{DSS} = 2 \text{ mA}$$

$$V_S = I_D R_S = 2 \times 10^{-3} \times 10^3 = 2 \text{ V}$$
$$V_{DS} = V_{DD} - V_S = 5 - 2 = 3 \text{ V}$$
$$V_{GD} = V_{SD} = -V_{DS} = -3 \text{ V}$$

22. **A** We know that

$$I_D = \frac{I_{DSS}}{V_P^2}(V_{GS} - V_p)^2$$

By the voltage divider action:

$$V_{GS} = \frac{R_2}{R_1 + R_2}(-10) = \frac{10 \times 10^3}{100 \times 10^3}(-10) = -1.0 \text{ V}$$

Then

$$I_D = \frac{18}{9}(-1 + 3)^2 = 8 \text{ mA}$$

The voltage across the drain and source is

$$V_{SD} = 10 - I_D R_D = 10 - 8 \text{ mA} \times 1 \text{ k}\Omega = 10 - 8 = 2 \text{ V}$$

The dissipation in the JFET is

$$P_{FET} = V_{DS}I_D = 2 \times 8 \times 10^{-3} = 16 \times 10^{-3}\ \text{W}$$

23. **A** The output current is

$$\begin{aligned} i_o &= i_i + g_m V_{BE} \\ &= i_i + g_m r_\pi i_i \\ &= i_i(1 + g_m r_\pi) \\ &= i_i(1+\beta) \end{aligned}$$

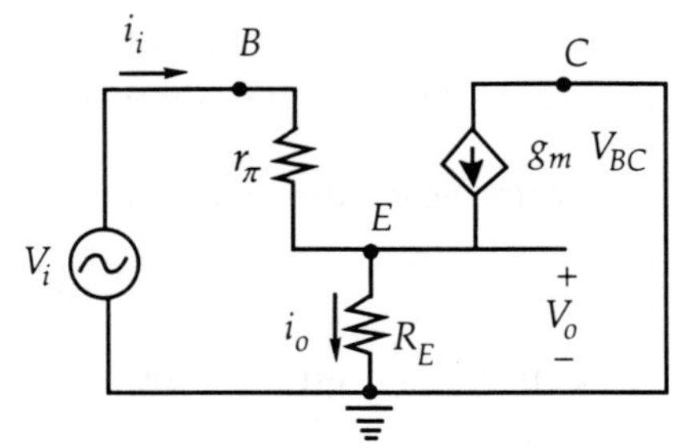

using $V_{BE} = i_i r_\pi$, $\beta = g_m r_\pi$.

small-signal (AC) equivalent circuit

Therefore, the current gain is

$$A_i = \frac{i_o}{i_i} = 1 + \beta = 1 + 100 = 101$$

24. **B** The output current is

$$\begin{aligned} i_o &= i_i + g_m V_{BE} \\ &= i_i + g_m i_i r_\pi \\ &= i_i(1 + g_m r_\pi) \\ &= i_i(1+\beta) \end{aligned}$$

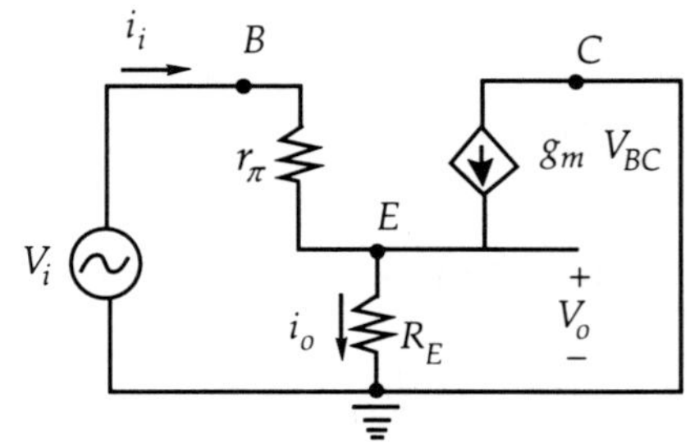

The output voltage is

$$V_o = i_o R_E$$

The input voltage is

$$\begin{aligned} V_i &= i_i r_\pi + V_o \\ &= i_i r_\pi + i_i(1+\beta)R_E = i_i[r_\pi + (1+\beta)R_E] \end{aligned}$$

The voltage gain is

$$\begin{aligned} A_V &= \frac{V_o}{V_i} = \frac{i_o R_E}{i_i[r_\pi + (1+\beta)R_E]} \\ &= \frac{R_E}{r_\pi + (1+\beta)R_E} A_i = \frac{R_E}{r_\pi + (1+\beta)R_E}(1+\beta) = \frac{10^3(101)}{2\times 10^3 + (101)\times 10^3} = 0.98 \end{aligned}$$

25. **D** $\dfrac{V_m}{V_i} = -\dfrac{Z_F}{Z_i}$ or $\left|\dfrac{V_m}{V_i}\right| = \dfrac{100 / R_i}{\sqrt{1 + (\omega^2 / 100)}}$

For a gain of 25 (maximum occurring at $\omega = 0$):

$$25 = \frac{100}{R_i}. \qquad \therefore R_i = 4\ \text{k}\Omega$$

26. **B** The bandwidth occurs at an amplitude of $1/\sqrt{2}$ (or 3dB in power, or 1/2 in power), which is where $\omega^2/100 = 1$ or $\omega = 10 = 2\pi f$. Therefore

$$f = 10/2\pi = 1.59\ \text{Hz}$$

27. **B** We design the two poles so that

$$\left|\frac{V_o}{V_i}\right| = \frac{K}{\left(\sqrt{1+\omega^2/100}\right)^2} = \frac{K}{1+\omega^2/100} = \frac{K}{1+(\omega R_F C_F)^2}$$

But $\tau = R_F C_F = 10^{-1}$ which requires a parallel $R-C$ with time constant $\tau = 10^{-1}$.

28. **C** $Z = \dfrac{j\omega 10}{j\omega + 10} = \dfrac{10}{1 - j10/\omega}$

$$\left|\frac{V_o}{V_m}\right| = \frac{10/10\times 10^3}{\sqrt{1+(10/\omega)^2}} \text{ and } \left|\frac{V_o}{V_i}\right| = \frac{(10^5/R_i)(10/10^4)}{\sqrt{1+(10/\omega)^2}\sqrt{1+(\omega/10)^2}}$$

This results in a bandpass with

$$\left|\frac{V_o}{V_i}\right| = \frac{100/1\ \text{k}\Omega}{2} = \frac{10^{-1}}{2} = 5\times 10^{-2}$$

centered at $\omega = 10$.

29. **A** For V_I positive, the upper configuration acts as a voltage follower (the diode can supply current in the forward direction): $V_o = V_I$ (for $V_I > 0$). For V_I negative, the lower configuration (inverting configuration) allows $V_o/V_I = -R_2/R_1$ to be positive (and the diode can supply current in the forward direction). If $R_2 = R_1$ then for either positive or negative $V_I, V_o = |V_I|$. $\therefore V_o = |V_I| =$ full wave rectifier.

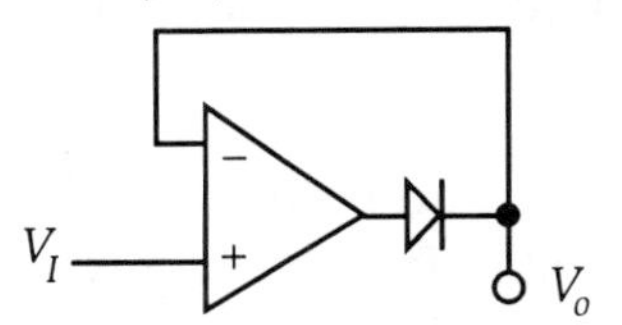

30. **B** If R_1 is open, the upper circuit appears as a voltage follower for positive voltage and no output for negative voltage (the diode is reverse biased for V_o negative). So this results in a half-wave rectifier.

31. **D** The input constraint (KVL) is

$$V_i = i_b R_b + 0.7$$

The output constraint (KLV) is

$$V_{ce} = V_o = -i_c R_c + 20$$

For $V_i = 1.7$

$$i_b R_b = 1.0 = i_b \times 2\times10^3. \quad \therefore i_b = 0.5 \text{ mA} = 5\times10^{-4} \text{ A}$$

Since $\beta = 100$, $i_c = \beta i_b = 5\times10^{-2}$ A. Then

$$V_o = -5\times10^{-2}\times100+20 = 15 \text{ V}$$

32. **B** See No. 31. For $V_i = 3.7$

$$i_b \times 2\times10^3 = 3. \quad \therefore i_b = 1.5 \text{ mA} = 1.5\times10^{-3} \text{ A}$$

Then with $\beta = 100$, $i_c = \beta i_b = 1.5\times10^{-1}$ and

$$V_o = -1.5\times10^{-1}\times100+20 = 5 \text{ V}$$

33. **C** See No. 31. For $V_i = 2.7$

$$i_b = 1.0 \text{ mA}, \; i_c = \beta i_b = 100 \text{ mA}, \text{ and } i_e = i_b + i_c = 101 \text{ mA}$$

34. **C** The dipole is created with two oppositely charged particles in close proximity. The net charge is zero. By Gauss' law:

$$\oint \varepsilon \mathbf{E}\cdot d\mathbf{A} = Q. \quad \therefore \oint \mathbf{E}\, d\mathbf{A} = 0$$

since no net charge is enclosed.

35. **D** The potential from which the dipole field is derived is

$$V = \frac{Qd\cos\theta}{4\pi\varepsilon_o r} = \frac{K\cos\theta}{r}$$

At $r = a$ (surface of the sphere) $V = K_2 \cos\theta$.

36. **B** The dipole field satisfies Maxwell's equations:

$$\nabla\times\mathbf{E} = 0$$
$$\nabla\cdot\varepsilon\mathbf{E} = \rho \quad \text{since } \rho = 0$$

However at $r = 0$, two charged particles very closely spaced do exist and become a singularity. Then $\nabla\times\mathbf{E} = \nabla\cdot\mathbf{E} = 0$ everywhere except at the origin.

37. **D** The surface charges can be replaced by point charges of magnitude $4\pi a\rho_o$. The potential from the equivalent point charge at (0,0,–a) is $-\left.\frac{4\pi a\rho_o}{4\pi\varepsilon_o r}\right|_{r=2a}$. The potential from (0,0,a) is $\left.\frac{4\pi a\rho_o}{4\pi\varepsilon_o r}\right|_{r=a}$. Inside the sphere:

$$V = -\frac{\rho_o}{2\varepsilon_o} + \frac{\rho_o}{\varepsilon_o} = \frac{\rho_o}{2\varepsilon_o}$$

38. **A** Since equal and opposite equivalent point charges are located at $(0,0,\pm a)$, the x-y plane is at zero potential.

39. **B** The electric field due to a point charge is

$$\mathbf{E} = \frac{Q}{4\pi\varepsilon_o r^2}\mathbf{a}_r$$

Taking care with the vector directions,

$$\mathbf{E} = \frac{Q(-\mathbf{a}_y - \mathbf{a}_z)/\sqrt{2}}{4\pi\varepsilon_o a^2(\sqrt{2})^2} + \frac{Q(\mathbf{a}_y - \mathbf{a}_z)/\sqrt{2}}{4\pi\varepsilon_o a^2(\sqrt{2})^2} = \frac{-2\mathbf{a}_z(4\pi a^2\rho_o)}{\sqrt{2}8\pi\varepsilon_o a^2}$$

$$\therefore |\mathbf{E}| = -\frac{\rho_o}{\sqrt{2}\,\varepsilon_o}$$

40. **D** At $t = 0$, $x(t)$ is a maximum, so

$$x(0) = K(2+4+2) = 8K$$

This can at most equal unity (for 100% modulation). Therefore,

$$K = \frac{1}{8} = 0.125$$

41. **C** Since

$$x_c(t) = A_c[1+\mu x(t)]\cos 2\pi f_c t$$

The frequencies transmitted are

$$f_c \pm f_m; \qquad f_c \pm 2f_m; \qquad f_c \pm 3f_m$$

Since $f_c = 10f_m$ the highest frequency carried is

$$f_c + 3f_m = 10f_m + 3f_m = 13f_m$$

42. **B** Using the signal

$$x_c(t) = A_c\left[1+\tfrac{1}{8}(2\cos\omega_m t + 4\cos 2\omega_m t + 2\cos 3\omega_m t)\right]\cos 10\omega_m t$$

with $A_c = 1$, a sketch of the spectrum provides the power in the side-bands:

$$2\left[2\left(\frac{1}{16}\right)^2 + 2\left(\frac{1}{8}\right)^2 + 2\left(\frac{1}{16}^2\right)\right] = \frac{3}{32} \cong 0.1 \text{ watts}$$

43. **D** This tone-modulated signal is given by

$$x_c(t) = A_c\left[1+\cos(6\pi\times 10^3 t)\right]\cos(600\pi\times 10^3 t)$$

The highest frequency is then 3 kHz + 300 kHz = 303 kHz

44. **A** The power dissipated can be obtained by considering the positive spectrum (phasor representation) of

$$x_c(t) = A_c\cos(600\pi\times 10^3 t) + \frac{A_c}{2}\cos(606\pi\times 10^3 t) + \frac{A_c}{2}\cos(594\pi\times 10^3 t)$$

The power dissipated is then

$$\langle P\rangle = \frac{1}{2}\left[A_c^{\,2} + \left(\frac{A_c}{2}\right)^2 \times 2\right] = \frac{3}{4}A_c^{\,2}$$

45. **A** The USSB-SC for the positive spectrum (phasor representation) is given by

$$z_{USSC-SC} = \frac{A_c}{2}\cos(606\pi \times 10^3 t)$$

46. **C** Since the current through an inductor cannot change instantaneously (unless an infinite voltage can be supported),

$$i(0^-) = i(0^+) = 0$$

47. **B** Since the current at $t = 0^+$ is initially zero, all the voltage V_s must initially drop across the inductor. But

$$v_L = L\frac{di}{dt} = V_s \text{ at } t = 0^+. \qquad \therefore \frac{V_s}{L} = i'(0^+)$$

48. **B** The roots for the RLC circuit (characteristic equation)

$$s_{1,2} = -\frac{R}{2L} \pm \sqrt{\left(\frac{R}{2L}\right)^2 - \frac{1}{LC}}$$

give overdamped (exponential decay) when

$$\left(\frac{R}{2L}\right)^2 > \frac{1}{LC}$$

49. **C** Since $i(0^+) = 0$ then $\varphi = \frac{\pi}{2}$. Note: this could also be $\phi = -\frac{\pi}{2}$. Since current starts at zero at $t = 0$ it builds up as a sine, but since K was not specified it could be negative.

50. **C** If the system is linear, which is not given, both a Norton and Thevenin are possible. Since it is not possible to evaluate nonlinearity by two measurements only answer **C** is acceptable.

51. **D** Given that the system is linear, a Thevenin is possible; this satisfies (A). Given the two measurements:

$$v_T = R_T\left(\frac{20}{2}\right) + 20$$

$$v_T = R_T\left(\frac{25}{5}\right) + 25$$

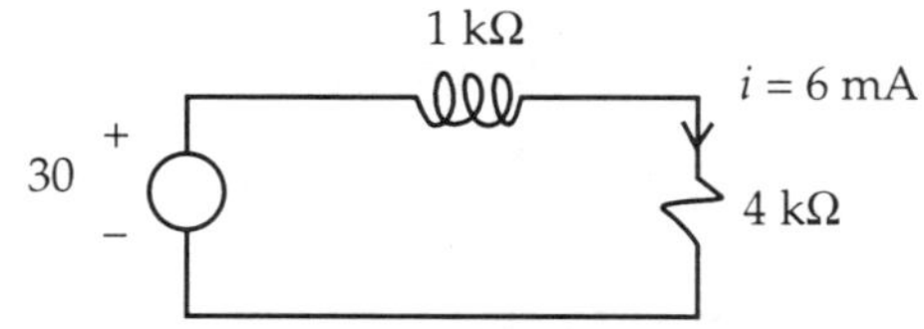

A Thevenin with $v_T = 30$ V and $R_T = 1$ kΩ results. A 4 kΩ resistor would result in power dissipated of

$$P = i^2 R = 36 \times 10^{-6} \times 4 \times 10^3 = 0.144 \text{ W}$$

Thus (B) is also acceptable. Since the system is linear a Norton is possible; this satisfies (C). Consequently (D) is the answer.

52. **C** Since the stack pointer is decremented on a load, it will decrease in value and hence the stack will fill up towards lower memory addresses.

53. **D** The SP will be decremented 10 times and hence be $C000_{16} - A_{16} = FFF6_{16}$.

54. **C** A stack is not effective for random access of data.

55. **D** The flip-flop icon indicates edge-sensitivity with the ∧ symbol and falling-edge with the O symbol.

56. **A** The flip-flop has the following next state table:

A	B	Q_n	Q_{n+1}
0	0	0	1
0	0	1	0
0	1	0	1
0	1	1	1
1	0	0	0
1	0	1	0
1	1	0	0
1	1	1	1

From this, a K-map can be formed to determine the characteristic equation:

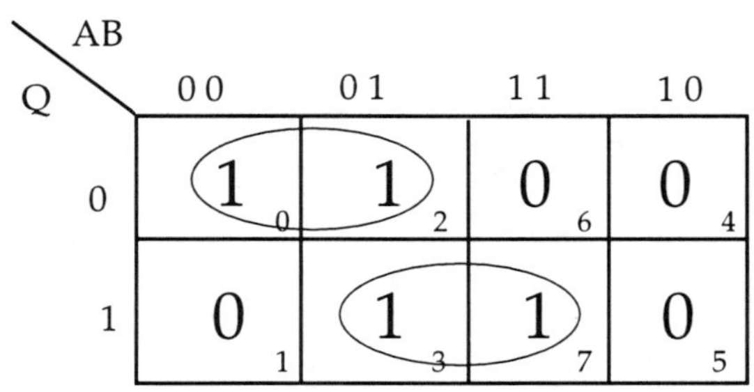

$$Q_{n+1} = \overline{A}Q_n + BQ_n$$

57. **B** This is found by using the truth table in the problem statement for evaluating all conditions on A and B which force the desired transition from Q_n to Q_{n+1}.

58. **A**

59. **C**

60. **D**

The D/S Review for Chemical Engineering

edited by M. C. Potter

Overview of the ChE/DS Topics

Sections

1. Chemical Thermodynamics .... *Wilkinson*
2. Chemical Reaction Engineering .... *Wilkinson*
3. Flow Reactors .... *Wilkinson*
4. Distillation .... *Wilkinson*
5. Mass Transfer .... *Wilkinson*
6. Convective Mass Transfer .... *Wilkinson*
7. Fluid Flow .... *Wilkinson*

What You Need to Know about the NCEES Handbook

Most of the problems on the Chemical Engineering test of the afternoon session can be worked using pages 82 to 85, 58 (Transport Phenomena), and 54 to 56 (Heat Transfer - see the Mechanical Engineering part of this review book) of the NCEES Handbook. Many of the problems, however, on the afternoon Chemical Engineering test will come from the material in pages 1 to 81 of the NCEES Handbook, material used for the General tests of the morning and afternoon sessions. Since all engineers must take the General morning exam, it is assumed that you are familiar with that material. You must realize, however, that you will have to make reference to equations, tables, and charts located in those first 81 pages of the NCEES Handbook when you are working problems in the afternoon Chemical Engineering session. The general engineering sections that you should be particularly familiar with are the sections on Mathematics, Chemistry, Thermodynamics (pages 45 and 46) and Fluid Mechanics (pages 35 to 42). A review of general engineering material is found in our *Fundamentals of Engineering*, a review book written specifically for the FE/EIT exam.

Do I Have to Take the DS Test?

If, after you quickly scan the pages of this Chemical Engineering review section, you feel that you could more easily pass the General exam of the afternoon session, you should register to take the afternoon General test. If you have already registered for the Chemical Engineering afternoon test, you may contact your state board and request a change. Make sure your state board recognizes a *pass* on the General afternoon test for your particular discipline before you make that decision; the vast majority of the state boards will recognize a pass on any of the afternoon tests by any registrant.

Why is the ChemE Review is so Brief?

Most review courses will only set aside one evening to remark on special concerns in the ChemE DS Test. The basic principles are concise in this ChemE, review although of course in practice and in homework they are expanded on greatly. Finally, the material the NCEES has determined to require ChemE DS Test takers to know is similarly quite brief. Our review takes all three factors into consideration. Allot a proportional amount of time to your ChemE DS preparation—but do not neglect study for the general subject a.m. session! (After all, it has twice as many questions: 120, compared to 60.)

This Review is Focused on the NCEES Handbook

In order to provide a very efficient, yet effective review, this DS Review has been designed to review only the material associated with the formulas and figures presented in the NCEES Handbook. The Handbook is the only material allowed in the exam. Thus it makes an efficient base to study around. It is assumed that you have a copy of the NCEES Handbook dated 1996 or later. If you do not, you can obtain one from Great Lakes Press by calling 1-800-837-0201.

GLP Offers a Concise A.M. Session Review & Other Study Aides

We also have an effective review for the General tests, both morning and afternoon, entitled *Fundamentals of Engineering*. It is used at hundreds of engineering schools across the nation. We also distribute the HP-48G and GX calculators, which are pre-programmed with many of the equations you'll need when working problems on the FE/EIT exam. Also, we've developed *Jump Start your HP-48G/X*, a 150-page book designed to efficiently help engineers get the most out of the HP48G/X. You can order these items by calling 1-800-837-0201.

Chemical Engineering

by Bruce Wilkinson

Chemical Engineering fundamentals often involve the chemical transformation of compounds but also involve the transport phenomena associated with the handling and purification of the products or raw materials. Basic to all of the operations involved is the chemical thermodynamics underlying the physical properties of the materials in question. Although many of the fundamentals and design operations encountered in Chemical Engineering are solvable by computer software, it is necessary for us to be able to solve problems in a more traditional manner to understand the principles and assumptions used so that the computer operations are correctly applied.

The examination questions are designed to evaluate the applicant's understanding of the fundamental principals involved in typical Chemical Engineering operations. The equations relevant to Chemical Engineering are found on pages 54, 55, 56, 58, and 82 to 85 of the NCEES Handbook. The thermodynamic fundamentals are summarized on pages 45 and 46 and fluid dynamics is summarized on pages 35 through 40. You should be familiar with the material outlined on these pages as well as the example questions given by the NCEES.

1. Chemical Thermodynamics

Fundamental to the separation of components by distillation or evaporation is an understanding of the vapor/liquid equilibrium characteristics of the system. When vapor and liquid phases coexist in thermodynamic equilibrium, the partial molal fugacity of each component $\bar{f}_i$ will be equal in each phase. that is: $\bar{f}_i^L = \bar{f}_i^V$. Since the partial molar fugacity cannot be measured directly, we express them in terms of the fugacities of the pure component f_i at the temperature and pressure involved and the activity coefficient γ_i of the component. Thus,

$$\gamma_i^L f_i^L x_i = \gamma_i^V f_i^V y_i$$

For low and moderate pressures with components which do not chemically associate with each other, the activity coefficients approach unity, and the pure component fugacities can be expressed in terms of a fugacity coefficient ν_i. Then, $f_i = \nu_i P_i$ where P_i is the partial pressure of the component. Equilibrium may then be represented as

$$v_i^V P_i y_i = v_i^L P_i^* x_i$$

where P_i^* is the vapor pressure of the component and P is the total system pressure. Vapor/liquid equilibrium at low and moderate pressures (where the fugacity coefficients approach unity) can then be represented by Raoult's Law:

$$P y_i = P_i^* x_i$$

Example 1.1 will illustrate.

One aspect of engineering that is unique to the chemical branch is the application to chemical reactions. An understanding of chemical equilibrium and chemical reaction kinetics is therefore critical for the chemical engineer. The equilibrium relationship that represents the relative composition of reactants and reaction products in a reaction can be used to predict the degree of conversion of reactants to desired products and the effects of changing the operating conditions on this equilibrium.

In general, the equilibrium relationship representing a reversible chemical reaction such as

$$aA + bB \leftrightarrow cC + dD$$

can be represented in terms of the change in standard Gibbs Free Energy by

$$-\Delta G^\circ = RT \ln K_a$$

where K_a = Equilibrium Constant for the reaction, and

$$K_a = a_C^c \times a_D^d \,/\, a_A^a \times a_B^b$$

where a_i is the activity of component i, defined as $a_i = \bar{f}_i / f_i^o$ with $\bar{f}_i$ being the partial molal fugacity of component and f_i^o the fugacity in the standard state. This relationship allows the determination of changes in chemical equilibrium which result from modification of the operating temperature of the reaction.

In general, for common compounds, the standard Gibbs Free Energy change can be determined from tabulated standard Free Energies of formation and the relationship

$$-\Delta G^\circ = c\Delta G_c^\circ + d\Delta G_d^\circ - a\Delta G_a^\circ - b\Delta G_b^\circ$$

For ideal gases, the partial molal fugacity can be expressed as the partial pressure of the component and the fugacity in the standard state is 1 atm. Therefore,

$$K_a = K_p = P_C^c \times P_D^d / P_A^a \times P_B^b$$

or, since, for ideal gases, Raoult's Law defines $P_i = y_i P$,

$$K_a = K_p = P^{c+d-a-b} \times y_C^c \times y_D^d \,/\, y_A^a \times y_B^b$$

Example 1.2 illustrates the applicability of these relationships.

Example 1.1

The vapor pressures of benzene (component 1) and toluene (component 2) are summarized below:

T, °C	P_1^*, bars	P_2^*, bars
88	1.285	0.508
90	1.361	0.542
94	1.526	0.616
98	1.705	0.698
100	1.800	0.742
104	2.004	0.836

1. What will be the mole fraction of benzene in the vapor phase in equilibrium with a 0.4 mole fraction benzene liquid at a total pressure of 1 bar?
   (A) 0.72
   (B) 0.62
   (C) 0.51
   (D) 0.61

2. If the temperature is 90°C and the pressure is 1 bar, what are the benzene liquid and vapor compositions in equilibrium?
   (A) 0.56, 0.76
   (B) 0.52, 0.65
   (C) 0.44, 0.76
   (D) 0.2, 0.34

3. At what pressure would a 0.2 mole fraction benzene liquid be in equilibrium with a 0.375 mole fraction benzene vapor?
   (A) 1.0 bar
   (B) 0.6 bar
   (C) 1.25 bar
   (D) 1.07 bar

4. What will be the temperature in question 3?
   (A) 104°C
   (B) 88°C
   (C) 100°C
   (D) 98°C

5. A gaseous mixture containing 0.4 mole fraction benzene and 0.6 mole fraction toluene is cooled to 100°C at 1 bar pressure. What fraction of the mixture is liquefied and what will be the liquid composition (mole fraction benzene)?
   (A) 0.41, 0.82
   (B) 0.6, 0.39
   (C) 0.2, 0.24
   (D) 0.91, 0.13

Solutions:

1. **B** Raoult's Law: $Py_i = P_i^* x_i$ for each component.

   For benzene: $y_1 = P_1 \times 0.4$, for toluene: $y_2 = P_2^* \times 0.6$

   Assume T = 94°C: $P_1 = 1.526$; $P_2 = 0.616$

   Then: $y_1 = 0.61$ $y_2 = 0.37$ Total 0.98

   Normalize to total of 1.00 (because total is close to unity):

   $y_1 = 0.62$ $y_2 = 0.38$

2. **A** $P_1^* = 1.361$ bar

   Assume x, calculate y's, check sum of y's.

   Assume $x_1 = 0.5$:

   Sum of y's $= 0.5 \times 1.361 + 0.5 \times 0.542 = 0.952$. Too low.

   Assume $x_1 = 0.6$:

   Sum of y's $= 0.6 \times 1.361 + 0.4 \times 0.542 = 1.03$. Too high.

   Assume $x_1 = 0.56$:

   Sum of y's $= 0.56 \times 1.361 + 0.44 \times 0.542 = 1.00$. OK.

3. **D** $P_i = x_i P_i^*$ Check sum of $P_i = P$.

   Assume T = 100°C:

   $P_1 = 0.2 \times 1.800 = 0.36$, $P_2 = 0.8 \times 0.742 = 0.594$ bar; sum = 0.954.

   $y_1 = 0.36/0.954 = 0.377$. $y_2 = 0.594/0.954 = 0.623$

   Value of y_1 is slightly higher than desired 0.375, choose higher T.

   Assume T = 104°C:

   $P_1 = 0.2 \times 2.004 = 0.401$, $P_2 = 0.8 \times 0.836 = 0.669$ bar; sum = 1.07.

   $y_1 = 0.401/1.07 = 0.375$. OK.

4. **A** From the above calculations, T = 104°C.

5. **C** $P_1^* = 1.800$ $P_2^* = 0.742$

   $P_1 = y_1 = 1.800x_1$ $P_2 = y_2 = 0.742x_2$

   Then, $1 - y_1 = 0.742(1 - x_1)$

   and, $1 - 1.800x_1 = 0.742(1 - x_1)$

   or, $x_1 = 0.244$ = benzene liquid composition

   Perform a material balance for L moles liquid and V moles vapor.

   Assuming a basis of one mole of feed:

   $0.244L + 1.800 \times 0.244(1 - L) = 0.4$

   $\therefore L = 0.2$ = fraction of feed that is liquid.

Example 1.2 Chemical Equilibrium

Pure ethane at 200°C and 5 atm pressure is stored for future use. There is some danger that the ethane will decompose to ethylene according to the equation:

$$C_2H_6(g) \rightarrow C_2H_4(g) + H_2(g)$$

Assume the reaction is first order in each component and that $\Delta G° = 10{,}000$ cal/gmol at 200°C. The gases behave ideally.

1. What fraction of the ethane will dehydrogenate in the long term?
   (A) 0
   (B) 0.12
   (C) 0.71
   (D) 0.002

2. If the pressure were lowered to 1 atm, what fraction of the ethane would dehydrogenate?
   (A) 0.00001
   (B) 0.5
   (C) 0.005
   (D) 0

Solution:

1. **D** $-\Delta G° = RT \ln K_a$

   $-10000 = 1.987 \times 473 \ln K_a$

   $K_a = 2.39 \times 10^{-5}$

   $K_a = K_p = P^{1+1-1}\left(y_p^2/y_r\right)$ where y_p represents the mole fraction of products and y_r the mole fraction of reactants. Since $y_r = 1 - 2y_p$,

   $$2.39 \times 10^{-5} = 5y_p^2/(1 - 2y_p)$$

   Solution by quadratic formula yields:

   y_p = mole fraction of ethylene in equilibrium = 0.0022

2. **A** Repeating the calculation with $P = 1.0$ rather than 5.0 yields $y_p = 0.0049$.

2. Chemical Reaction Engineering

The fundamentals of chemical reaction kinetics are given on page 82 of the NCEES Handbook. Since chemical reactions often form the basis for chemical engineering practice, it is important to be able to determine the order of chemical reactions based on experimental information and to be able to evaluate reaction vessel sizing. The most commonly encountered situations involve homogeneous reactions and these are the ones most easily analyzed. However, heterogeneous reaction systems are also used in industry and their analysis is usually more complex. For all systems, the rate of reaction is usually defined as

$$r_A = (1/V)dN_A/dt$$

where r_A represents the moles of A disappearing per unit time per unit volume. If the volume of the system is a constant,

$$r_A = dC_A/dt$$

where C_A is the concentration of A, moles per liter.

The rate of reaction is defined in terms of the concentrations of reactants and a reaction rate constant which is a function of the temperature of the system. Thus,

$$-r_A = k(C_A C_B \cdots) \quad \text{at constant volume}$$

where

$$k = Ae^{-E_a/RT}$$

with E_a being the activation energy for the reaction. Example 2.1 illustrates this relationship.

In analyzing the order of a specific reaction, it is often convenient to express the fractional conversion of some reactant A, as

$$X_A = \frac{C_{Ao} - C_A}{C_{Ao}}$$

Then, the overall order of the reaction is related to the exponents on the individual concentration terms in the rate equation as

$$-r_A = kC_A^x C_B^y$$

where the overall rate is $n = x + y$ (for isothermal reactions in which k is constant and the reactor volume is fixed).

For zero order reactions, the rate of the reaction is independent of the concentration of reactants or the degree of conversion of reactants. Thus,

$$-r_A = k \quad \text{or} \quad -dC_A/dt = k$$

Integration results in $C_A = C_{Ao} - kt$ or $X_A = kt/C_{Ao}$.

Alternatively, for first order reactions, with $-r = kC_A$ there results

$$C_A = C_{Ao}e^{-kt} \quad \text{or} \quad X_A = 1 - e^{-kt}$$

Similarly, for second order reactions we have $-r_A = kC_A{}^2$ resulting in

$$\frac{1}{C_A} - \frac{1}{C_{Ao}} = kt$$

Example 2.2 illustrates these relationships.

Example 2.1

The pyrolysis of ethane proceeds with an activation energy of about 70 000 cal/gmol. How much faster will the reaction proceed at 700°C than at 600°C?

(A) 20 times
(B) 60 times
(C) 2 times
(D) 0.5 times

Solution:

$$k = Ae^{-E/RT}$$

$$\frac{k_{700}}{k_{600}} = \frac{e^{(-70\,000/1.987\times 973)}}{e^{(-70\,000/1.987\times 873)}}$$

$$= 63.3 \text{ times faster}$$

The answer is **B.**

Example 2.2

For the following aqueous reaction, experimental batch data have been measured at 90°C:

$$A \rightarrow B + C$$

Given: $C_A = 0.5$ moles/liter at the beginning of the reaction and $C_A = 0.2$ moles/liter at 10 minutes with the rate equation:

$$-dC_A/dt = kC_A^2$$

1. What is the overall order of the reaction?
(A) 0
(B) 2
(C) 0.5
(D) 1

2. What are the units of the rate constant?
(A) moles/liter · min
(B) liters/mole · min
(C) min^{-1}
(D) dimensionless

3. What is the value of the rate constant?
(A) 0.03 (B) 0.1 (C) 1.0 (D) 0.003

4. What is the total reaction time to achieve 90% conversion of A?
(A) 20 minutes
(B) 25 minutes
(C) 16 minutes
(D) 60 minutes

Solutions:

1. **B** Overall order is 2 (exponent on concentration)

2. **B** $-dC_A/dt = kC_A^2$. The units are moles/liter · min $= k\,(\text{moles/liter})^2$

   $k =$ liters/mole · min

3. **A** The equation is

$$-dC_A/dt = kC_A^2$$

   Rearrange:

$$-dC_A/C_A^2 = kdt$$

   Integrate:

$$1/C_A - 1/C_{Ao} = kt. \quad \text{Then } 1/0.2 - 1/0.5 = k(10)$$

   $\therefore k = 0.03$ liters/mole · min

4. **D** For 90% conversion, $C_A = 0.5$ moles/liter. Using the equation from No. 3:

$$1/0.5 - 1/5 = 0.03t$$

   $\therefore t = 60$ minutes

3. Flow Reactors

Most chemical reactions are carried out on a flow basis in order to provide a continuous process. Plug flow and a continuous stirred tank are the most common processes used, and you should be familiar with the analysis of each type. Plug flow reactors, as the name implies, allow the reaction to proceed as the reactants/products flow through a pipe with no longitudinal mixing. Such a situation might be expected when flow is turbulent so that the radial velocity distribution of flow is negligible. For the plug flow reactor case, applying a differential analysis of a section of the reactor yields

$$F_{Ao}dX_A = -r_A dV_{PFR}$$

where V_{PFR} is the reactor volume. Integrating results in

$$V_{PFR} = -F_{Ao}\int_0^{X_A} \frac{dX_A}{r_A}$$

where F_{Ao} is the moles of A fed per unit time.

Similarly, for a Continuous Stirred Tank Reactor, which is defined as one to which feed is continuously fed and a mixture of product and unreacted feed continuously removed with a well- mixed reactor volume being constant,

$$F_{Ao}\,X_A = -r_A\,V_{CSTR}$$

where r_A is evaluated at the exit concentration from the reactor. A comparison of plug flow and CSTR is given in Example 3.2.

Example 3.1 Plug Flow Reactor

In a homogeneous gas phase reaction, phosphine decomposes as

$$4PH_3(g) \rightarrow P_4(g) + 6H_2(g) \qquad @\,600°C$$

The reaction rate is $-r_{PH_3} = (10/\text{hr})\, C_{PH_3}$.

1. Since the volume changes with reaction, what is the value of ε, the constant reflecting this volume change?
   (A) 0
   (B) 0.2
   (C) 0.5
   (D) 0.75

2. What would be the size of a plug flow reactor operating at 600°C and 4.6 atm and producing 80% conversion of a feed consisting of 40 gmol of phosphine per hour?
   (A) 55 L
   (B) 850 L
   (C) 2000 L
   (D) 138 L

Solutions:

1. **D** From Stoichiometry,

$$\varepsilon = (7-4)/4 = 0.75$$

2. **D** Referring to page 83,

$$V_{\text{PFR}} = F_{A_0} \int_0^{X_A} dX_A/(-r)$$

where A represents phosphine. Substituting for r and C_A in terms of X_A yields

$$V_{PFR} = \frac{F_{A0}}{10C_{A0}} \int_0^{X_A} \frac{1+\varepsilon X_A}{1-X_A} dX_A$$

which, upon integration, results in

$$V_{\text{PFR}} = \frac{F_{A0}}{10C_{A0}} \left[(1+\varepsilon)\ln\frac{1}{1-X_A} - \varepsilon X_A \right]$$

But

$$C_{A0} = \frac{4.6 \text{ atm}}{(0.08205 \text{ atm/gmol}\cdot\text{K})(873\text{K})}$$
$$= 0.0642 \text{ gmol/liter}$$

and $X_A = 0.8$. Then

$$V_{PFR} = \frac{40}{10 \times 0.0642} \left[(1+0.75)\ln\frac{1}{1-0.8} - 0.75 \times 0.8 \right] = 138 \text{ L}$$

Example 3.2 Continuous Stirred Tank Reactor

For the case discussed in Example 3.1, what would be the volume of a CSTR to achieve the same results?

(A) 55 L
(B) 5000 L
(C) 200 L
(D) 400 L

Solution:

From the equation on page 83,

$$V_{\text{CSTR}} = F_{A0}X_A/(-r)$$

where r is evaluated at the outlet conditions.

At 80% conversion, the product stream is

Phosphine	40×0.2	8 gmol
P_4	$40 \times 0.8/4$	8 gmol
Hydrogen	$(40 \times 0.8) \times 6/4$	48 gmol
Total		64 gmol

Volume (using perfect gas law):

$$V = \frac{0.08205 \times 64 \times 873}{4.6} = 997 \text{ liters}$$

Then, $C_A = 8 \text{ gmol}/997 \text{ liters} = 0.0080 \text{ moles/liter}$

So that

$$-r_A = 10 \times 0.0080 = 0.08 \text{ gmol/liter} \cdot \text{hour}$$

Then:

$$V = \frac{40 \times 0.8}{0.080} = 400 \text{ L}$$

The answer is **D**.

4. Distillation

The separation of miscible liquids by boiling, to take advantage of the relative volatility of one or more components, is an important operation in many chemical engineering plants. An understanding of the principles involved gives us the ability to solve such separation problems as may be encountered in practice. Most distillation calculations involve the simultaneous application of material (and sometimes energy) balances with phase equilibria to simulate the operation of a theoretical column. Extension to apply practical information on plate efficiencies enables the results of the theoretical calculations to be extrapolated to an actual column design.

Distillation operations can be conducted on a batch or continuous cycle. The original distillation systems involved the simple boiling of a batch of raw material to remove some of the more volatile component and thus produce a product richer in this component than the feed stock. The degree of purification possible under these circumstances is limited by the relative volatilities of the bottoms (waste) stream as compared to the overhead (product) stream. A more sophisticated separation can be achieved by recycling (or refluxing) part of the overhead stream to the top of a fractionating column in which a succession of boiling and condensation steps takes place. In this way, the overhead stream can be significantly richer in the volatile components than would be expected by the analysis of the bottoms stream and the relative volatilities.

The simplest design calculations involve binary systems, and the principles can be illustrated by the analysis of such a system. Application of the same procedures to multicomponent mixtures can be achieved and computer programs are available to simplify this rather tedious task.

The analysis of simple batch (or flash) distillation involves the solution of component and overall material balances and, finally, the application of vapor-liquid equilibria relationships between the components. The material balances result in a relationship between the amount of material unevaporated W and the original charge to the still W_0 as given by

$$\ln\frac{W}{W_0} = \int_{x_0}^{x} \frac{dx}{y-x}$$

where x and y are liquid and vapor compositions.

It should be noted that during such a simple distillation the vapor and liquid compositions are continuously changing. Solution of this integral with a constant relative volatility α, defined as $y = \alpha x$, results in

$$\ln\frac{W}{W_0} = \frac{1}{\alpha-1}\ln\frac{x(1-x_0)}{x_0(1-x)} + \ln\frac{1-x_0}{1-x}$$

An example of this practice is given in Example 4.1.

Continuous distillation involves the return (or reflux) of a portion of the overhead product stream to the top of a column consisting of a number of stages on which successive boiling and condensation takes place. This produces increased concentration of the more volatile components toward the top of the column. The feed to such a unit is near the center of the column, while products are taken off overhead and at the bottoms (and sometimes at intermediate locations in the case of multicomponent separations). The analysis of such a system involves stagewise calculation of material, energy, and equilibrium balances to calculate the number of theoretical stages required to achieve the desired separation.

For binary systems, the simplest analysis can be applied when an assumption of constant molal overflow can be made. This case reflects the situation when the molal heats of vaporization of the components being separated are equal, heats of mixing can be neglected, and the column can be assumed to have no significant heat losses. For this case, the graphical McCabe-Thiele method can be used to analyze the system requirements. On a plot of vapor mole fraction vs. liquid mole fraction, the equilibrium relationship and the desired overhead and bottoms compositions to be produced from a given feed composition can be located. In the top (or rectifying) section of the column, the slope of an operating or material balance line can be given as

$$y = \frac{L}{V}x + \frac{D}{V}x_D$$

where L, D, V are the liquid, overhead product, and vapor molal flow rates; and x_D is the desired overhead product composition. A similar relationship,

$$y = \frac{L}{V}x - \frac{B}{V}x_B$$

can be derived for the lower (stripping) section with L and V representing the flows below the feed plate, and x_B the bottoms composition with a flow rate of B. The intersection of the two operating lines represents the feed point whose location is determined by a line drawn through the $x = y$ point at the feed composition and whose slope is defined by a quantity $q/(q-1)$, where q is calculated by determining the heat required to convert one mole of feed to a saturated vapor divided by the molal heat of vaporization of the feed stream.

Using the known feed composition and thermal conditions, as well as the desired overhead and bottoms product composition, it is possible to determine the number of theoretical stages required to achieve the desired separation when various reflux ratios (defined as $L/(V-L)$ or L/D) are assumed. Increasing the reflux ratio decreases the number of theoretical stages required but increases the amount of heat that must be added to the base of the column to maintain the desired quantity of products. An economic balance between energy and equipment requirements determines the least-cost system.

The conversion from theoretical stages for a given separation to the actual stages required involves the application of empirical values for the Murphree Plate Efficiency. This is defined as the actual vapor change produced by a stage as a fraction of the theoretical vapor change.

Example 4.2 illustrates the graphical solution of a simple binary system analysis.

Example 4.1 Flash Distillation

Assume a 18 mole % ethanol/water mixture is to be separated into an overhead product and a bottoms stream of 6 mole % ethanol. The relative volatility over the composition range is approximately 6.

1. What is the composition of ethanol in the first drop of vapor condensed?
   (A) 0.57
   (B) 0.7
   (C) 0.3
   (D) 0.9

2. If the process is continued until the re-boiler composition is 6 mole %, what fraction of the feed will have been removed?
   (A) 0.32
   (B) 0.2
   (C) 0.61
   (D) 0.68

3. What will be the average distillate composition?
   (A) 0.2
   (B) 0.9
   (C) 0.44
   (D) 0.3

4. What will be the composition of the last drop of distillate produced?
   (A) 0.45
   (B) 0.20
   (C) 0.28
   (D) 0.08

Solutions:

1. **A** Referring to the equations on page 83:

$$y = \frac{\alpha x}{1+(\alpha-1)x}$$

where α is relative volatility. At $x = 0.18$,

$$y = \frac{6\times 0.18}{1+5\times 0.18} = 0.57$$

2. **C** $$\ln\left(\frac{W}{W_0}\right) = \frac{1}{\alpha-1}\ln\left[\frac{x(1-x_0)}{x_0(1-x)}\right] + \ln\left[\frac{1-x_0}{1-x}\right] = \frac{1}{5}\ln\left[\frac{0.06(0.82)}{0.18(0.94)}\right] + \ln\left[\frac{0.82}{0.94}\right] = -0.3836$$

$$\therefore \frac{W}{W_0} = 0.681$$

The fraction removed is then = 1.000 – 0.681 = 0.319

3. **B** By material balance on ethanol:

$$0.18 \times 1 = 0.681 \times 0.06 + 0.319 \times y$$

$$\therefore y = 0.436$$

4. **C** Calculation of y when $x = 0.06$: $y = 0.277$

Example 4.2 Continuous Distillation

A fractionating column operating at 1 atm pressure is supplied with 1000 gmol/hr of a 60 mole % component A, 40 mole % component B feed. The feed is 25 % vapor and 75% liquid. The bottoms are to be 5 mole % A. The overhead product is to be 90 mole % A as a saturated liquid. The column is operated at an external reflux ratio equal to 1.5 times the minimum L/D. The equilibrium data are given on the diagram shown:

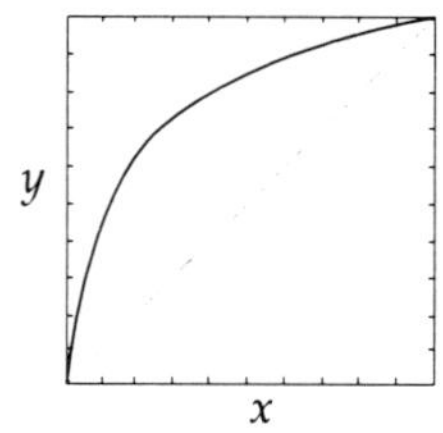

1. What is the minimum internal reflux ratio (L/V)?
   (A) 0.35
   (B) 0.20
   (C) 0.54
   (D) 0.9

2. What is the operating internal reflux ratio $(L/V)_{OP}$?
   (A) 0.27
   (B) 0.45
   (C) 0.60
   (D) 0.55

3. What is the minimum number of theoretical stages necessary to achieve the separation?
   (A) 2.5
   (B) 3.5
   (C) 4.5
   (D) 5.0

4. What is the actual number of theoretical stages required at the operating reflux ratio?
   (A) 9
   (B) 8
   (C) 6
   (D) 8.2

5. If the Murphree efficiency of the column is 0.7, how many actual stages are required?
   (A) 6
   (B) 7.1
   (C) 8.6
   (D) 11.2

6. If the molal heat of vaporization is 10 000 cal/gmol, what is the reboiler duty in cal/hr?
   (A) 5.1×10^6
   (B) 6.4×10^6
   (C) 9.2×10^6
   (D) 3×10^6

Solutions:

Locate Feed, Product and Bottoms on the diagram shown:

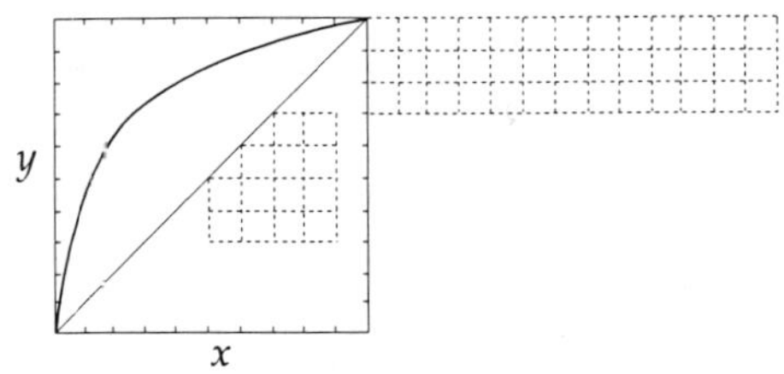

1. **B** $q = 0.75$ (25% vapor). Slope of q line $= q/(q-1) = 0.75/(-0.25) = -3$

   Locate q line on grid at feed.

   Locate line through product and intersection of q line with equilibrium.

   Slope $= (L/V)_{MIN} = 0.2$.

2. **A** Using this value, $(L/D)_{MIN} = 0.25$.

$$(L/D)_{OP} = 1.5 \times 0.25 = 0.375 \text{ or } (L/V)_{OP} = 0.27$$

   (by material balance around top of column)

3. **B** Using a 45° line, step off minimum number of theoretical stages from product to bottoms:

$$\text{stages} = 3.5$$

4. **C** Using operating reflux ratio, construct enriching and stripping section operating lines and step off number of theoretical stages:

$$\text{stages} = 6$$

5. **C** Number of actual stages: $6/0.7 = 8.6$

6. **B** Material balances:

$$0.6F = 0.9D + 0.05B$$

$$600 = 0.9(1000 - B) + 0.05B$$

$\therefore B = 353$ gmol/hr and $D = 647$ gmol/hr.

Then

$$L\text{ (above feed)} = 0.375 \times 647 = 243$$

$$V\text{ (above feed)} = 647 + 243 = 890\text{ gmol/hr}$$

$$V\text{ (below feed)} = 890 - 250 = 640\text{ gmol / hr}$$

$$= \text{boil up rate in reboiler}$$

$$\text{Heat duty on boiler} = (640\text{ gmol/hr}) \times 10\,000$$

$$= 6.4 \times 10^6\text{ cal/hr}$$

5. Mass Transfer

Mass transfer or the tendency of components in a system to migrate from a region of high to low concentration is commonly encountered in chemical engineering practice. The movement of molecules can be accomplished by molecular diffusion (due to their random walk) or to convective forces resulting from fluid motion.

Molecular diffusion can be through a stagnant film in one direction only (unidirectional) or by counterdiffusion of two molecules (equimolar counter diffusion). Both are encountered in practice. Mass flux of component *A* through static component *B* via unidirectional diffusion due to a pressure gradient in the *z* direction can be analyzed by using the diffusivity D_m and

$$\frac{N_A}{A} = \frac{D_m P}{RT p_{B,lm}} \frac{p_{A2} - p_{A1}}{z_2 - z_1}$$

Similarly, equimolar counterdiffusion under similar circumstances can be expressed by

$$\frac{N_A}{A} = \frac{D_m}{RT} \frac{p_{A2} - p_{A1}}{z_2 - z_1}$$

Examples 5.1 and 5.2 demonstrate the principles involving simple diffusional mass transfer.

Example 5.1

A thin layer of palladium metal is highly permeable to hydrogen but not to other gases such as nitrogen. Thus, hydrogen can be separated from a mixture of 40 mole percent hydrogen in nitrogen by passing the mixture over a palladium surface. On the other side of the palladium, the hydrogen pressure is adjusted so that, at the surface of the palladium next to the gas mixture, a mole fraction of 0.2 hydrogen is maintained. The resistance to mass transfer is represented by a stagnant film 0.1 cm thick. The temperature is 20°C and the pressure is 10 atm. The diffusivity of hydrogen in nitrogen is 0.764 cm^2/sec. What will be the flux of hydrogen through the palladium in gmol/sec·cm^2?

(A) 12×10^{-5}

(B) 0.02

(C) 0.004

(D) 9×10^{-4}

Solution:

Diffusion of hydrogen through stagnant nitrogen:

$$\frac{N_A}{A} = \frac{D_m P}{RT p_{B,lm}} \frac{p_{A2} - p_{A1}}{z_2 - z_1}$$

where $p_{A1} = 0.2 \times 10 = 2$ atm; $p_{A2} = 0.4 \times 10 = 4$ atm; $p_{B1} = 8$ atm; $p_{B2} = 6$ atm;

$$p_{B,lm} = 6.95.$$

Then,

$$\frac{N_A}{A} = \frac{0.764 \times 10}{0.08205 \times 293 \times 6.95} \frac{2}{0.1 \times 1000}$$
$$= 9.14 \times 10^{-4} \text{ gmol / sec} \cdot \text{cm}^2$$

The answer is **D**.

Example 5.2

Pure chlorine gas is contained in a 500 mL bulb from which a 0.2-cm inside diameter, 20-cm-long tube leads to the atmosphere. If the temperature is 20°C, the absolute pressure is 1 atm and the diffusivity of chlorine in air is 0.124 cm^2/sec, what is the initial diffusion rate of chlorine from the bulb in gmol/hour?

(A) 4.2×10^{-3}

(B) 2.9×10^{-5}

(B) 6.1×10^{-5}

(D) 9.2×10^{-7}

Solution:

This is a case of equimolar counter diffusion of chlorine and air:

$$N_A = \left(\frac{D_m A}{RT}\right) \frac{p_{A1} - p_{A2}}{z_2 - z_1}$$
$$= \frac{0.124 \times 1 \times \pi \times 0.2^2/4}{0.08205 \times 293 \times 20 \times 1000}$$
$$= 8.1 \times 10^{-9} \text{ gmol / sec} \quad \text{or} \quad 2.9 \times 10^{-5} \text{ gmol / hr}$$

The 1000 in the denominator is needed to convert the units.

The answer is **B**.

6. Convective Mass Transfer

The forced diffusion of a component through two or more other components by means of fluid flow or agitation determines the height of absorption, distillation or extraction columns. The analysis of these systems is often applied to packed towers (as contrasted to stage-wise units), and the height of a transfer unit times the corresponding number of transfer units yields the height of the separator to be used. The basic equations applicable to such a system are given on page 84 of the NCEES Handbook.

Convective mass transfer flux can be calculated using individual mass transfer coefficients analogous to heat transfer coefficients and concentration driving forces analogous to temperature driving forces. Thus

$$\frac{N_A}{A} = k_G'(p_{AG} - p_{Ai})$$

$$= k_L'(C_{Ai} - C_{AL})$$

$$= K_G'(p_{GA} - p_A^*)$$

$$= K_L'(C_A^* - C_{AL})$$

where the subscript i refers to interface and the superscript * refers to equilibrium pressures and concentrations relative to the opposing phases. Overall mass transfer coefficients can be related to individual ones by

$$\frac{1}{K_G'} = \frac{1}{k_G'} + \frac{H}{k_L'}$$

$$\frac{1}{K_L'} = \frac{1}{H k_G'} + \frac{1}{k_L'}$$

where H is the Henry's Law constant for the system.

For mass transfer in circular conduits under turbulent flow conditions, the individual inside mass transfer coefficient may be calculated using the Sherwood equation:

$$\frac{k_m D}{D_m} = 0.023\left(\frac{DV\rho}{\mu}\right)^{0.8}\left(\frac{\mu}{\rho D_m}\right)^{1/3}$$

For an absorption tower, the height of the tower can be calculated in terms of the number of transfer units times the height of a corresponding transfer unit. The height of a transfer unit is influenced by the characteristics of the materials being handled, the molal velocity of fluid flow (which affects turbulence) and the interfacial contact area (a) of the medium (packing, plates, etc.). Specifically,

$$H_{OG} = \frac{G}{K_G a}, \qquad H_{OL} = \frac{L}{K_L a}$$

Similarly, the number of transfer units required for a given separation is given by

$$N_{OG} = \int_{y_1}^{y_2} \frac{dy}{y - y^*} + \frac{1}{2}\ln\frac{1-y_2}{1-y_1}$$

$$N_{OL} = \int_{x_1}^{x_2} \frac{dx}{x - x^*} + \frac{1}{2}\ln\frac{1-x_1}{1-x_2}$$

Example 6.1

A cylindrical tube with an inside diameter of 1 cm is cast of benzoic acid. Water is passed through the tube at velocity of 20 cm/sec. The saturation concentration of benzoic acid in water at 57°F is 2.36 grams/liter. The Schmidt number is 1870 and the viscosity of water is 1.17 $\text{gm/cm}\cdot\text{sec}$.

1. How long must the tube be if the solution leaving the tube is to be 1% saturated?
   (A) 21 meters
   (B) 58 meters
   (C) 32 meters
   (D) 115 meters

2. What is the mean mass transfer coefficient for this system based on the arithmetic average driving force?
   (A) 0.0004 cm/sec
   (B) 0.01 cm/sec
   (C) 0.1 cm/sec
   (D) 0.0001 cm/sec

Solutions:

1. **D** First, calculate the Reynolds number:

$$\text{Re} = VD\rho / \mu$$

$$= \frac{20 \times 1 \times 1}{0.00117} = 17\,094$$

$\therefore$ Turbulent. Using Sherwood's Dimensionless Group Equation:

$$D_m = \frac{\mu}{\rho S_c} = \frac{0.00117}{1 \times 1870} = 6.25 \times 10^{-7}\ \text{cm}^2/\text{sec}$$

$$\frac{kD}{D_m} = 0.023\,(\text{Re})^{0.8}(\text{Sc})^{0.33} = 689$$

Thus, k = 0.00043 cm/sec. Then,

$$N_A = k(\rho_{As} - \rho_{Ab})\pi DL = 20(\rho_{A2} - \rho_{A1})\pi D^2 / 4$$
$$= 0.0004(2.36 - 0.0168)\frac{L}{1000} = 20\frac{(2.36 - 0.0236)}{4 \times 1000}$$

$\therefore L = 11566$ cm or 116 meters.

2. **A** Mass transfer coefficient = k = 0.00043 cm/sec.

Example 6.2

The waste product from a process contains 1 mole % of component A. The concentration of A is to be reduced to 100 ppm by volume in an absorption tower using a dilute alkaline solution in water as the solvent at a rate 1.25 times the minimum. For this system, the equilibrium data can be represented as $p = 5x$ in atmospheres. The liquid side resistance H_L is 0.5 m and the gas phase resistance is H_G = 1.2 m. The tower operates at 1 atm and 20°C with a Schmidt No. of 1 in the gas phase. The gas rate of 20 000 gmol/hr is used with 1 cm Raschig Rings.

1. What is the number of overall gas phase transfer units?
   (A) 15.6
   (B) 4.3
   (C) 8.2
   (D) 12.9

2. What is the height of the column?
   (A) 21 m
   (B) 25 m
   (C) 43 m
   (D) 18 m

Solutions:

1. **A** At the top of the column, $y_1 = 0.0001$ and $x = 0$.

   At the bottom of the column, $y_2 = 0.01$ and $x_2 = ?$

   At minimum L/G, $x_2 = x^* = 0.002$.

   Then, at minimum, $L(0.002 - 0) = G(0.01 - 0.0001)$ so that

$$(L/G)_{\text{MIN}} = 4.95$$
$$(L/G)_{\text{OP}} = 1.25 \times 4.95 = 6.19$$

Then,

$$L = 6.19 \times 20\,000 = 123\,700 \text{ gmol / hr}.$$

Material balance:

$$(0.01 - 0.0001) = 123700(x_2 - 0). \quad \therefore x_2 = 0.0016$$

Finally,

$$N_{\text{OG}} = \frac{y_2 - y_1}{(y - y^*)_{\text{LM}}}$$

But,

$$(y - y^*)_2 = 0.01 - 5 \times 0.0016 = 0.002$$
$$(y - y^*)_1 = 0.0001 - 0 = 0.0001$$

so that

$$(y - y^*)_{\text{LM}} = \frac{0.002 - 0.0001}{\ln(0.0019 / 0.0001)} = 0.00063$$

and

$$N_{\text{OG}} = \frac{0.01 - 0.0001}{0.00063} = 15.6$$

2. **B** The height of the column is found as follows:

$$H_{\text{OG}} = H_{\text{G}} + \frac{mG}{L}$$
$$= 1.2 + \frac{5 \times 0.5}{6.19} = 1.6 \text{ m}$$
$$\therefore z = H_{\text{OG}} N_{\text{OG}} = 15.6 \times 1.6 = 25 \text{ m}$$

7. Fluid Flow

The flow of fluids being used in a chemical process must be analyzed in many applications. The most common flow involves transfer of fluids from one process step to another, or raw materials/products to their corresponding destinations. In most cases, circular pipes are involved but, with some of the more exotic fluids (plastics, suspensions, etc.), noncircular cross sections are also encountered during processing.

For incompressible fluid flow in circular cross sections, the evaluation of momentum transfer resulting in friction losses is achieved by considering the velocity distribution in the pipe. At low Reynolds Numbers ($VD\rho/\mu < 2100$), a parabolic velocity profile is exhibited in the fluid and momentum transfer occurs primarily between layers of fluid so that the surface roughness of the pipe has little effect on the friction loss. At higher Reynolds Numbers, the velocity profile is nearly flat, and momentum transfer occurs at the wall of the pipe so that surface roughness dominates the friction loss phenomenon. Friction losses can be evaluated by

$$h_f = f\frac{L}{D}\frac{V^2}{2g}$$

where f is a friction factor dependent only on the Reynolds Number when the flow is laminar but also on surface roughness in turbulent flow conditions; the frition factor is found on the Moody diagram.

In addition to friction losses, the other factors that affect pressure differences along a pipe are variations in flow geometry, contractions or expansions, elevation changes, and pipe fittings. The combination of these can be achieved by applying the energy equation:

$$\frac{P_1}{\rho g} + z_1 + \frac{V_1^2}{2g} = \frac{P_2}{\rho g} + z_2 + \frac{V_2^2}{2g} + h_f + h_{f,fitting}$$

Values for $h_{f,fitting}$ can be determined as a function of the geometry and fluid velocity. An additional term can be included in the above balance to account for mechanical work done by or on the system.

Example 7.1

An oil (specific gravity = 0.9, viscosity = 10 cP) flows out of a constant pressure tank through 200 m of horizontal 1-cm (actual I.D.) pipe at 1 liter/min. To increase the flow, a second 1-cm pipe is connected to the tank and runs in parallel with the first pipe for 100 m where it joins the first pipe. Neglect the effects of fittings, entrance, etc. What will the total flow rate be after the addition of the pipe?

(A) 1.5 L/s

(B) 1.3 L/s

(C) 2.0 L/s

(D) 1.1 L/s

Solution:

For original case:

$$V = \frac{Q}{A} = \frac{0.001}{\pi \times 0.01^2 / 4} = 12.7 \text{ m / min}$$

$$\text{Re} = \frac{VD\rho}{\mu} = \frac{(12.7/60) \times 0.01 \times 900}{10 \times 0.001} = 190. \quad \therefore \text{ laminar flow}$$

The pressure drop (see p. 37 of the Handbook) is given by

$$\Delta p_o = \frac{8\mu QL}{\pi R^4}$$

For case 1:

$$\Delta p_o = \frac{8\mu Q_1 \times 200}{\pi R^4} = 200AQ_1 \text{ where } A = \text{Const}$$

For case 2, first 100 m:

$$\Delta p_A = 100AQ_A$$

For case 2, second 100 m:

$$\Delta p_B = 100AQ_B$$

But $2Q_A = Q_B$. Thus:

$$200AQ_1 = 100AQ_A + 100A(2Q_A). \quad \therefore Q_1 = 1.5Q_A = 0.75Q_B$$

Or $Q_B = 1.33Q_1$. $\therefore$ The flow under case 2 = 1.33 times the original flow = 1.33 L/min.

The answer is **B**.

Example 7.2

An oil having a density of 0.9 gm/cm^3 and a viscosity of 8 cP runs from a main to a shower head located 20 m above the main through a 1-inch wrought-iron pipe (I.D. = 2.66 cm). The pressure drop through the shower head is 0.7 atm. The desired flow is 1 L/sec. Contraction and enlargement losses are negligible. What must the pressure be in the main?

(A) 1.52 atm
(B) 3.26 atm
(C) 2.6 atm
(D) 11.1 atm

Solution:

First, calculate the Reynolds number:

$$V = Q / A = \frac{1000}{\pi \times 2.66^2 / 4} = 180 \text{ cm / s}$$

$$\text{Re} = \frac{180 \times 2.66 \times 0.9}{0.08} = 5385. \quad \therefore \text{ turbulent flow}$$

For 1-inch wrought-iron pipe, e/D =0.0018 (Moody diagram) so that at Re = 5385 we find f = 0.025. Then, the head loss is

$$h = f\frac{L}{D}\frac{V^2}{2g} = 0.025\frac{2000}{2.66}\frac{180^2}{2\times 980} = 310 \text{ cm}$$

Energy balance:

$$\frac{\Delta p}{\rho g} + \Delta z + \Delta\frac{V^2}{2g} + h_f + \left(\frac{\Delta p}{\rho g}\right)_{shower} = 0$$

$$\therefore\ \Delta p = (20 + 0 + 3.10 + 0.7 \times 100\,000 / 9800) \times 9800 = 296\,000 \text{ Pa} \text{ or } 2.96 \text{ atm}$$

The answer is **C**.

Equation Summaries

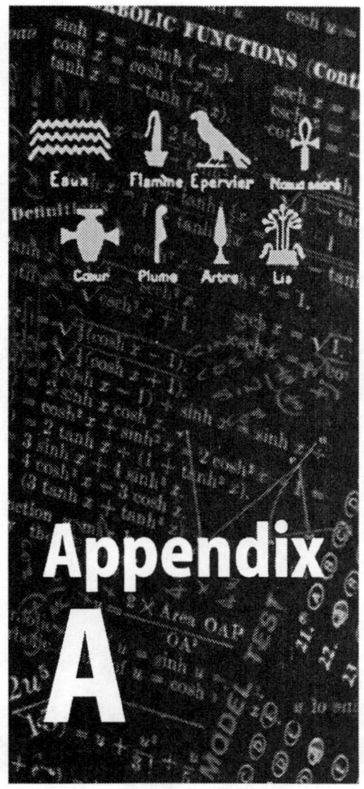

These summaries are useful for study

The following pages are Equation Summary Sheets of the FE/EIT subjects which rely most on equations. They are intended to be used for quick overviews, handy problem solving—and as part of a special strategy for preparing for the exam in its new format.

The selected equations are presented in the same format and nomenclature as is to be found in the NCEES Reference Handbook, a newsprint booklet which is given to all exam applicants. Be prepared! You may find some of the nomenclature to be different from what you are used to using, as the NCEES has apparently obtained some of it from older or obscure texts.

They can help you anticipate and prepare for NCEES obstacles

The current edition of the NCEES Reference Handbook has many extraneous equations which will in all likelihood be of little aid in solving exam problems. As discussed earlier, an excellent exam preparation strategy is to identify the equations you find to be most useful, then highlight them and become acquainted with their position in the Handbook, so that you may quickly access them in the 'clean' Handbook during the exam itself. This strategy should maximize your ability to perform as well as possible, given the present exam format.

Calculators are a good way to access equations during the exam

A programmable calculator may be pre-programmed to solve many of the problems on the FE exam, e.g., problems involving matrices, permutations, standard deviations, Mohr's circle problems, etc. Or you may choose to buy the HP-48G, which has hundreds of equations and constants preprogrammed. (We offer this superior calculator at a significant discount.) Not all states allow the use of the HP-48G, so be sure you check with your state board.

Mathematics

—Selected Equations from the NCEES Reference Handbook—

Straight Line:

$y = mx + b$ (slope - intercept form)

$y - y_1 = m(x - x_1)$ (point - slope form)

$m = \dfrac{y_2 - y_1}{x_2 - x_1}$ (slope)

$m_1 = -\dfrac{1}{m_2}$ (two perpendicular lines)

Quadratic Equation: $ax^2 + bx + c = 0$ $\quad$ roots $= \dfrac{-b \pm \sqrt{b^2 - 4ac}}{2a}$

Conic Sections:

	General Form	$h = k = 0$
Parabola:	$(y-k)^2 = 2p(x-h)$	$y^2 = 2px$ Focus: $(p/2, 0)$ Directrix: $x = -p/2$
Ellipse:	$\dfrac{(x-h)^2}{a^2} + \dfrac{(y-k)^2}{b^2} = 1$	$\dfrac{x^2}{a^2} + \dfrac{y^2}{b^2} = 1$ Focus: $\left(\sqrt{a^2 - b^2},\ 0\right)$
Hyperbola:	$\dfrac{(x-h)^2}{a^2} - \dfrac{(y-k)^2}{b^2} = 1$	$\dfrac{x^2}{a^2} - \dfrac{y^2}{b^2} = 1$ Focus: $\left(\sqrt{a^2 + b^2},\ 0\right)$
Circle:	$(x-h)^2 + (y-k)^2 = r^2$	$x^2 + y^2 = r^2$

Logarithms:

$\ln x = 2.3026 \log x$ $\quad$ $\log xy = \log x + \log y$ $\quad$ $\log x/y = \log x - \log y$

$\log_b b^n = n$ $\quad$ $\log_b b = 1$ $\quad$ If $b^c = x$, then $\log_b x = c$

$\log x^c = c \log x$ $\quad$ $\log 1 = 0$

Trigonometry:

$\sin \theta = y / r$ $\quad$ $\cos \theta = x / r$

$\tan \theta = y / x$ $\quad$ $\cot \theta = x / y$

$\csc \theta = r / y$ $\quad$ $\sec \theta = r / x$

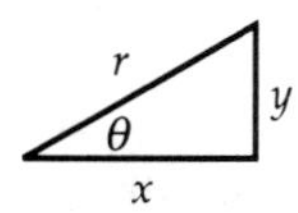

Law of Sines: $\dfrac{a}{\sin A} = \dfrac{b}{\sin B} = \dfrac{c}{\sin C}$

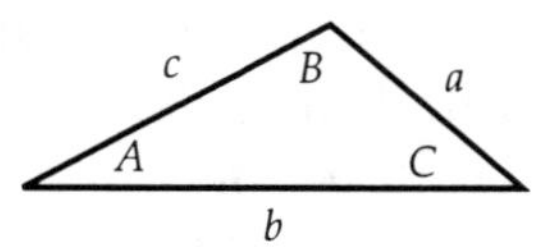

Law of Cosines: $a^2 = b^2 + c^2 - 2bc \cos A$

Identities:

$\tan \theta = \sin \theta / \cos \theta$

$\sin^2 \theta + \cos^2 \theta = 1$

$\sin(\alpha + \beta) = \sin \alpha \cos \beta + \cos \alpha \sin \beta$

$\cos(\alpha + \beta) = \cos \alpha \cos \beta - \sin \alpha \sin \beta$

$\sin 2\alpha = 2 \sin \alpha \cos \alpha$

$\cos 2\alpha = \cos^2 \alpha - \sin^2 \alpha$

$= 2\cos^2 \alpha - 1$

$= 1 - 2\sin^2 \alpha$

Complex Numbers:

$$i = \sqrt{-1} \qquad x + iy = re^{i\theta} \qquad \cos\theta = \frac{e^{i\theta} + e^{-i\theta}}{2}$$

$$r = \sqrt{x^2 + y^2} \qquad e^{i\theta} = \cos\theta + i\sin\theta \qquad \sin\theta = \frac{e^{i\theta} - e^{-i\theta}}{2i}$$

$$(x + iy)^n = r^n(\cos n\theta + i\sin n\theta)$$

Matrices:

Transpose: $\mathbf{B} = \mathbf{A}^T$ if $b_{ji} = a_{ij}$

Inverse: $\mathbf{A}^{-1} = \dfrac{\text{adj}(\mathbf{A})}{|\mathbf{A}|}$

Adjoint: $\text{adj}(\mathbf{A}) =$ matrix formed by replacing $\mathbf{A}^T$ elements with their cofactors

Cofactor: cofactor = minor $\times (-1)^{h+k}$ where h = column, k = row

Minor: minor = determinant that remains after the common row and column are struck out

Vectors:

$$\mathbf{A}\cdot\mathbf{B} = a_x b_x - a_y b_y + a_z b_z = |\mathbf{A}||\mathbf{B}|\cos\theta = \mathbf{B}\cdot\mathbf{A}$$

$$\mathbf{A}\times\mathbf{B} = \begin{vmatrix} \mathbf{i} & \mathbf{j} & \mathbf{k} \\ a_x & a_y & a_z \\ b_x & b_y & b_z \end{vmatrix} = |\mathbf{A}||\mathbf{B}|\mathbf{n}\sin\theta = -\mathbf{B}\times\mathbf{A}$$

where $\mathbf{n}$ is $\perp$ plane of $\mathbf{A}$ and $\mathbf{B}$

$$\mathbf{i}\cdot\mathbf{i} = \mathbf{j}\cdot\mathbf{j} = \mathbf{k}\cdot\mathbf{k} = 1 \qquad \mathbf{i}\times\mathbf{j} = \mathbf{k}, \quad \mathbf{j}\times\mathbf{k} = \mathbf{i}, \quad \mathbf{k}\times\mathbf{i} = \mathbf{j}$$

Taylor Series:

$$f(x) = f(a) + \frac{f'(a)}{1!}(x - a) + \frac{f''(a)}{2!}(x - a)^2 + \cdots$$

Maclaurin Series: a Taylor series with $a = 0$

Probability and Statistics:

$$P(n,r) = \frac{n!}{(n-r)!}$$ (permutation of n things taken r at a time)

$$C(n,r) = \frac{P(n,r)}{r!} = \frac{n!}{r!(n-r)!}$$ (combination of n things taken r at a time)

$$\bar{x} = \frac{x_1 - x_2 + \cdots + x_n}{n}$$ (arithmetic mean)

$$\sigma^2 = \frac{\sum (x_i - \bar{x})^2}{n - 1}$$ (variance)

$$\sigma = \sqrt{\text{variance}}$$ (sample standard deviation)

$$\text{median} = \begin{cases} \text{middle value if odd number of items} \\ \tfrac{1}{2}(\text{sum of middle two values}) \text{ if even number of items} \end{cases}$$

mode = value that occurs most often

Calculus: $f'(x) = 0 \begin{cases} \text{maximum} & \text{if } f''(x) < 0 \\ \text{minimum} & \text{if } f''(x) > 0 \end{cases}$

L'Hospital's Rule: $\lim_{x \to a} \frac{f(x)}{g(x)} = \lim_{x \to a} \frac{f'(x)}{g'(x)}$ if $\frac{f(a)}{g(a)} = \frac{0}{0}$ or $\frac{\infty}{\infty}$

$$\frac{d}{dx}(uv) = u\frac{dv}{dx} + v\frac{du}{dx} \qquad \frac{d}{dx}(\ln u) = \frac{1}{u}\frac{du}{dx} \qquad \frac{d}{dx}(\sin u) = \cos u\frac{du}{dx}$$

$$\frac{d}{dx}\left(\frac{u}{v}\right) = \frac{v\,du/dx - u\,dv/dx}{v^2} \qquad \frac{d}{dx}\left(e^u\right) = e^u\frac{du}{dx} \qquad \frac{d}{dx}(\cos u) = -\sin u\frac{du}{dx}$$

$$\frac{d}{dx}\left(u^n\right) = nu^{n-1}\frac{du}{dx}$$

$$\int x^n dx = \frac{x^{n+1}}{n+1} \quad n \neq -1 \qquad \int \sin x\,dx = -\cos x \qquad \int \sin^2 x\,dx = \frac{x}{2} - \frac{\sin 2x}{4}$$

$$\int \frac{dx}{ax+b} = \frac{1}{a}\ln|ax+b| \qquad \int \cos x\,dx = \sin x \qquad \int \cos^2 x\,dx = \frac{x}{2} + \frac{\sin 2x}{4}$$

$$\int e^{ax} dx = \frac{1}{a}e^{ax}$$

Differential Equations: $y'' + 2ay' + by = f(x)$ (linear, 2nd order, constant coefficient, nonhomogeneous)

Homogeneous solution:

$$y_h(x) = C_1 e^{r_1 x} + C_2 e^{r_2 x} \quad \text{if } r_1 \neq r_2 \quad \text{where } r^2 + 2ar + b = 0$$
$$= (C_1 + C_2 x)e^{r_1 x} \quad \text{if } r_1 = r_2$$
$$= e^{-ax}(C_1 \cos \beta x + C_2 \sin \beta x) \quad \text{if } a^2 < b. \quad \beta = \sqrt{b - a^2}$$

Particular solution:

$$y_p = B \quad \text{if } f(x) = A$$
$$= Be^{\alpha x} \quad \text{if } f(x) = Ae^{\alpha x}$$
$$= B_1 \sin \omega x + B_2 \cos \omega x \quad \text{if } f(x) = A_1 \sin \omega x + A_2 \cos \omega x$$

General solution: $y(x) = y_h(x) + y_p(x)$

Mechanics of Materials

—Selected equations from the NCEES Reference Handbook—

Definitions:

$$\sigma = \varepsilon E$$
$$\tau = \gamma G$$
$$E = 2G(1+\nu)$$
$$\nu = -\frac{\varepsilon_{lateral}}{\varepsilon_{longitudinal}}$$

E = modulus of elasticity
G = shear modulus
σ and τ = normal and shear stress
ε and γ = normal and shear strain
ν = Poisson's ratio

Uniaxial Loading:

$$\left.\begin{aligned}\sigma &= \frac{P}{A}\\ \varepsilon &= \frac{\delta}{L}\end{aligned}\right\} \quad \delta = \frac{PL}{AE}$$

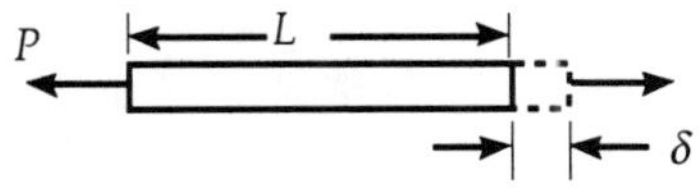

Thermal Deformation:

$$\delta_t = \alpha L(T - T_o)$$

α = coefficient of thermal expansion

Thin-walled Pressure Vessel:

$$\sigma_t = \frac{pD}{2t}$$ hoop (circumferential) stress

$$\sigma_a = \frac{pD}{4t}$$ axial (longitudinal) stress

t = cylinder thickness
D = cylinder diameter
p = pressure

Stress and Strain:

Stress Condition

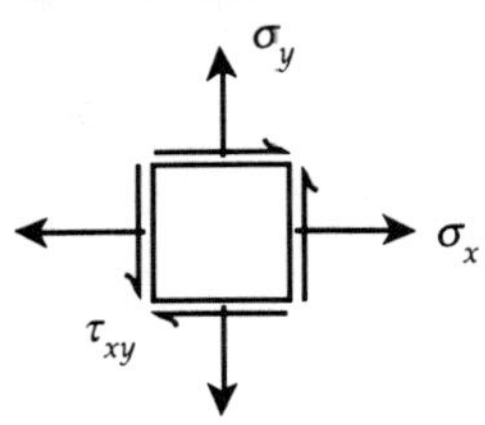

Mohr's Circle

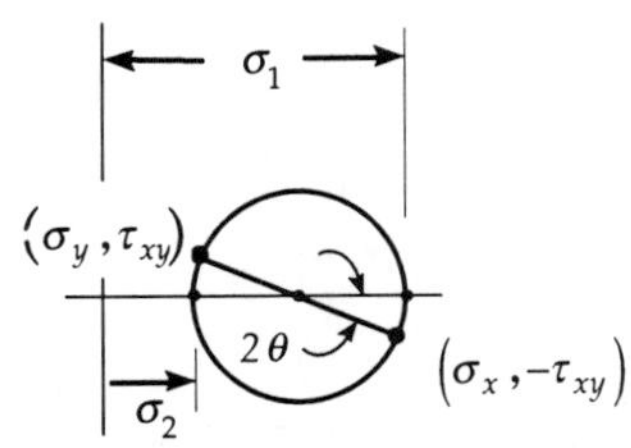

Maximum and Minimum Stresses

$$\sigma_1 = \sigma_{max} = \frac{\sigma_x + \sigma_y}{2} + \left[\left(\sigma_x - \sigma_y\right)^2/4 + \tau_{xy}^2\right]^{1/2}$$

$$\sigma_2 = \sigma_{min} = \frac{\sigma_x + \sigma_y}{2} - \left[\left(\sigma_x - \sigma_y\right)^2/4 + \tau_{xy}^2\right]^{1/2}$$

$$\tau_{max} = \frac{\sigma_1 - \sigma_2}{2}$$ = radius of Mohr's circle

3-D Strain:

$$\varepsilon_x = \frac{1}{E}\left[\sigma_x - \nu\left(\sigma_y + \sigma_z\right)\right]$$
$$\gamma_{xy} = \frac{\tau_{xy}}{G}$$

Torsion:

$$\tau = \frac{Tr}{J}$$ (shear stress)

$$\phi = \frac{TL}{JG}$$ (angle of twist)

J = polar moment of inertia
$= \pi r^4/2$ for a circle

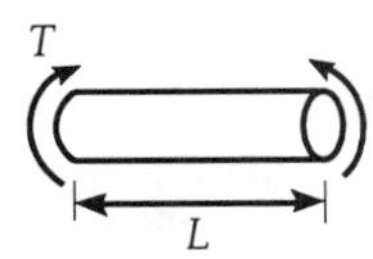

Beams:

$$V = \frac{dM}{dx}$$ V = vertical shear force, M = bending moment

$$\sigma = -\frac{My}{I}$$

I = centroidal moment of inertia
$= bh^3/12$ for a rectangle
$= \pi r^4/4$ for a circle

y = distance from neutral axis

$$\tau = \frac{VQ}{Ib}$$

Q = moment of area between y-position and top or bottom

$$EIy'' = M$$ differential equation of deflection curve

Columns:

$$P_{cr} = \frac{\pi^2 EI}{k^2 L^2}$$

$$k = \begin{cases} 1 & \text{ends pinned} \\ 0.5 & \text{ends fixed} \\ 0.7 & \text{one pinned, one fixed} \\ 2 & \text{one fixed, one free} \end{cases}$$

Dynamics

—Selected equations from the NCEES Reference Handbook—

Kinematics (motion only)

Tangential and Normal Components:

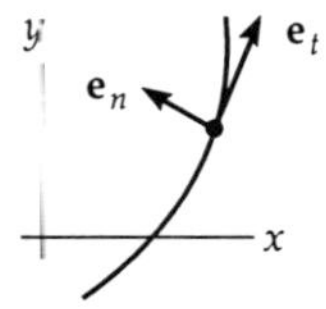

$$\mathbf{a} = \frac{dv_t}{dt}\mathbf{e_t} + \frac{v_t^2}{\rho}\mathbf{e_n}$$

$$\mathbf{v} = v_t \mathbf{e_t}$$

ρ = radius of curvature

Plane Circular Motion:

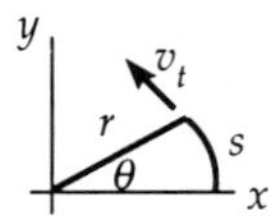

$$\mathbf{e_r} = -\mathbf{e_n} \qquad \mathbf{e_\theta} = \mathbf{e_t}$$

$$\omega = \dot{\theta} = \frac{v_t}{r} \qquad a_t = r\alpha$$

$$\alpha = \dot{\omega} = \ddot{\theta} = \frac{a_t}{r} \qquad a_n = \frac{v_t^2}{r} = r\omega^2$$

$$v_t = r\omega \qquad s = r\theta$$

Straight Line Motion:

$$s = s_o + v_o t + a_o t^2/2$$
$$v = v_o + a_o t$$
$$v^2 = v_o^2 + 2a_o(s - s_o)$$

Projectile Motion:

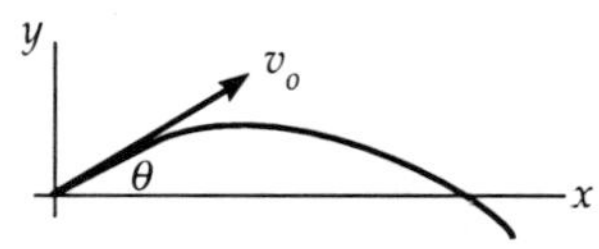

$$a_x = 0, \quad a_y = -g$$
$$v_x = v_o \cos\theta \qquad x = v_o t \cos\theta$$
$$v_y = v_o \sin\theta - gt \qquad y = v_o t \sin\theta - \tfrac{1}{2}gt^2$$

Kinematics (forces and motion)

$$\sum \mathbf{F} = \frac{d}{dt}(m\mathbf{v}), \qquad \sum F_t = ma_t = m\frac{dv_t}{dt}, \qquad \sum F_n = ma_n = m\frac{v_t^2}{\rho}$$

Impulse and Momentum: $m[v_x(t) - v_x(0)] = \int_0^t F_x(t)dt$ or change in momentum = impulse

Work and Energy: $PE_1 + KE_1 + W_{1\to2} = PE_2 + KE_2$ where

$$KE = \tfrac{1}{2}mv^2$$
$$PE = mgh \quad \text{(gravity)}$$
$$= \tfrac{1}{2}kx^2 \quad \text{(spring)}$$
$$W_{1\to2} = \text{friction force work}$$

Impact:

$$m_1v_1 + m_2v_2 = m_1v_1' + m_2v_2'$$

$$e = -\frac{v_{1n}' - v_{2n}'}{v_{1n} - v_{2n}} = \begin{cases}1 & \text{elastic}\\ 0 & \text{plastic}\end{cases}$$

v_1, v_2 = velocities before impact
v_1', v_2' = velocities after impact

Rotation: $I_o\alpha = \sum M_o$ where $I_o = \int(x^2 + y^2)dm$, rotation about O.

constant M:

$$\alpha = \frac{M}{I}$$
$$\omega = \omega_o + \frac{M}{I}t$$
$$\theta = \theta_o + \omega_o t + \frac{M}{2I}t^2$$

work and energy: $I_o\frac{\omega^2}{2} - I_o\frac{\omega_o^2}{2} = \int_{\theta_o}^{\theta} M d\theta$

Banking of Curves: $\tan\theta = \frac{v^2}{rg}$ where r = radius of curvature
θ = angle between surface and horizontal

Electric Circuits

—Selected Equations from the NCEES Reference Handbook—

Electrostatics:

$F_2 = \dfrac{Q_1 Q_2}{4\pi\varepsilon r^2}$ (force on charge 2 due to charge 1) ε = permittivity — $\mathrm{C^2/N \cdot m^2 = F/m}$

$= 8.85 \times 10^{-12}$ for air or free space

$E = \dfrac{Q}{4\pi\varepsilon r^2}$ (electric field intensity due to point charge Q — C)

$E_L = \dfrac{\rho_L}{2\pi\varepsilon r}$ (radial field due to line charge ρ_L — C/m)

$E_s = \dfrac{\rho_s}{2\varepsilon}$ (plane field due to sheet charge ρ_s — $\mathrm{C/m^2}$)

$Q = \oint \varepsilon \mathbf{E} \cdot d\mathbf{A}$ (enclosed charge — C)

$E = \dfrac{V}{d}$ (electric field between plates with potential difference V separated by the distance d)

$H = \dfrac{I}{2\pi r}$ (magnetic field strength due to current in long wire)

$B = \mu H$ (magnetic flux density)

$\mathbf{F} = I\mathbf{L} \times \mathbf{B}$ (force on conductor) $\mathbf{L}$ = length vector of conductor

DC Circuits:

Resistors:

$V = IR$ (Ohm's law) $R_T = R_1 + R_2 + \cdots$ (series)

$P = VI = \dfrac{V^2}{R} = I^2 R$ (power) $R_T = \left[\frac{1}{R_1} + \frac{1}{R_2} + \cdots\right]^{-1}$ (parallel)

Capacitors:

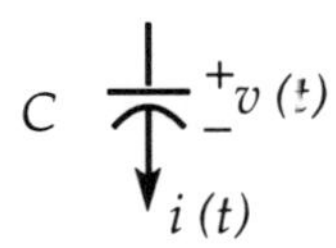

$i = C\dfrac{dv}{dt}$ $\text{energy stored} = \frac{1}{2}Cv^2$ $C_{eq} = C_1 + C_2 + \cdots$ (parallel)

$v = \dfrac{1}{C}\int i\,dt$ $C_{eq} = \left[\frac{1}{C_1} + \frac{1}{C_2} + \cdots\right]^{-1}$ (series)

Inductors:

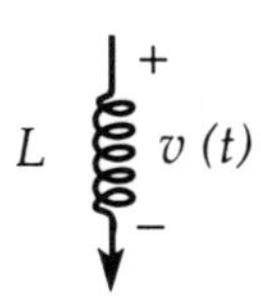

$i = \dfrac{1}{L}\int v\,dt$ $\text{energy stored} = \frac{1}{2}Li^2$ $L_{eq} = L_1 + L_2 + \cdots$ (series)

$v = L\dfrac{di}{dt}$ $L_{eq} = \left[\frac{1}{L_1} + \frac{1}{L_2} + \cdots\right]^{-1}$ (parallel)

Kirchhoff Voltage Law (KVL): $\sum V_{rises} = \sum V_{drops} = 0$

Kirchhoff Current Law (KCL): $\sum I_{in} = \sum I_{out}$

Thevenin equivalent circuit:

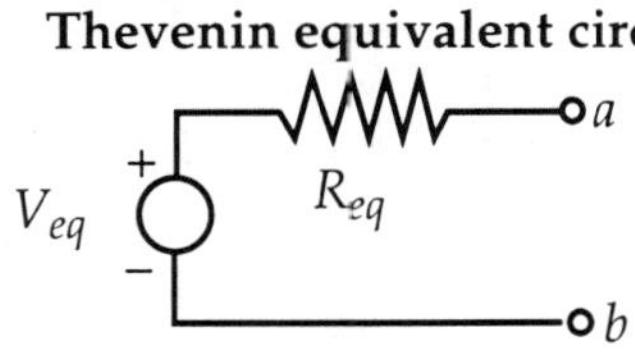

$R_{eq} = \dfrac{V_{eq}}{I_{sc}}$ I_{sc} = short circuit current

V_{eq} = open circuit voltage

RC Transients:

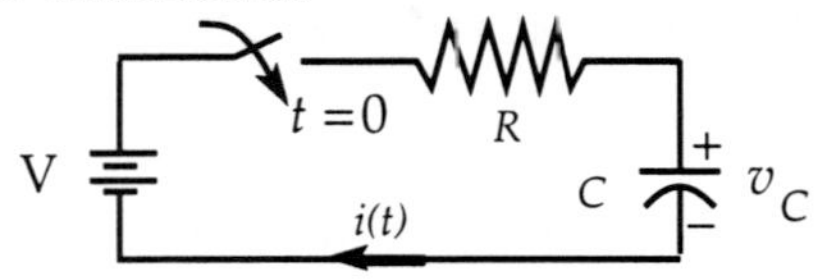

$$v_C(t) = v_C(0)e^{-t/RC} + V\left(1 - e^{-t/RC}\right)$$

$$i(t) = \{[V - v_C(0)]/R\}e^{-t/RC}$$

RL Transients:

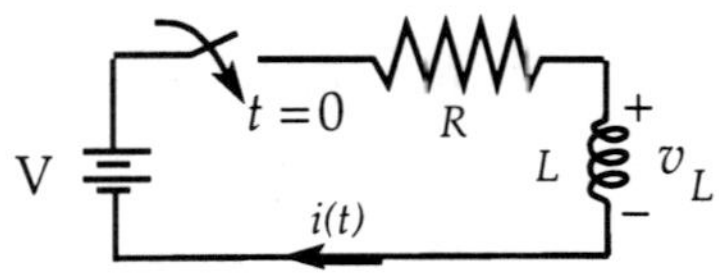

$$v_L(t) = -i(0)Re^{-Rt/L} + Ve^{-Rt/L}$$

$$i(t) = i(0)e^{-Rt/L} + \frac{V}{R}\left(1 - e^{-Rt/L}\right)$$

Operational Amplifiers:

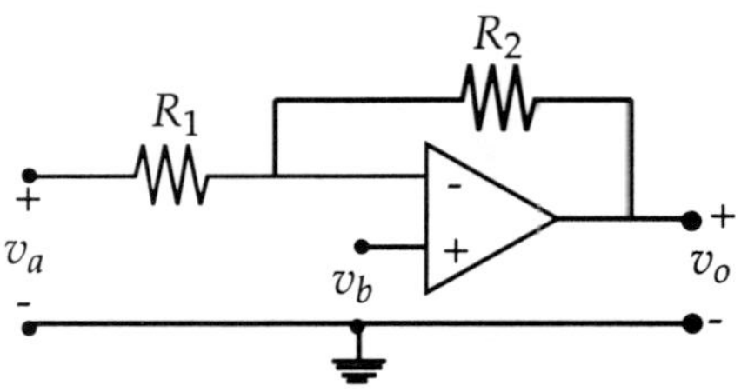

$$v_o = -\frac{R_2}{R_1}v_a + \left(1 + \frac{R_2}{R_1}\right)v_b$$

inverting if $v_b = 0$

non - inverting if $v_a = 0$

AC Circuits: (single phase)

$$f = \frac{1}{T} = \frac{\omega}{2\pi}$$

f = frequency (Hz)
T = period (sec)
ω = angular frequency (rad / s)

$V_{avg} = \frac{2}{\pi}V_{max}$ (full-wave rectified sine wave)

$V_{avg} = \frac{1}{\pi}V_{max}$ (half-wave rectified sine wave)

$V_{rms} = \frac{1}{\sqrt{2}}V_{max}$ (full-wave rectified sine wave)

$V_{rms} = \frac{1}{2}V_{max}$ (half-wave rectified sine wave)

Resistor: $Z = R$ — Z = Impedance

Capacitor: $Z = -\frac{j}{\omega C} = -jX$ — X = Reactance

Inductor: $Z = j\omega L = jX$

$V = IZ$

$P = \frac{1}{2}V_{max}I_{max}\cos\theta = V_{rms}I_{rms}\cos\theta$ (real power) ($\theta = 0$ for resistors)

$Q = \frac{1}{2}V_{max}I_{max}\sin\theta = V_{rms}I_{rms}\sin\theta$ (reactive power)

p.f. $= \cos\theta$ (power factor)

Resonance: $f = \frac{1}{2\pi\sqrt{LC}}$ (resonant frequency for series and parallel circuits)

Fluid Mechanics

—Selected Equations from the NCEES Reference Handbook—

Properties:

$\rho = \frac{m}{V\!\!\!/}$ (density)

$\gamma = \rho g$ (specific weight)

$v = \frac{\mu}{\rho}$ (kinematic viscosity)

$\tau_n = -p$ (normal stress)

$\tau_t = \mu \frac{dv}{dy}$ (tangential stress)

μ = dynamic viscosity

Statics:

$p_2 - p_1 = -\gamma h$ (h is vertical upward)

$F_{\text{buoyant}} = \gamma V\!\!\!/_{\text{displaced}}$ (Archimedes' principle)

$F = \gamma h_C A$

$z^* = \frac{I_C}{A Z_C}$

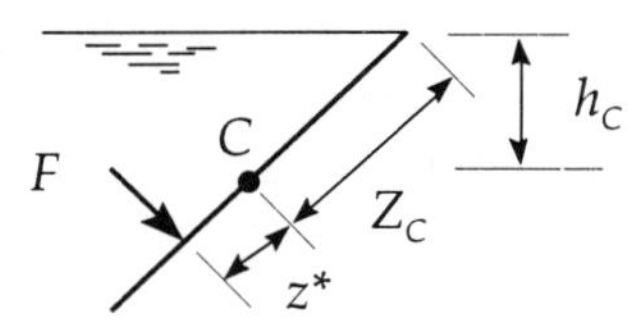

One-Dimensional Flows:

$A_1 V_1 = A_2 V_2$ (continuity equation)

$Q = AV$ (flow rate)

$\dot{m} = \rho A V$ (mass flow rate)

$$-\frac{\dot{W}_S}{\gamma Q} + \frac{p_1}{\gamma} + \frac{V_1^2}{2g} + z_1 = \frac{p_2}{\gamma} + \frac{V_2^2}{2g} + z_2 + h_f$$ (Energy Equation — if $h_f = \dot{W}_s = 0$, then Bernoulli Eq.)

$$h_f = f \frac{L}{D} \frac{V^2}{2g}$$ (Darcy's Equation — find f on Moody Diagram)

$$\text{Re} = \frac{VD\rho}{\mu}$$ (Reynolds Number)

$$h_{f,\,\text{fitting}} = C \frac{V^2}{2g}$$ (minor losses — C is loss coefficient)

$$\sum \mathbf{F} = \rho Q (\mathbf{V}_2 - \mathbf{V}_1)$$ (Momentum equation)

Perfect Gas:

$p = \rho R T$ (perfect gas law)

$c = \sqrt{kRT}$ (speed of sound)

$\text{M} = \frac{V}{c}$ (Mach number)

Similitude: If viscous effects dominate (internal flows) then Reynolds numbers on prototype and model are equated:

$$(\text{Re})_p = (\text{Re})_m \quad \text{or} \quad \left(\frac{V\ell\rho}{\mu}\right)_p = \left(\frac{V\ell\rho}{\mu}\right)_m$$

If gravity dominates (dams, weirs, ships) then Froude numbers are equated:

$$(\text{Fr})_p = (\text{Fr})_m \quad \text{or} \quad \left(\frac{V^2}{\ell g}\right)_p = \left(\frac{V^2}{\ell g}\right)_m$$

Open Channel:

$$Q = \frac{C}{n} A R^{2/3} S^{1/2} \quad \text{where} \quad R = \frac{A}{P_{wetted}}$$

$$C = \begin{cases} 1.0 & \text{metric} \\ 1.49 & \text{english} \end{cases}$$

Thermodynamics

—Selected Equations from the NCEES Reference Handbook—

Properties:

P	(absolute pressure, kPa or lbf/in^2)
$v = \dfrac{V}{m}$	(specific volume, m^3/kg or ft^3/lbm)
u	(internal energy, kJ/kg or Btu/lbm)
$h = u + Pv$	(enthalpy, kJ/kg or Btu/lbm)
s	(entropy, kJ/kg·K or Btu/lbm-°R)
c_p	(constant pressure specific heat, kJ/kg·K or Btu/lbm-°R)
c_v	(constant volume specific heat, kJ/kg·K or Btu/lbm-°R)
$x = \dfrac{m_v}{m_{total}}$	(quality)

Two phase system:

$v = v_f + xv_{fg}$ where $v_{fg} = v_g - v_f$

$h = h_f + xh_{fg}$

v_f = saturated liquid value

v_g = saturated vapor value

Ideal gas:

$Pv = RT, \quad PV = mRT$ where $R = \dfrac{\overline{R}}{M}, \quad \overline{R} = 8.314\dfrac{\text{kJ}}{\text{kmol}\cdot\text{K}}$ or $1545\dfrac{\text{ft - lbf}}{\text{lbmol - °R}}$

$\Delta u = c_v \Delta T, \qquad \Delta h = c_p \Delta T$

$$\Delta s = c_p \ln\frac{T_2}{T_1} - R\ln\frac{P_2}{P_1} = c_v \ln\frac{T_2}{T_1} + R\ln\frac{v_2}{v_1}$$

$$\frac{T_2}{T_1} = \left(\frac{P_2}{P_1}\right)^{\frac{k-1}{k}} = \left(\frac{v_1}{v_2}\right)^{k-1}, \qquad P_2 v_2^k = P_1 v_1^k, \quad k = c_p/c_v \quad \text{(constant entropy process)}$$

First law (system):

$q - w = \Delta u$ where $w = \int P dv$

$$= RT\ln\frac{v_2}{v_1} = RT\ln\frac{P_1}{P_2}$$ (isothermal process with ideal gas)

First law (control volume):

$h_i + V_i^2/2 = h_e + V_e^2/2$	(nozzles, diffusers)	i = inlet
$h_i = h_e + w$	(turbine, compressor)	e = exit
$h_i = h_e$	(throttling device, valve)	
$h_i + q = h_e$	(boilers, condensers, evaporators)	

Cycles:

$$\eta = \frac{W}{Q_H} = \frac{Q_H - Q_L}{Q_H} \quad \text{(efficiency)} \qquad COP = \frac{Q_H}{W} \quad \text{(heat pump)}$$

$$= 1 - \frac{T_L}{T_H} \quad \text{(Carnot cycle)} \qquad = \frac{Q_L}{W} \quad \text{(refrigerator)}$$

Second Law: No engine can produce work while transferring heat with a single reservoir. (Kelvin-Planck)

No refrigerator can operate without a work input. (Clausius)

$$\Delta S \geq \int \frac{\delta Q}{T} \qquad \Delta S = \frac{Q}{T} \text{ (reservoir or } T = \text{const)}$$

$$\Delta S_{total} = \Delta S_{surr} + \Delta S_{system} \geq 0 \qquad \Delta S = C_p \ln \frac{T_2}{T_1} \quad \text{(solid or liquid)}$$

Heat Transfer:

$$q = -kA\frac{dT}{dx} \quad \text{(conduction)} \qquad k = \text{conductivity}$$

$$= -kA\frac{T_2 - T_1}{L} \quad \text{(through a wall)} \qquad R = \frac{L}{kA} \quad \text{(resistance factor)}$$

$$q = hA(T_1 - T_2) \quad \text{(convection)} \qquad R = \frac{1}{hA} \quad \text{(resistance factor)}$$

$$= \varepsilon\sigma A\left(T_1^4 - T_2^4\right)F_{12} \quad \text{(radiation)} \qquad h = \text{convection coefficient}$$

$$\varepsilon = 1 \text{ for black body} \quad \text{(emissivity)}$$

$$\sigma = 5.67 \times 10^{-8} \frac{\text{W}}{\text{m}^2 \cdot \text{K}^4} \quad \text{(Stefan - Boltzmann constant)}$$

$$F_{12} = 1 \text{ if one body encloses the other} \quad \text{(shape factor)}$$

English and SI Units

The following tables present the SI (Systems International) units and the conversion of English units to SI units, along with some of the more common conversion factors.

SI Prefixes

Multiplication Factor	*Prefix*	*Symbol*
10^{15}	peta	P
10^{12}	tera	T
10^{9}	giga	G
10^{6}	mega	M
10^{3}	kilo	k
10^{-1}	deci	d
10^{-2}	centi	c
10^{-3}	mili	m
10^{-6}	micro	μ
10^{-9}	nano	n
10^{-12}	pico	p
10^{-15}	femto	f

SI Base Units

Quanitiy	*Name*	*Symbol*
length	meter	m
mass	kilogram	kg
time	second	s
electric current	ampere	A
temperature	kelvin	K
amount of substance	mole	mol
luminous intensity	candela	cd

SI Derived Units

Quantity	*Name*	*Symbol*	*In Terms of Other Units*
area	square meter		m^2
volume	cubic meter		m^3
velocity	meter per second		m/s
acceleration	meter per second squared		m/s^2
density	kilogram per cubic meter		kg/m^3
specific volume	cubic meter per kilogram		m^3/kg
frequency	hertz	Hz	s^{-1}
force	newton	N	$m \cdot kg/s^2$
pressure, stress	pascal	Pa	$kg/(m \cdot s^2)$
energy, work, heat	joule	J	$N \cdot m$
power	watt	W	J/s
electric charge	coulomb	C	A s
electric potential	volt	V	W/A
capacitance	farad	F	C/V
electric resistance	ohm	Ω	V/A
conductance	siemens	S	A/V
magnetic flux	weber	Wb	$V \cdot s$
inductance	henry	H	Wb/A
viscosity	pascal second		$Pa \cdot s$
moment (torque)	meter newton		$N \cdot m$
heat flux	watt per square meter		W/m^2
entropy	joule per kelvin		J/K
specific heat	joule per kilogram-kelvin		$J/(kg \cdot K)$
conductivity	watt per meter-kelvin		$W/(m \cdot K)$

Conversion Factors to SI Units

English	SI	SI Symbol	To Convert from English to SI Multiply by
Area			
square inch	square centimeter	cm^2	6.452
square foot	square meter	m^2	0.09290
acre	hectare	ha	0.4047
Length			
inch	centimeter	cm	2.54
foot	meter	m	0.3048
mile	kilometer	km	1.6093
Volume			
cubic inch	cubic centimeter	cm^3	16.387
cubic foot	cubic meter	m^3	0.02832
gallon	cubic meter	m^3	0.003785
gallon	liter	L	3.785
Mass			
pound mass	kilogram	kg	0.4536
slug	kilogram	kg	14.59
Force			
pound	newton	N	4.448
kip (1000 lb)	newton	N	4448
Density			
pound/cubic foot	kilogram/cubic meter	kg/m^3	16.02
pound/cubic foot	grams/liter	g/L	16.02
Work, Energy, Heat			
foot-pound	joule	J	1.356
BTU	joule	J	1055
BTU	kilowatt-hour	kWh	0.000293
therm	kilowatt-hour	kWh	29.3

Conversion Factors to SI Units (continued)

English	*SI*	*SI Symbol*	*To Convert from English to SI Multiply by*
	Power, Heat, Rate		
horsepower	watt	W	745.7
foot pound/sec	watt	W	1.356
BTU/hour	watt	W	0.2931
BTU/hour-ft^2-°F	watt/meter squared-°C	$W/m^2 \cdot °C$	5.678
tons of refrig.	kilowatts	kW	3.517
	Pressure		
pound/square inch	kilopascal	kPa	6.895
pound/square foot	kilopascal	kPa	0.04788
inches of H_20	kilopascal	kPa	0.2486
inches of Hg	kilopascal	kPa	3.374
one atmosphere	kilopascal	kPa	101.3
	Temperature		
Fahrenheit	Celsius	°C	5/9 (°F-32)
Fahrenheit	kelvin	K	5/9 (°F+460)
	Velocity		
foot/second	meter/second	m/s	0.3048
mile/hour	meter/second	m/s	0.4470
mile/hour	kilometer/hour	km/h	1.609
	Acceleration		
foot/second squared	meter/second squared	m/s^2	0.3048
	Torque		
pound-foot	newton-meter	$N \cdot m$	1.356
pound-inch	newton-meter	$N \cdot m$	0.1130
	Viscosity, Kinematic Viscosity		
pound-sec/square foot	newton-sec/square meter	$N \cdot s/m^2$	47.88
square foot/second	square meter/second	m^2/s	0.09290
	Flow Rate		
cubic foot/minute	cubic meter/second	m^3/s	0.0004719
cubic foot/minute	liter/second	L/s	0.4719
	Frequency		
cycles/second	hertz	Hz	1.00

Conversion Factors

Length

1 cm = 0.3937 in
1 m = 3.281 ft
1 yd = 3 ft
1 mi = 5280 ft
1 mi = 1760 yd
1 km = 3281 ft

Area

1 cm^2 = 0.155 in^2
1 m^2 = 10.76 ft^2
1 ha = 10^4 m^2
1 acre = 100 m^2
1 acre = 4047 m^2
1 acre = 43560 ft^2

Volume

1 ft^3 = 28.32 L
1 L = 0.03531 ft^3
1 L = 0.2642 gal
1 m^3 = 264.2 gal
1 ft^3 = 7.481 gal
1 m^3 = 35.31 ft^3
1 acre-ft = 43,560 ft^3
1 m^3 = 1000 L

Velocity

1 m/s = 3.281 ft/s
1 mph = 1.467 ft/s
1 mph = 0.8684 knot
1 knot = 1.688 ft/s
1 km/h = 0.2778 m/s
1 km/h = 0.6214 mph

Force

1 lb = 4.448 x 10^5 dyne
1 lb = 32.17 pdl
1 lb = 0.4536 kg
1 N = 10^5 dyne
1 N = 0.2248 lb
1 kip = 1000 lb

Mass

1 oz = 28.35 g
1 lb = 0.4536 kg
1 kg = 2.205 lb
1 slug = 14.59 kg
1 slug = 32.17 lb

Work and Heat

1 BTU = 778.2 ft-lb
1 BTU = 1055 J
1 Cal = 3.088 ft-lb
1 J = 10^7 ergs
1 kJ = 0.9478 ft-lb
1 BTU = 0.2929W · hr
1 ton = 12,000 BTU/hr
1 kWh = 3414 BTU
1 quad = 10^{15} BTU
1 therm= 10^5 BTU

Power

1 Hp = 550 ft-lb/s
1 HP = 33,000 ft-lb/min
1 Hp = 0.7067 BTU/s
1 Hp = 2545 BTU/hr
1 Hp = 745.7 W
1 W = 3.414 BTU/hr
1 kW = 1.341 Hp

Volume Flow Rate

1 cfm = 7.481 gal/min
1 cfm = 0.4719 L/s
1 m^3/s = 35.31 ft^3/s
1 m^3/s = 2119 cfm
1 gal/min= 0.1337 cfm

Torque

1 N · m = 10^7 dyne · cm
1 N · m = 0.7376 lb-ft
1 N · m = 10 197 g · cm
1 lb-ft = 1.356 N · m

Viscosity

1 lb-s/ft^2 = 478 poise
1 poise = 1 g/cm · s
1 N · s/m^2 = 0.02089 lb-s/ft^2

Pressure

1 atm = 14.7 psi
1 atm = 29.92 in Hg
1 atm = 33.93 ft H_2O
1 atm = 1.013 bar
1 atm = 1.033 kg/cm^2
1 atm = 101.3 kPa
1 psi = 2.036 in Hg
1 psi = 6.895 kPa
1 psi = 68 950 dyne/cm^2
1 ft H_2O= 0.4331 psi
1 kPa = 0.145 psi

Reader Remarks & Rewards Survey!

Yes, you! ...Your suggestions, comments and errata help us continuously to improve this review. Just fill out, tear, fold, tape and mail this prepaid survey! *Thank you!*

Name: ______________________________

Address: ______________________________

Phone: ______________ Position: ______________

"You can tell them I said so!" (check here) ☐

Comments

____

____

____

fold ▶

Any subjects we missed? Where?: ______________________________

Quick Survey

1. Do you think this book is:
(A) Too sketchy
(B) Just about right
(C) Too much material

2. Do you think our problems are:
(A) Too easy
(B) Just about right
(C) Too difficult

3. Do you think our solutions are:
(A) Too sketchy
(B) Just about right
(C) Too much explanation

4. Did you participate in a review course?
(A) Yes Course location/name: ______________
(B) No

5. Did you use other material in your preparation? If so, please list:
____

6. How long has it been since your undergraduate college graduation?
_____ years _____ recently _____ haven't yet

7. Rank factors in order of influence on your initial appraisal of this book:
_____ Price
_____ Reputation
_____ Written by professors
_____ Presentation of material
_____ Depth/Amount of Material
_____ Other ______________

tear

Rewards for Errata!

Find mistakes? We offer up to $3 for each error found in *"FE/EIT Discipline Review"* (2nd ed.), not already discovered, depending on significance.

error	proposed correction	pg. #	prob. or ex. #

fold ▶

____

Send info to my friend...

I think my friend would like information about your publications. Please send to the following address a catalog and other information relating to the following subjects in engineering: FE/EIT ☐ PE ☐ Mechanical ☐ Civil ☐ Electrical ☐

Name: ______________________________

Address: ______________________________

______________________________ FEDSF9

fold

Just fill out, fold & drop this survey card into any mailbox!

NO POSTAGE NECESSARY IF MAILED IN THE UNITED STATES

BUSINESS REPLY MAIL
FIRST-CLASS MAIL PERMIT NO 71 GROVER, MO

POSTAGE WILL BE PAID BY ADDRESSEE

GREAT LAKES PRESS
PO BOX 550
GROVER, MO 63040-9919

Cut or tear here—fold, tape and mail!

fold

Catalog of Titles by Great Lakes Press

Editor and Principal Author: Dr. Merle C. Potter, Ph.D., P.E.

Fundamentals of Engineering Review, new 7th edition, $52.95
The most effective review for the FE/EIT exam. Written by nine PhD teaching professors for the A.M. and General P.M. sessions. Contains all you need to pass —no more, no less! •Includes hardcover reference book which reviews all 12 general subjects; tables and charts; 1100 problems and full solutions; two 8-hour practice exams; NCEES equation summaries. 615 pages.

FE Discipline Review, 2nd edition, $34.95
Reviews the four major discipline P.M. areas: civil, mechanical, electrical and chemical. Covers the most important equations in the *NCEES Reference Handbook*. Includes reviews, practice problems and solutions—plus full 4-hr practice exams for CE, ME & EE. Written by fifteen PhD teaching professors. 362 pages.

Principles & Practice of Mechanical Engineering Review, 2nd ed., $69.95
Full review for the Principles & Practice of Engineering (PE) Mechanical Exam. •The only review written by PhD professors—nine experts in their fields. •Includes reviews of all 15 major subject areas tested; hundreds of excellent exam-simulating practice problems and solved examples. The only concise review. 591 pages. (Practice problems solutions manual, $19.95, 120 pp.)

Principles & Practice of Civil Engineering Review, 2nd edition, $69.95
Full review for the Principles & Practice of Engineering (PE) Civil Exam. •Written by ten PhD professors. •Covers all 16 major subject areas; hundreds of practice problems with solutions. The only concise review. 693 pages. (Practice problems solutions manual, $19.95, 130 pp.)

Principles & Practice of Electrical Engineering Review, $69.95
Full review for the Principles & Practice of Engineering (PE) Electrical Exam. •Written by ten PhD professors. •Concisely covers all major subject areas. 600 pages. (Practice problems solutions manual, $19.95, 110 pp.)

Jump Start the HP-48G/GX, $19.95
Unique user manual written expressly for engineers. Superior to HP manual. Get the most out of your calculator! Makes passing tests much easier! Covers: Using Equations, Solved Problems, Special Applications, Writing Programs, Basic Statistics, File Transfer, and more. 170 pages.

GRE, GMAT & GRE/Engineering Time•Savers™, $17.95
The most authoritative guides to these tests—authors are PhD teaching professors from major universities. •Study same subjects as in other reviews...in *half* the time! •Includes 3 practice tests with solutions and charts to identify your strengths and weaknesses •Complete reviews of all subjects.

QUICK & EASY ORDER FORM

Name (print): ____________________

Street Address: ____________________

____

____

Day ph.: ____________________

Credit Card: VISA MC (circle one)

Number: ____________________

Exp.: ____________

Name of Title	Price	Quan.	Multi-Discount!	Subtotal
Fundamentals of Engr	$52.95 x		–	=
FE/Discipline Review	$34.95 x		–	=
PE Review ME/CE/EE	$69.95 x		–	=
Solutions ME/CE/EE	$19.95 x		–	=
JumpStart the HP-48	$19.95 x		–	=
GRE Time•Saver	$17.95 x		–	=
GMAT Time•Saver	$17.95 x		–	=
GRE/Engr Review	$17.95 x		–	=
Shipping	$4 per item x		– N.A.	=
(MI & MO residents add 6% sales tax.) TOTAL				$

MULTI-DISCOUNT! Take 15% off each book on orders of 3 or more!

30-day money-back satisfaction guarantee on all GLP titles.
Mail this order form, credit info, & check or money order to:
Great Lakes Press POB 550 Grover, MO 63040

Call 1-800-837-0201 / fax 1-314-273-6086
www.glpbooks.com / custserv@glpbooks.com

FE7-97